The First Partner

HILLARY

RODHAM

CLINTON

ALSO BY JOYCE MILTON

The Rosenberg File
(with Ronald Radosh)

*The Yellow Kids: Foreign Correspondents in the
Heyday of Yellow Journalism*

*Loss of Eden: A Biography of Charles and
Anne Morrow Lindbergh*

Tramp: The Life of Charlie Chaplin

The First Partner

HILLARY

RODHAM

CLINTON

A Biography

JOYCE MILTON

Perennial
An Imprint of HarperCollinsPublishers

A hardcover edition of this book was published in 1999 by William Morrow and Company, Inc.

THE FIRST PARTNER—HILLARY RODHAM CLINTON. Copyright © 1999 by Joyce Milton. Epilogue copyright © 2000 by Joyce Milton. All rights reserved. Printed in the United States of America. No part of this book may be used or reproduced in any manner whatsoever without written permission except in the case of brief quotations embodied in critical articles and reviews. For information address HarperCollins Publishers Inc., 10 East 53rd Street, New York, NY 10022.

HarperCollins books may be purchased for educational, business, or sales promotional use. For information please write: Special Markets Department, HarperCollins Publishers Inc., 10 East 53rd Street, New York, NY 10022.

Designed by Jo Anne Metsch

First Perennial edition published 2000

The Library of Congress has catalogued the hardcover edition as follows:

Milton, Joyce.
The first partner—Hillary Rodham Clinton / Joyce
Milton.—1st ed.
p.   cm.
Includes bibliographical references and index.
ISBN 0-688-15501-4
1. Clinton, Hillary Rodham.   2. Presidents' spouses—United
States—Biography.   3. Clinton, Bill, 1946–   .   I. Title.
E887.C55M545 1999
937.929'092—dc21
[b]   99-12610

ISBN 0-688-17772-7 (pbk.)

00 01 02 03 04 RRD 10 9 8 7 6 5 4 3 2 1

Contents

Introduction

It all seems so long ago, that night of November 3, 1992, when my friends and I celebrated the victory of William Jefferson Clinton. For the first time, the White House would be occupied by members of my own generation, and more than that, people who knew people I knew. Like many women my age, I felt I could identify with Hillary Rodham Clinton. As it happened, her father, like mine, was a staunch Methodist, a World War II veteran and the product of one of those blue-collar Pennsylvania towns where the men earned their hard-scrabble existence in the mines and the mills. Her mother, like mine, was a homemaker who had put off her own education to raise a family, then signed up for college courses after her children were on their own. And Hillary was also the first in her family to "go east" to school, arriving just in time to be part of the tumultuous years of civil rights and antiwar protests.

Most important, Hillary Clinton was a product of the first wave of postwar feminism. The early 1970s were a time when enormous opportunities were opening up in previously male-dominated professions. To take advantage of them, a woman did not necessarily have to be intellectually brilliant, but she did need nerve and a thick skin. Women in those years were traveling into the future without a road map, and most of us were a lot less confident than we seemed. Decisions that may seem trivial in retrospect took on enormous symbolic importance. What name should we use after marriage? Could a feminist wear makeup? Was it acceptable to hire another woman to do our housework? We were haunted by the fear that we were—horrors!—just a few compromises away from becoming like our mothers. Hillary Clinton had come through all that, and to all appearances she was one of the lucky ones who managed not only to have it all but to hold on to what she had. She was the wife of a very successful man, a mother, a corporate attorney and a social activist in her own right.

Like so many other women of our generation, I was curious to see how Hillary—her first name seemed to define her better than either surname—would transform the role of First Lady of the United States. The First Ladies I remembered, from Jackie Kennedy on, fit the mold of the ideal corporate wife. They were women who devoted all their efforts to enhancing their husbands' careers without having any visible agendas of their own. But there weren't many women like this around anymore, and the change transcended party lines. After all, Marilyn Quayle was also a lawyer-wife, and we could expect to see many more professional women in the White House in the future, regardless of whether the Republicans or the Democrats were in power. Hillary herself seemed to recognize this, and she talked with apparent insight about her situation, insisting in one interview, "What I represent is generational change. It's not about me."

When a publisher asked me to write a small book introducing Hillary to elementary school readers, I was happy to say yes. Mine was one of several children's books about Hillary published early in the Clinton administration. One of them—not mine—bore the subtitle *A New Kind of First Lady*. But new in what way? By the time I handed in my manuscript in September 1993, my initial optimism about Hillary's potential had faded. Her health care initiative was already in trouble, and the administration was beset by embarrassing problems—failed Cabinet appointments, the firing of the White House travel office staff and the botched handling of the standoff between the Bureau of Alcohol, Tobacco, and Firearms and the

Branch Davidians. Rather than carving out a niche for herself, Hillary appeared to be intent on doing everything at once—leading the health care initiative, acting as her husband's shadow chief of staff, choosing china patterns for formal dinners, conducting a philosophical conversation about what she perceived as America's crisis of "meaning" and, meanwhile, posing for glam shots for *Vogue* magazine.

There were other, more pedestrian signs of trouble. In my experience, letters and calls to the White House were always answered. The replies might not always be responsive, but at least there was a reply. After the Clintons moved in, phone messages and letters often seemed to disappear down a rabbit hole. The First Lady let it be known that she was upset about inaccuracies in magazine articles about her early life, yet when I called her office no one was prepared to correct them. Had Hillary really appeared on the TV quiz show *College Bowl,* as reported? People called with questions like that all the time, I was told by one impatient staffer, who nevertheless didn't know the answer.

In 1992, Skip Rutherford, a longtime supporter from Arkansas who is now spearheading the construction of Bill Clinton's presidential library, predicted, "The White House will be more personable, more open, more inclusive, more fun than at any other time in American history." But the opposite proved to be true. Hillary Clinton, in particular, was obviously having a hard time coping with being so much in the public eye. She hated having her psyche analyzed by an endless parade of journalists and biographers—an understandable reaction but also an unrealistic one, especially as the fortunes of the Clinton administration increasingly came to revolve around the psychodrama of Bill and Hillary's tension-filled marriage.

A self-confessed policy wonk and proud of it, Hillary wanted to be judged by her ideas. That's fair enough. But the more one looked at Hillary Clinton's ideas, the more one saw that despite her reputation as a very smart woman, she is not a clear thinker, perhaps because she trusts her intellectual rationalizations a lot more than she trusts her emotions.

No one can doubt that Hillary Clinton is a very hard worker who strives to do good, consistent with her own ambitions. But she has also been a victim of that great delusion of the 1960s—namely, that it's possible to continually reinvent oneself, rewriting the rules to suit whatever role one happens to be playing at the moment.

Even more than Bill Clinton, Hillary can be said to have earned the nickname "the comeback kid." Her reputation for having a grasp of policy issues has survived despite her association with failed programs, including the health care debacle. Although she told us in so many words that she was not just "a little woman standing by her man," her efforts to help her husband cover up his history of sexual indiscretions are seen in exactly those terms, as an expression of wifely loyalty. Her proclamation of a "vast right-wing conspiracy" reverberated with echoes of Joe McCarthy's "conspiracy so immense," yet she is seen as a victim, not a demagogue. Whatever her shortcomings, no one can deny that Hillary Rodham Clinton is resilient. Doubtless she will be with us, testing out some new role or other, long after Bill Clinton has faded from the scene.

ONE

The First Victim

W H E N a woman with servants spends the weekend cleaning out her closets, it usually is not a good sign. And when Hillary Rodham Clinton told reporters that closet cleaning and hearing a good sermon at church had been the highlights of the past few days she was, by her standards, baring her soul. That Saturday, January 17, 1997, her husband had given a six-hour deposition to lawyers representing Paula Corbin Jones in her sexual harassment case.

Although the Clintons did their best to put up a show of unconcern, anyone who knew William Jefferson Clinton realized that for him to testify under oath about his sexual history was a very bad idea indeed. How this no-win situation was allowed to come about ranks as the greatest mystery of the political partnership of Bill and Hillary Rodham Clinton, which had proved in so many other ways to be a resounding success.

Certainly, it didn't take a Yale-trained lawyer to recognize that there were other options. Even a young Pentagon employee named Monica Lewinsky, who would never be accused of being politically astute, recognized that the Jones case cried out for a settlement, regardless of its merits. In the months before Clinton's deposition was taken, Lewinsky and her Pentagon colleague Linda Tripp thrashed out scenarios that would lead to an out-of-court resolution, thus solving both the President's problem and theirs. As Lewinsky saw it, the imperative was clear. "The American people elected him," she reminded Tripp, "so let him do his stupid job. You know?"

One "game plan" laid out by Lewinsky assigned a key role to Hillary Rodham Clinton: The First Lady would go on *Larry King Live,* having let it be known that she was prepared to take a question about the Jones case. When asked, Hillary "would respond emotionally. It's hard to see something we don't see from her often," Lewinsky mused. Hillary would then say, "The country is being robbed of its time that the President spends on other issues. They wish it would simply be settled. It's been hard on our family. I would like nothing more than for this to be a non-issue in our lives and in the lives of the American people."

His wife having cleared the way, Bill Clinton could appear the next morning with press spokesman Mike McCurry at his side and make a brief announcement, saying that for the sake of his family and the country he had decided to give Paula Jones the apology she was demanding and settle the lawsuit. It would be "sort of a gallant statement," Monica thought, and given Bill Clinton's high poll ratings, "a two week story," maybe "a three week story" at most.[1]

This was not, however, the scenario the Clintons chose to follow. The Jones deposition might be a minefield, salted with booby traps, but they had negotiated treacherous territory before and survived. Several women who had indicated that they might be prepared to cooperate with Paula Jones's attorneys had already reneged. Notably, Kathleen Willey, an attractive widow appointed by the President to the United Service Organization's Board of Governors, and her friend Julie Steele had backed off from a story earlier reported in *Newsweek* that the President had fondled Willey and placed her hand on his genitals when she visited the Oval Office one day in November 1993 to ask him for a job.

There were still a few witnesses who might pose problems for Bill Clinton, among them Linda Tripp, who had told *Newsweek* reporter Mike Isikoff that she saw Willey emerge from the Oval Office that day, disheveled and apparently "joyful" over being the object of the

President's advances. Like numerous other female career employees in the West Wing of the White House, Tripp resented the way jobs had been doled out to women who caught the President's eye, and she was furious that Willey, who hadn't been too proud to take the appointment offered her, would be presented to the world as a victim. From Tripp's point of view, her statement to Isikoff had not only been the truth, it happened to defend Bill Clinton against the charge that he was a sexual harasser.

This, of course, was not the way the President's allies saw it. Soon after the *Newsweek* story appeared, Tripp heard from Norma Asnes, a New York producer and personal friend of Hillary Clinton. Tripp knew Asnes slightly, having met her at an official Pentagon function a few years earlier. Suddenly, however, Asnes had become very chummy. She invited Tripp to spend a few days at her luxurious Fifth Avenue apartment and asked her to come along on a chartered yacht cruise planned for the following summer. And in November, after Tripp's named appeared on the witness list for the Paula Jones suit, Asnes talked of introducing her to executives who could help her find a better-paying job outside of government. Tripp was flattered at first but also suspicious of Asnes's motives. Referring to Asnes's closeness to the First Lady, she told Lewinsky, "Let's not forget whose friend she is." Tripp also expressed her reservations to Asnes, who told her, "I like to enjoy the people I'm with. I like them to be articulate, bright, and mentally stimulating. They don't have to be at my level. It doesn't matter if they're not millionaires."

Far from finding this reassuring, Tripp was insulted. "So I'm one of the plebeians," she told Lewinsky. "I hate to sound like a skeptic. But—why? I mean, we don't know each other that well."[2]

The question of what—or, more to the point, who—inspired Norma Asnes's desire to play Pygmalion would come to interest Independent Counsel Kenneth Starr. His investigators later interviewed Asnes, but in the push to deliver a timely report to Congress, the role played by the First Lady's friend was just one more lead that was never developed. But for Tripp, who had already been called a liar by the President's attorney Robert Bennett, Asnes's approaches were evidence that the occupants of the White House were taking an intense interest in her deposition. Knowing that she would undoubtedly be asked if she was aware of other women besides Willey whom the President had approached, Tripp decided to protect herself with evidence.

The first hint that Paula Jones's attorneys might have a few surprises in store for the President came a week before Christmas when Lewinsky, an ex–White House intern, told the President's friend Vernon Jordan that she, too, had been subpoenaed. Worse news yet, unlike some plain vanilla subpoenas served on other women connected with the case, Monica's summons specifically mentioned certain gifts she had received from Bill Clinton, including a brooch, a hat pin and a book. This was an obvious signal that the Jones lawyers had inside information about Clinton's relationship with Lewinsky, and if ever there was a time to panic, this was it. But the President still wasn't ready to tell Bob Bennett that the time had come to settle the case. By January 17, 1998, the day of the deposition, Lewinsky had signed an affidavit denying that she'd had sexual relations with the President, and the gifts in question had been returned to Clinton's private secretary, Betty Currie. Even if Lewinsky told the Jones lawyers a different story, it would seem that she was now a tainted witness, with nothing to back up any charges she might make.

After the deposition, Bill and Hillary had planned to show the world a united front by going out to dinner at a Washington restaurant. But the day proved to be a lot tougher than the President had expected. The questions the Jones attorneys asked left no doubt that not only did they know about the gifts the President had given Lewinsky, they knew that Lewinsky—referred to for the purposes of the lawsuit as "Jane Doe #6"—had visited the White House more than three dozen times after her transfer to a job at the Pentagon, ostensibly to see Betty Currie, and they knew that U.S. Representative to the United Nations Bill Richardson had offered her a job in New York.

The Clintons did not dine out on Saturday evening. And by the time they retired for the night, there was more bad news. The Drudge Report, the Internet gossip sheet loathed but avidly followed at the White House, was reporting that *Newsweek* had the intern story but had decided to spike it just minutes before its deadline. Drudge did not disclose Lewinsky's name, but he mentioned the existence of tapes of "intimate phone conversations." This can only have sent a shudder through Clinton, who'd had phone sex with Lewinsky on several occasions.

But what about Hillary Rodham Clinton? Had she known about Monica Lewinsky?

The President's dalliance was not exactly unknown inside the walls of the White House. Members of the Secret Service had recognized Lewinsky as the President's mistress and took bets on the timing of her visits to the Oval Office, and Hillary's own deputy chief of staff, Evelyn Lieberman, had been worried enough about Lewinsky's knack for getting close to Bill Clinton to have her transferred from the White House to the Pentagon. Lewinsky, devastated that Clinton had "changed the rules" of their relationship, believed his promise that he would find her another White House job after the 1996 elections, but Linda Tripp heard from a friend who worked for the National Security Council that Lewinsky was persona non grata at the White House and would never be allowed to come back.

Still, it is possible that Hillary was unaware of all this at the time. As she would tell it, the first she heard of the Monica Lewinsky situation was on Wednesday morning, January 21, the day the story broke in *The Washington Post,* when her husband woke her from a sound sleep and told her, "You're not going to believe this, but—I want to tell you what's in the newspapers." Washington insiders had been following Matt Drudge's bulletins on the breaking story for four days, but to Hillary "this came as a very big surprise."

It is hard to imagine that Mrs. Clinton had literally never heard of Monica Lewinsky—who was, after all, on the Jones case witness list—but there was, it seems, substantial truth to the First Lady's statement that her husband had kept the details from her as long as possible. In fact, one White House source claimed that it was Hillary's aide, Roberta Green, not the President, who was the first to brief her on the breaking story. At any rate, it was only later that morning, on the train to Baltimore, where she was to make an appearance at Goucher

College, that Hillary was told that one of the gifts her husband had given Lewinsky was a copy of Walt Whitman's *Leaves of Grass*. "He gave me that same book after our second date," she told an aide, the pain evident in her voice.[3]

With hindsight, Hillary had two reasons to be angry with her husband. The first was that he had allowed the affair to happen at all, courting a woman young enough to be his daughter with the same poetry he had used to attract her. The second was that out of shame and fear of the possible repercussions he had tried to handle the aftermath of the affair himself. Clinton had continued to string Lewinsky along for months with the promise that he would find a way to bring her back to the White House. This may have been his clumsy idea of letting her down gently, but the confused signals he was giving out only made her more hurt, agitated and, ultimately, demanding. By the time Clinton turned to Vernon Jordan and Bill Richardson in a final effort to find Lewinsky a job outside Washington, it was too late. Linda Tripp already had her on tape. Last of all, when the deposition was already on record, he turned to Hillary.

This was a familiar pattern in the Clinton marriage. And strangely, Hillary seemed to thrive on it. By the time she returned to Washington from her speech at Goucher, she had regained her shattered composure. Addressing a crowd of reporters who met her train at Union Station, she acknowledged that it was "difficult and painful" to see her husband attacked so unfairly, "but I also have now lived with this for, gosh, more than six years. I have seen these charges and accusations evaporate and disappear, if they're ever given the light of day."

Hillary had already called her daughter, Chelsea, at Stanford to tell her that the allegations were untrue, and she now got on the phone and summoned old friends Mickey Kantor, Harold Ickes and Hollywood TV producer Harry Thomason to the White House. Hillary thought Bill's initial denial of the Lewinsky story, in an interview with National Public Radio, had been "too vacillating." Over the weekend, Thomason spent hours coaching the President, and on Monday morning, January 26, Clinton interrupted an announcement about day-care programs in the Roosevelt Room to wag his finger at the TV cameras and declare, "I want to say one thing to the American people. I'm going to say this again. I did not have sexual relations with that woman, Miss Lewinsky."

Unlike her husband, Hillary needed no coaching to carry on. Standing beside him in the Roosevelt Room, radiant in a lemon-yellow ensemble, she had never looked better. Even her hairstyle, previously just ever so slightly unflattering, was now perfect. When the child-care event ended, Hillary continued with her busy schedule, leaving for New York, where she paid a visit to a child-care center in Harlem, her arrival cheered by an admiring crowd.

On Tuesday morning, looking tired but still feisty, the First Lady was interviewed on the *Today* show by Matt Lauer. She defended her husband so vigorously that it is necessary to read the transcript closely to assure oneself that she never actually said that the allegations of an affair weren't true. What she did say was that her husband might possibly have given Monica Lewinsky gifts but his actions needed to be "put into context":

> Anyone who knows my husband knows that he is an extremely generous person to people he knows, to strangers, to anybody who is around him. And I think that, you know, his behavior, his treatment of people, will certainly explain all of this. . . . I mean, I've seen him take his tie off and hand it to somebody.

Hillary went on to raise an issue that she often discussed in private. "I don't know what it is about my husband that generates such hostility. But I have seen it for twenty-five years."

She went on to speculate that the hostility was a reaction to Bill Clinton's "personality, his kind of gregariousness," as well as to his "political ideas," and she went on to blame his current problems on "this vast right-wing conspiracy that has been conspiring against my husband since the day he announced for president." She predicted that "when all of this is put into context . . . some people are going to have a lot to answer for."

This incendiary language, which Hillary had crafted with the help of White House aide Sidney Blumenthal, and reportedly used without the approval of top presidential advisers, was widely ridiculed at first, yet it would prove to have a potent appeal for energizing the left wing of the Democratic Party. Ironically, the segment of the party that had always been cool to Bill Clinton and his policies found it easy to rally around Hillary and her charge that his public embarrassment was all the fault of a right-wing cabal.[4]

In the 1998 election campaign the First Lady would emerge as the Democrats' most potent weapon, inspiring the party faithful who might otherwise have been too discouraged to go to the polls. Even Democrats who could not quite bring themselves to see Bill Clinton's actions as just gregariousness gone awry were impressed by Hillary's staunch loyalty and unruffled dignity. She was, after all, a victim twice over—not only of her husband's indiscretions but, from her supporters' point of view, of his hostile, scandal-obsessed enemies. When Mrs. Clinton appeared at a "Get Out the Vote" rally in Chicago, Jesse Jackson no doubt spoke for many in the crowd when he told her, "Hillary, you've come through rain and you're not wet. You've walked through fire and there's not a singe on your clothing."

Although one would imagine that this was a time of severe emotional stress for Hillary, she appeared to be unusually calm, displaying an equanimity that amazed both personal friends and her staff. Shortly after the *Today* appearance, when Nancy Bekavac, who knew the Clintons from Yale, called to say that she'd be unable to make a planned trip to Washington because her secretary had quit unexpectedly, Hillary returned the call and commiserated with Bekavac about her personnel problem as if she had nothing much on her mind at the moment. One White House aide delicately ascribed the First Lady's coping strategy to an ability to escape from sordid facts and the "politics of defamation" by "moving to this large environment of ideas." Hillary, the aide noted, was reframing the issue in her own mind as a symptom of a larger social problem—the breakdown of "civil society."[5]

In Hillary's terms, it seems, this breakdown was the fault of consumerism, the tabloid press and negative political tactics, which led the public to be mesmerized by scandal when they ought to be concentrating on more elevated matters like social policy. She elaborated on these views in early February, when she flew to Davos, Switzerland, to address 2,000 international businessmen at the World Economic Forum. After suggesting that the Clinton administration might yet revive its plan for government-mandated health care, she made a plea for a return to civil society, which she defined as "the stuff of family life" and of "religious belief and spirituality. . . . There is no doubt that we are creating a consumer-driven culture that undermines both capitalism and democracy. There is the same relentless pressure and instantaneous rush to judgment."

The trouble, of course, was that Hillary's belief in a "vast right-wing conspiracy" hardly contributed to a more civil level of political discourse. The President was now committed to a firm denial that left him no wriggle room. Inside the White House, the Clintons were trying out the theory that Monica Lewinsky was a "stalker," an obsessive young woman who fantasized a relationship with the President of the United States. According to Sid Blumenthal's later testimony to Ken Starr's grand jury, Clinton told him that Monica Lewinsky "came at me and made a sexual demand on me." But he had refused to comply, telling Lewinsky, "I've been down that road before. I've caused a lot of pain for a lot of people. And I'm not going to do that again." Yet when Clinton's former pollster and adviser Dick Morris called, offering to make a public attack on Lewinsky's credibility, the President warned him off. The problem, he explained, was that he couldn't be sure that Lewinsky was cooperating with Independent Counsel Ken Starr. While early reports that Lewinsky had physical evidence of a sexual encounter in the form of a semen-stained dress had been publicly discounted, Clinton had reason to be wary.

During this period the President confided to Sid Blumenthal, "I feel like a character in a novel. I feel like somebody who is surrounded by an oppressive force that is creating a lie about me and I can't get the truth out. I feel like the character in *Darkness at Noon*." It would be hard to think of a more apt literary reference. Arthur Koestler's great novel is the story of a veteran Bolshevik who finds himself accused during the purges of terrible crimes against the party to which he has devoted his life. As a good revolutionary, Nicolai Rubashov has always believed that the end justifies the means. Now that he has become expendable, he is doomed by his own principles. Koestler, interestingly enough, describes Rubashov as entrapped by his habitual abuse of language, which he calls the "grammatical fiction"—Rubashov's "Truth" has been a lie all along.

While Bill Clinton's comparison of himself to Rubashov suggests a painfully acute self-knowledge, Hillary appeared to have risen above the contradictions of her situation. Even as her husband faced the ignominy of being the first elected President to be impeached by the House of Representatives and then tried by the United States Senate, Hillary was more popular than ever. Her poll numbers soaring, she was being courted by the Democratic leadership of New York State as a future Senate candidate. And she appeared on the cover of *Vogue,* looking regal in a black velvet dress by Oscar de la Renta. *Vogue* editor Anna Wintour, who elsewhere had pronounced Mrs. Clinton "brilliant," now said admiringly, "I think she psyches herself into this battle mode and goes forward, not looking right or left. She told us she doesn't even read the newspapers."[6]

This was a strange place for Hillary Rodham Clinton to find herself. The onetime campus activist, who scorned appearances in favor of intellectual substance, had come a long way to reinvent herself as a *Vogue* cover girl. And the former spokesperson for the generation of the 1960s, which prided itself on exposing hypocrisy and "telling it like it is," was now admired by many of her peers for her ability to close her mind to unpleasant realities. One is tempted, like Bill Clinton, to draw a lesson from Nicolai Rubashov, who cried out in his moment of greatest anguish, "What a mess we have made of our golden generation!"

TWO

Rugged Individualist

C HILDREN are not rugged individualists." This is the first sentence of Hillary Rodham Clinton's book *It Takes a Village,* and it's a startling one because Hillary Rodham *was* a rugged individualist as a child. Moreover, that was exactly what her parents raised her to be.

Hillary's parents belonged to the generation that grew up poor during the Great Depression and came to young adulthood during World War II. Descended from generations of men who worked in the lace mills, first in Northumberland, England, and then in eastern Pennsylvania, Hugh Rodham worked in a coal mine and stacked boxes in a factory to put himself through Penn State, where he played football and majored in physical education. By 1937, Hugh was a curtain salesman with the Columbia Lace Company, and he met Dorothy Howell when she showed up to apply for a secretarial job.

Unlike Hugh, who came from a close-knit, feisty family, Dorothy Howell was born in Chicago to a fifteen-year-old mother and a seventeen-year-old father. At the age of eight, her parents having split up, Dorothy and her three-year-old sister were shipped off to relatives in Alhambra, California. The girls traveled across the country unescorted, with Dorothy doing her best to look after her little sister. In California, they were taken in by their hypercritical grandmother and distant, emotionally withdrawn grandfather. At fourteen, Dorothy escaped from the house by becoming a live-in baby-sitter for a family that treated her kindly. After high school, she left California with no regrets and moved back to Chicago, where she supported herself by doing secretarial work.

During World War II, Hugh served in the Navy, became a chief petty officer, and was spared overseas duty when he was assigned to teach physical education as part of the Gene Tunney Program. He and Dorothy married in 1942, five years after they had met. They waited another five years to have their first child, Hillary Diane, born on October 26, 1947. A son, Hugh junior, followed in 1950. Hugh senior had left his sales job and was setting up his own custom drapery business. He solicited orders from hotels, corporations and airlines, printed the fabric himself and did his own sewing. Dorothy kept the books. Hugh didn't believe in credit; he would die in 1993 without ever having owned a credit card. He and Dorothy became prodigious savers, living in a one-bedroom apartment on the North Side of Chicago until they had saved up enough money for a house.

Not long after Hugh junior's birth they were able to purchase a home on the corner of Wisner and Elm in suburban Park Ridge, northwest of the city. Here, Anthony, their second son, would arrive in 1954.

The yellow brick Georgian house, which the Rodhams sold in the late 1980s for a sum reported to be over $400,000, was purchased for cash, without a mortgage. Then in the throes of a postwar growth spurt, Park Ridge was attracting lots of young families looking for good schools and a safe environment for their children. Even compared to the surrounding towns, which were generally bastions of the Republican Party, Park Ridge had a definite conservative cast. The John Birch Society was active there in the early 1950s. Today, it is the hometown of Representative Henry Hyde, the pro-life conservative and chairman of the House Judiciary Committee.

The neighborhood where the Rodhams chose to live was described by one local as a "country club area," but the Rodhams were not a country club family. Hugh and Dorothy socialized with some of their neighbors at backyard cookouts, but they mostly kept to themselves. Hillary would recall that none of her parents' friends were professionals, even though Park Ridge was a town filled with "doctors, lawyers and Indian chiefs." Dorothy Rodham taught Sunday school, but Hugh didn't even attend church, preferring to do his praying at home. By no means social climbers, the Rodhams were proud people who tended to avoid situations where they might feel the sting of being thought inferior.

Hillary was just old enough at the time of the move to face having to fight for acceptance in the new neighborhood. Every time she got dressed up and went out to play, she would recall, "I'd get beaten up and kicked around so bad" that there was nothing to do but run back to the house in tears. One girl in particular made Hillary's life miserable; she had been accepted by the boys on the block and wasn't about to let another girl into the gang. Finally, Dorothy told her daughter, "There's no room in this house for cowards. . . . The next time she hits you, I want you to hit her back." Hillary took her mother's advice and soon came home crowing, "I can play with the boys now."

The dominant personality in the Rodham family was Hillary's father, Hugh, a man inevitably described as "gruff." He chewed tobacco and barked out his opinions, ever ready for a good political argument with anyone brave enough to disagree with him. "My father was confrontational, completely and utterly so," Hugh junior said of him. He was also sparing of praise for his three children. Hughie junior became quarterback of his high school football team and won the conference championship, throwing ten completed passes out of eleven. His father's verdict when he saw Hughie after the game was, "I got nothing to say to you, except that you should have completed the other one." Hillary's straight-A report cards evoked only a grumbled "You must go to a pretty easy school."

Hugh senior was also notoriously tight with a dollar. The only status symbol he allowed himself was his late-model Cadillac. Otherwise, the family scrimped. Other Park Ridge families had summer cottages in Michigan or Wisconsin or took vacations in Florida;

the Rodhams vacationed at a family cabin on Lake Winona, near Scranton, where Hugh and the boys fished while Hillary and her mom groused about the lack of indoor plumbing. Other kids in town got new wardrobes when the fashions changed; Hillary and her brothers were expected to wear out their clothes, and it could take some hard arguing to convince Dad that a new pair of shoes was in order. In the summer, the children were sometimes rewarded with pennies for pulling dandelions from the lawn, but Hugh couldn't see why they should be paid for working around the house, and when he had a big order to finish, the boys went into the shop to help out, again without receiving wages for their labor. Informed that the neighbors gave their children regular allowances, Hugh said of his brood, "They eat and sleep for free. We're not going to pay them for it as well!"

Roger Morris, writing of Hillary's childhood in *Partners in Power,* tells us, "Among both relatives and friends, many thought Hugh Rodham's treatment of his daughter and sons amounted to the kind of psychological abuse that might have crushed some children."[1] Maybe so, but Rodham would have been a familiar type in any Pennsylvania mining or mill town. A man who has earned his living in a coal mine, even briefly, knows that the world can be a brutally hard place, and his great fear is likely to be that his children, growing up in affluence unimaginable by the standards of his own childhood, will turn out soft and selfish. Like so many children of Depression poverty, both Hugh and Dorothy worried a lot about this. On family trips to Chicago, they made a point of driving through skid row neighborhoods in order to show the children how they might be living but for their good fortune. On one of their summer vacations in Pennsylvania, Hugh even took them to see a coal mine. But for Hillary and her brothers, a coal mine was as relevant to their lives as a tepee. They measured themselves against the neighborhood kids, and while Park Ridge wasn't an ostentatious community, its children took certain luxuries for granted.

To an extent that his children probably found impossible to comprehend, Hugh senior had his reasons for pinching pennies. He was, after all, a self-employed businessman, with three children to educate and no pension or job security. The drapery company's only employee was hired after Hugh arrived for work one morning in 1958 and found him sleeping off a bender in the doorway of the shop; rather than rousting the drunk back to the streets, Hugh bought him a meal, helped him get cleaned up and put him to work. Otherwise, the business was destined to remain a one-man operation, dependent on Hugh's hard physical labor as well as his ability to hustle new orders.

For all his grouchiness, Hugh was devoted to his family. In the morning he sat at the kitchen table helping the children with their math homework, and he played endless games of pinochle with them in the evenings. He went out to watch Hughie play football, planting himself in a folding lawn chair on the sidelines instead of joining other parents in the bleachers. He taught Hillary to read the stock tables; and when she fretted about her low batting average in school softball games, he took her out to the park and threw her one pitch after another until she learned to connect with a curveball.

Hillary and her father often seemed to be involved in a contest of wills. She did everything he asked, and he would respond by raising the bar a few inches higher. Hillary's brothers had no doubt that she was their father's favorite, a daddy's girl who could do no wrong in his eyes. This may have been far from obvious to Hillary. In later years, however, Hillary would insist that her father's behavior was "empowering." As she told *Glamour* magazine in a 1992 interview, "My father would come home and say, 'You did well, but couldn't you do better?' "

Dorothy's nickname for her husband was "Mr. Difficult." After a lonely and difficult childhood, Dorothy had set aside her dream of higher education to marry a man who was a good provider but often irascible. She kept a neat home whose only flourishes were a few treasured antiques; chauffeured the children to softball, football and soccer games, to dancing lessons and to movies; and used any time left over to help Hillary with projects like organizing a neighborhood

"Olympics" to raise money for charity. Hugh was a Goldwater conservative and loved to argue politics. Dorothy quietly cast her votes for Democrats.

In later years, Dorothy would strike many who met her as stern, even slightly bitter. At Hillary's forty-sixth birthday party, to which the other women came dressed as Southern belles and femmes fatales, Dorothy appeared as a nun. But Dorothy was no victim. "I loved having my children at home," she said later. And when the last of the children entered college, she became the first woman in the neighborhood to go back to school, signing up for college courses in philosophy and Spanish. For a long time, however, her dreams were invested in her daughter. Long before Hillary started first grade, Dorothy began telling her that school would be a "great adventure. . . . She was going to learn great things, live new passions. I motivated her in a way that she wasn't resigned to go to school. I wanted her to be excited by the idea."

Though we look back on suburban life in the 1950s as a more innocent time, even Park Ridge had its dangers. Hillary would recall "the occasional stranger who exposed himself to us or tried to get one of us to go for a ride." Once, while she and her friends were playing at a building site, a man threatened them with a butcher knife. Another time, an older boy threw Hillary to the ground, jumped on top of her and kissed her until she struggled free and ran away. When she got home, Dorothy sat her down and they talked about how to avoid such incidents in the future.

Discussing these events in *It Takes a Village,* Hillary would observe that the greater feeling of security that surrounded children in the 1950s was not just because the world was inherently safer. Children felt safe because the adults around them were braver and more composed—there was "nothing like the pervasive anxiety about safety that has seeped into every corner of our country's psyche." On the very next page, Hillary tells how she recently used a kit created by the Mazique Parent-Child Center to organize a "Baby Safety Shower" for a friend who had just adopted a baby, in order to raise her consciousness about such menaces as household poisons, overly soft mattresses, poorly designed cribs, dripping hot water faucets, lead-based paint, riding bicycles even a brief distance without a helmet and streets without speed bumps. She even recounts the story of a pair of children, ages two and four, who let themselves into an unlocked car and suffocated there. Of course, parents in the fifties did worry about their children's safety. But having survived hard times, as Dorothy Rodham had, they were not unduly anxious about the risks of everyday life. Hillary, on the other hand, seems unable to imagine that there are still parents around who trust their own instincts, as her mother did.

As Dorothy hoped, Hillary loved school. At Eugene Field Elementary, she was the perfect student. Her fifth-grade teacher, Mrs. King, was so fond of teaching Hillary, it was said, that she followed her to the sixth grade so that she could have her in class for a second year. As a Girl Scout, Hillary amassed an eye-popping collection of badges, and in a time when many girls avoided sports she was a better than fair athlete, playing softball, soccer, tennis and Ping-Pong. As she got older, she earned her junior life saving certificate and became a wading-pool lifeguard, learned canoeing and water safety, and still found time to earn spending money by babysitting.

At the competitive Maine East School, however, Hillary was no longer the top student in her class, not even in the top ten. She earned her place on the honor roll by working hard, once even submitting a history term paper that ran to seventy-five pages. Articulate and well prepared but not prone to challenge authority, she was the kind of student who makes teaching a pleasure. In classroom debates, she could state her opinions in complete sentences, reeling off the main points of her argument in logical order. Hillary took part in so many activities that it was a wonder she found time to shoehorn them all into her schedule, but many of her extracurricular interests involved talking and planning rather than doing—she was a student council member, a member of the "cultural values" committee, a member of the prom committee, a gym leader.

Despite a reputation as a teacher's pet, Hillary was outgoing and sociable, if more than a little bossy. While still a student at Eugene Field, she attracted her first admirer, a boy named Jim Yrigoyen. There was a fad for wearing dog tags at the time, and Hillary wore his dog tag for a while. But Yrigoyen's most memorable encounter with Hillary came some time later, when she assigned him to watch over some baby rabbits she had found in her backyard. A boy from the block came by and demanded one of the rabbits, and Yrigoyen let him have one. A few minutes later, Hillary returned, found one of the babies missing and dealt with Yrigoyen by punching him in the nose.

Hillary had a gift for making lasting friendships. Betsy Johnson Ebeling met Hillary in the sixth grade and for a time took piano lessons from the same teacher, an eccentric named Margaret-Lucy Lessard who was mad about Pomeranians. Lessard had her deceased pets stuffed and mounted when they died and kept them on display in a glass case in her living room, where her students found them understandably distracting. Ebeling remains close to Hillary today, one of eight lifetime friends who got together a few years ago to give her a silver friendship charm bracelet.

Boys, however, were sharply divided in their opinions about Hillary. Except for a slight overbite, she was a pretty girl, who could look sleek and sophisticated when she took the trouble to dress up, but this happened rarely. "Guys didn't think she was attractive," one man who went to school with Hillary recalled. "They liked girls who were 'girlish.' But Hillary was 'womanish.'" Hillary's reputation for being aloof and overserious was a turnoff for many—her high school newspaper once predicted that she would become a nun, taking the name "Sister Frigidaire"—but some boys, especially older ones, saw her composure as a challenge. Hillary had no great difficulty lining up dates when she wanted them, but she was too busy to go out often, and when she did she was far from sure of herself. Set to double-date with Ebeling for the prom, Hillary worried that her father wouldn't spring for the price of a new dress. When he did after all, she promptly decided that her new chiffon gown was too plain and wondered, somewhat illogically, if she got it only because her mom and dad had agreed to act as chaperons and didn't want to be embarrassed by her.

Hillary and Betsy Ebeling became part of the in crowd at Maine East, a conservative school even by the standards of the times. Drugs were simply not an issue, and Hillary and her girlfriends didn't drink or smoke. Maine East girls who were feeling rebellious held ear-piercing parties, but Hillary didn't go for that idea, either. Still, Hillary's girlfriends regarded her as fun-loving, and she had enough of a sense of humor about herself to play on her "Sister Frigidaire" image by taking the part of Carry Nation in a school skit, delivering a pro-temperance rant in mock defense of Park Ridge's purity as a "dry" community.

Those outside of the school in crowd were less likely to be Hillary fans. Penny Pullen, later an Illinois state legislator and now a pro-life activist and lecturer, was a member of Hillary's senior class but did not know Hillary well. "She was part of the popular crowd, and I was not," Pullen recalls. Once, however, during their senior year, Pullen happened to overhear Hillary talking to some of her friends about how she could increase her chances of winning the DAR award given to the most outstanding girl in the class. Hillary went on to speculate that winning the award would help her get a certain scholarship. Pullen was mildly appalled by this display of ambition, noting, "The DAR award wasn't something you campaigned for, it was given to the girl who earned it."[2]

Another classmate, who preferred not to give her name, told journalist Cynthia Hanson: "Hillary was so take-charge, so determined, so involved in every single activity that you'd think, 'Why don't you chill out a bit? Why don't you give somebody else a chance?' I always felt Hillary thought she knew what was best, and that's what everybody should do."[3]

Despite her seemingly ironclad self-confidence, Hillary could be shaken. Once, while playing in a junior high school soccer game, she turned to the goalie of the opposite team, a working-class girl from an Eastern European background, and commented, "Boy, it's really cold."

"I wish people like you would freeze," the goalie snapped.

"But you don't even know me," Hillary protested.

"I don't have to know you to know I hate you," came the retort.

In 1997, Hillary would recount this conversation in a speech to 11,000 New England teenagers delivered in connection with President Clinton's national "conversation" on race relations, citing the goalie's put-down of her as an example of her personal experience of prejudice, since, presumably, the goalie had sized her up as an "uppity, wealthy whitebread" kind of girl. Park Ridge must have been a genteel environment indeed if Hillary could be hurt enough to carry around the memory of this routine exercise in playground trash talk for four decades.

Hillary's next major encounter with working-class ethnics came in 1964, when overcrowding at Maine East High School caused students from her neighborhood to be transferred to Maine South, a brand-new building whose student body included Poles, Slavs and Italians. Hillary would speak of Maine South as a school riven by ethnic rivalries. "There was a lot of tension because of that," she recalled. "Some corridors you couldn't walk down without feeling unsafe, uncomfortable." Others who attended the school rate this as a considerable exaggeration.[4]

From a psychological point of view, Hillary Rodham's character was not complicated. She was a daddy's girl, striving to wrest some expression of approval from a father who had trouble demonstrating his pride and affection. And she was also the only daughter of a very intelligent woman who had put her own ambitions on hold but saw her little girl's future as limitless. Dorothy urged Hillary to aim high, suggesting that she might even become the first woman Supreme Court justice some day. But if the Rodhams put pressure on all their offspring to succeed, her younger brothers managed to put up a successful resistance. Hughie and Tony were known around town as happy jocks, average students who liked a good time. Hughie would develop into a typical 1970s college student; he kept a messy room, revered the pop icons of leftism without being an activist himself and brooded a lot about dropping out without actually doing it. Tony was less interested in academics and more gregarious.

Far from pushing Hillary to accumulate honors, her parents were a little startled by her competitiveness. At one high school awards day, Dorothy would recall, she was embarrassed by her daughter's repeated trips to the stage. She even wished sometimes that Hillary would show some interest in makeup and clothes. It is doubtful, however, that she mentioned this to Hillary. Appearances ranked low on the Rodham scale of values; hard work, tenacity and self-reliance were everything.

In 1961, a new influence came into Hillary's life, driving a 1959 red Chevrolet Impala convertible. Fresh out of seminary and enamored of the existentialist theology of Paul Tillich, the Reverend Don Jones had just been hired as the youth minister of the Rodhams' church, First Methodist. "I was somewhat flamboyant," says Jones. "I decided I was going to be myself and not conform to the traditional style of Methodist ministers."

Determined to make personal salvation relevant to his young charges, Jones renamed the youth group the "University of Life" and incorporated into its programs the music of Bob Dylan, poems by e.e. cummings and T. S. Eliot, and the François Truffaut film *The 400 Blows*. He drew parallels between Marxism and Christianity, pointing out that both systems of thought had their version of Eden (or in Marx's view, pre-capitalist society) and both looked toward a heaven (the dictatorship of the proletariat) with a lot of struggle in between.

First Methodist was a large, wealthy congregation; in addition to the senior minister and Reverend Jones, it employed a minister in charge of pastoral visitations, a music director and a Christian education specialist. Moreover, the church had a well-established outreach ministry, which belied the image of Park Ridge as a suburban enclave, indifferent to the problems of the poor and minorities. Dorothy Rodham helped organize a summer Bible school project that worked with the Mexican migrant workers who were bused in to nearby truck farms every year. During the summer after Jones arrived, Hillary, just fourteen, organized groups of older students from the

Bible school to provide a baby-sitting service for the farm workers' children, who were left to fend for themselves all day while their parents worked in the fields. Hillary and her group served cold drinks and cupcakes, and some of the girls sewed items of children's clothing and doll clothes. It fell to Jones to ferry Hillary and the other girls to the migrant camp in his Impala, a situation that made some parents in the congregation nervous.

Jones was oblivious to these fears at the time—"there was no flirting or anything like that," he says gravely. But while Jones proved to be no threat to the teenage girls of First Methodist, he was a part of a trend that would drive a wedge between Methodist congregants and the church hierarchy. Influenced by such theologians as Tillich and Dietrich Bonhoeffer, whose ideas were formed in response to the threat of Nazi Germany, the seminaries were emphasizing the necessity of expressing one's faith through commitment to social change, a message that at times had disconcerting repercussions.

On April 15, 1962, Jones took his University of Life to Orchestra Hall in Chicago to hear Martin Luther King, Jr., deliver a sermon entitled "Remaining Awake Through a Revolution." King began by saying that too many Americans were like Rip Van Winkle, so fast asleep that they weren't even aware of the great changes taking place in the country. This was precisely how Jones saw all-white Park Ridge, as "a kind of a cocoon" where his parishioners could avoid engagement with the moral issues of the day. From time to time Jones would be called on to preach, and he delivered a few sermons that he now sees as "fairly prophetic about criticizing the status quo."

Some parishioners, indeed, were shocked by Jones's message, but considering the conservative mood of Park Ridge, there was less controversy than one might think over the tack Jones was taking with the youth group. A program on teen pregnancy evoked a flurry of objections, but there was no major controversy when Jones took the University of Life to Chicago to hear Saul Alinsky, the radical organizer who used tactics like picketing executives' homes and staging "fart-ins" to wring concessions from corporations. Hillary would be so impressed by Alinsky that she wrote her college thesis on him.

However well intentioned, some of Jones's programs, with hindsight, hint at the pitfalls of striving for "relevance." Jones arranged visits to recreation centers in Chicago, where the youth of First Methodist mixed with teenagers from inner-city neighborhoods, some of them gang members. During one such session Jones held up a reproduction of Picasso's *Guernica* and went around the circle, asking the young people to relate what they saw in the painting to their own experiences. Not surprisingly, the inner-city kids excelled at this assignment, coming up with emotionally charged accounts of how violence had touched their lives, stories far more dramatic than the contributions of the white, middle-class kids from the suburbs. The technique worked to get the group talking, but the unspoken lesson was that the troubled inner-city teens led lives that, however violence-scarred, were also more authentic. Being victims of the system endowed them with a gloss of moral authority.

The long-term social effects of this attitude, which grew out of the existentialist emphasis on the uniqueness of the self and a romanticized view of alienation, would prove to be disastrous. But the immediate impact of Jones's programs was exhilarating. "It was a time when a lot of the idealism that was going to fuel the sixties and early seventies was becoming known about," recalled Hillary's friend Sherry Heiden. "We believed in the incredible social changes that can happen if you change your perspective." Another University of Life member, Rick Ricketts, said of the sessions, "Don presented the material in a very thoughtful way and in a way that was very comfortable to us. Don could tie all these things together." But it was Hillary who became Jones's most devoted student: "She seemed to be on a quest for transcendence," he recalls. Even after she left Park Ridge to attend Wellesley, Hillary continued to see Jones as a mentor, writing him long, painfully earnest letters filled with discussions of philosophy and talk of her search for ways to express her faith through social action.

One of these letters, written during her sophomore year in college, informed Jones that she had reread *Catcher in the Rye,* a book he had lent her when she was fourteen. Hillary confessed that back then the book had gone over her head. "She probably couldn't understand the alienation," surmises Jones, "because she'd never experienced it." On second reading, however, Hillary had been struck by similarities between the protagonist and her brother Hughie—both had the same "reverence for life." "I hope the world is kinder to him than it was to Holden Caulfield," she added.

"There is a kind of eerie coincidence," Jones observes, that Holden was a saver of children and Hillary dedicated her life to being a child saver, too. Although he has no reason to believe that Salinger's novel influenced Hillary's course in life, he wouldn't find it surprising. "It's a very moral book."[5]

But, surely, this is wrong. *Catcher in the Rye* is a masterful portrait of a depressed adolescent. But Holden Caulfield, who ends the novel in a mental institution, is incapable of saving anyone. He can't even save himself. Holden's fantasy of becoming a "catcher" of children, rescuing them before they fall off "some crazy cliff" that presumably symbolizes the loss of innocence, is sheer escapism. He sentimentalizes childhood and condemns grown-ups as hypocrites because he wants an excuse to avoid adult responsibilities and the difficult choices that come with being part of an imperfect world.

Jones's interpretation of *Catcher* was in tune with the spirit of the 1960s, however. It was a time when disaffected young people felt great confidence in their ability to lead and when the older generation, conscience-stricken over its own shortcomings, was often more than ready to let them.

THREE

The Art of the Impossible

Y mind exploded when I got to Wellesley," Hillary Rodham Clinton would recall three decades later. Perhaps only members of her generation who "went East" to college during the mid-1960s can appreciate what she meant by this statement. Propelled by high SAT scores and credentials as National Merit scholarship finalists, the first wave of the baby boomers headed off to elite campuses armed with supreme self-confidence that the world was theirs for the taking. Often, they were the first in their families to attend college or, like Hillary, the first to attend an elite institution hundreds of miles from home.

Filled with high expectations, the boomers were leaving home during interesting times as defined by the old Chinese curse. The spring semester of Hillary's senior year in high school coincided with Martin Luther King's campaign in Selma, Alabama, and his triumphal march into Montgomery, where he delivered a speech calling the protest a "shining moment in the conscience of man." Just weeks later, in April 1965, came the first march on Washington to protest American involvement in Vietnam. Twenty-five thousand protesters turned out, far more than the organizers expected, to hear Student Nonviolent Coordinating Committee activist Robert Parris, also known as Robert Moses, proclaim that the killing of South Vietnamese peasants was the moral equivalent of the murder of civil rights organizers in Alabama and Mississippi. August brought riots in the Watts district of Los Angeles. Barry McGuire's rendition of a song called "The Eve of Destruction" topped the pop charts.

For idealistic white students, these developments were traumatic. The civil rights movement was the burning moral issue of the decade, and now black activists, turning from the goal of integration to calls for black power, were telling them that they had no place in the struggle. Stranded without a cause, white activists searched for other models of revolution; and impressionable students, just months away from high school pep rallies and family barbecues, were soon being bombarded with the message that their whole way of life was tainted by the sins of hypocrisy, capitalist greed and imperialism.

During this period, the favorite reading matter of Hillary Rodham, that nice middle-class girl from Park Ridge and now a Wellesley freshman, was a magazine called *motive,* subsidized by the United Methodist Church. *motive* featured articles by SDS theoretician Carl Oglesby, who asked rhetorically, "What would be so wrong about a Viet Nam run by Ho Chi Minh, a Cuba by Castro?" According to Oglesby, it was morally wrong to condemn terrorists abroad or rioters in Los Angeles because "their violence is reactive and provoked, and it remains culturally beyond guilt." The editors of *motive* followed the trends, moving from an embrace of terrorism to advocating lesbianism and the rites of wicca until, in 1969, the Methodists finally withdrew their support. Hillary, meanwhile, pored over every issue and saved them in a file, which she still had many years later.

One of the Seven Sisters, Wellesley in 1965 was if anything even more isolated from the events of Selma and Watts than Park Ridge had been. Students were expected to abide by strict curfews, which made allowances for the occasional weekend date ending at midnight. Women changed into skirts for dinner and were subject to room checks, and everyone signed an honor code, known as the "vow," promising to abide by the regulations in the Wellesley student handbook, the so-called Grey Book. Wellesleyans were still expected to aspire to a future as a "well-adjusted housewife," and in Hillary's class of more than 400 there were only 6 blacks. And yet, there were already prophets of change among the faculty. Hillary's classmate Eleanor Acheson would recall that the college had recently recruited "a ton of younger assistant professors and lecturers and instructors [with] a lot of new ideas and some very aggressive advocacy-oriented ways of teaching."

Hillary was not at Wellesley very long before she felt an urgent need to make some gesture expressing her stand on race relations. She invited a black student to attend Sunday services at the Methodist church in town, then wondered if she had done the right thing. "I was testing me as much as I was testing the church," she wrote Don Jones.

This was a rare moment of self-doubt. Hillary emerged almost immediately as a leader and was soon in the forefront of every student battle. She campaigned to eliminate the honors system, having lapel buttons made up at her own expense with the slogan BREAK THE VOW. She led the fight against curfews and fought for fewer required courses and a pass-fail grading option. She protested the absence of minority representation on the faculty and pushed for the admission of more black students. She called for the establishment of a summer Outward Bound program on campus and served on a committee to revise the system for checking out library books.

For all this whirlwind of activity, Hillary was in no way a campus radical. In a different decade she could just as easily have been running the sorority pledge drive. The real radicals on campus tended to be suspicious of her eagerness to remain in the good graces of the administration, and the college newspaper, the *Wellesley News,* occasionally took potshots at her. Complaining about an essay contest that was intended to suggest ways to increase student participation, the paper mockingly suggested that anyone with ideas on "how to perk up school spirit" send them directly to Hillary Rodham.

Hillary was moving away from her parents' politics, though by the standards of the time her ideological journey was tortoiselike. Arriving at Wellesley as a Goldwater conservative, she was elected president of the Young Republicans Club during her freshman year and adopted the politics of the other club members, who were overwhelmingly Nelson Rockefeller liberals. She was excited about John Lindsay's campaign to become mayor of New York, an enthusiasm that led her to confess to Don Jones that she was developing leftist tendencies. By her junior year she had completed her transformation into a Eugene McCarthy Democrat. Even so, she kept some ties to the Republicans. Chosen that summer to take part in a college-sponsored internship program, she was assigned to work as a researcher for the Republican Conference on Capitol Hill, chaired by Melvin Laird, who would serve as Defense Secretary under Nixon. Laird assigned her to the Republican House member from Park Ridge, where she spent most of the summer studying revenue sharing. And in August, she went down to Miami for the Republican Party convention, where she worked for the nomination of Nelson Rockefeller. Eugene McCarthy having dropped out of the race, some liberal students saw Rockefeller as their best chance to nominate an antiwar presidential candidate in 1968.

A junior-year tradition that survived at Wellesley was the election of one woman from each dormitory as a "Vil Junior," a semiofficial adviser to young students. Hillary became chairman of the council of Vil Juniors, and a year later, as a senior, she defeated two other candidates for the presidency of the student government. Her duties included a weekly meeting with Wellesley's president, Ruth Adams, who found her "liberal in her attitudes, but . . . definitely not radical. She was, as a number of her generation were, interested in effecting change, but from within rather than outside the system. They were not a group that wanted to go out and riot and burn things. They wanted to go to law school, get good degrees and change from within."

Indeed, the most notable feature of Hillary's Wellesley career was her tendency to befriend students who were destined for success in later life. For a girl of no particular background from the Midwest, she almost instantly became part of the in group, which included Eleanor "Eldie" Acheson, the granddaughter of Truman's secretary of state, Dean Acheson; Betsy Griffith, later headmistress of the Madeira

School; Susan Roosevelt, later married to Massachusetts governor William Weld; and Jan Piercy, a future banker.

As one classmate later told Hillary biographer Donnie Radcliffe, "If you were somebody from Scarsdale High School, I'm sure you found it easier to be around her than if you were from someplace like Enid, Oklahoma. It could be very intimidating." This comment is interesting, since class and family background, far more than race or Vietnam, was the immediate issue on the Wellesley campus. It is doubtful that the girl from Enid guessed that Hillary Rodham's father cut drapery fabric with his own hands to pay her tuition.

Hillary was clubby, perhaps too much so. An editorial in the *Wellesley News* once accused her of elitism for ignoring democratic procedures and placing her friends on committees and in other positions of responsibility. "The habit of appointing friends and members of the in-group should be halted immediately in order that knowing people in power does not become a prerequisite for officeholding," the *News* warned.

Among her own circle of friends, Hillary wasn't always so serious. She danced to Motown and the Beatles and had a slightly ribald sense of humor. For a couple of years she dated a Harvard man named Geoff Shields, enjoying occasional hiking trips to Vermont or outings to Harvard football games. Still, excursions to parties organized by Shields's Harvard dorm typically ended with the couple in intense discussions of the race issue, often joined by Shields's black roommate.

Hillary and her Wellesley friends were self-confessed "wonks," the kind of students who lingered in the dining hall for hours debating the issues of the day. Hillary was invariably well informed, and others looked to her to fill them in on the facts. Still, she seems to have been driven less by political philosophy or ideology than a fervent desire to be on the "right" side, defined as the side taken by other people she considered moral and good. "She was very much into debating the moral basis for decisions," Shields recalls.

Whether she realized it or not, Hillary was probably far less influenced by New Left politics, or by political theory of any stripe, than by the Third Force or humanist psychology, which had already been thoroughly assimilated by educators, Protestant theologians and ministers, counselors and the media. The leaders of this school were Carl Rogers, an eminent child psychologist who exerted a profound influence on educational psychology, and Abraham Maslow, a theoretician who had for years spread his ideas through a correspondence circle made up of intellectuals from many disciplines.

Freud believed that repression was necessary for the sake of civilization. Humanist psychology taught that human beings were innately good and would progress toward self-actualization as long as they were not thwarted by the malign influence of our "screwed up" culture. Carl Rogers warned that teachers and parents must act not as authority figures but as "facilitators," allowing young people to follow their own paths to knowledge and morality. Maslow, whose books became runaway bestsellers in the mid-1960s, argued that rebellion against authority was a path to becoming a self-actualized personality. In his view, capitalism was bad because it promoted wasteful competition, and hierarchical structures were innately evil because they stifled personal growth. Curiously enough, Maslow, whose politics were vaguely socialist, made an exception for our national government. The New Deal and federal enforcement of civil rights statutes seemed to show that Washington could take the lead in breaking the power of more traditional institutions like the family, organized religion, and state and local power structures, paving the way for the utopian society that Maslow called Eupsychia.

Abbie Hoffman, a student of Maslow's at Brandeis who for a time dated his daughter, was already putting the theory of self-actualization into practice, demonstrating how it was possible to reinvent oneself as a revolutionary without bothering to have an organization or even a program. The Yippies anticipated the Seinfeld generation by being a party about nothing.

Freelance writer Betty Friedan, dissatisfied with Freud's theories about female psychology, also adopted Maslow's notion of self-actualization, writing *The Feminine Mystique,* the book credited with inspiring the modern feminist movement. Ironically, the prophet of personal growth himself hadn't given much thought to how his theories applied to women. Challenged by Friedan to come up with examples of self-actualized women, Maslow could think of only one—Eleanor Roosevelt.

The model of Eleanor Roosevelt, a woman who never had to earn a living and came out of the tradition of society women who strove to do something useful, may not have been terribly relevant for young women in the mid-1960s, but then it was hard to think of many better ones. Hillary and her friends expected to work, but just how they would manage to meld careers and home life was far from clear. Reading an article in *The New York Times* about Ted Sorensen, Hillary was horrified by the statement that Sorensen was so busy with the 1968 presidential campaign that he went home just once in three weeks, and then to pick up a supply of clean shirts. She and classmate Betsy Griffith assured each other that they'd never settle for marriage to a man like that.

Even at Wellesley, the larger issues of racism and the Vietnam War eventually began to crowd out concerns like eliminating the dress code, and by 1968, there was a sense of events spinning out of control. When the news of Martin Luther King's assassination reached the campus, Hillary returned to her room weeping tears of sorrow and rage and threw her book bag with so much force that it slammed into the wall, her roommate at the time would recall. "I can't stand it any more! I can't take any more!" she shouted.

Students saw King's death as not just a national tragedy but a personal assault, and at a meeting in the college chapel, grief and anger over his murder mutated into a sense of roiling frustration. There were calls for a student strike. Economics professor Marshall Goldman suggested that a memorial to Dr. King be held on a weekend, since giving up classes was not exactly a sacrifice, only to be put down by Hillary, who said tartly, "I'll give up my date Saturday night, Mr. Goldman, but I don't think that's the point. Individual consciences are fine, but individual consciences have to be made manifest." In the end, the meeting settled on a two-day teach-in, which took a month to plan and was poorly attended. Most students passed on the symbolic "strike" to attend their regular classes.

★   ★   ★

IN the spring of 1969, with commencement day fast approaching, members of Hillary's class led by Eldie Acheson began to talk of staging a counter-commencement. The administration planned to confer an honorary degree on Massachusetts senator Edward Brooke, a liberal black Republican once supported by Hillary and the Young Republicans Club but now tainted by his allegiance to Richard Nixon. Acting as mediator, Hillary took her friends' concerns to President Adams, who reluctantly agreed to let Hillary, as student government president, say a few words at the regular commencement exercises. Adams extracted a promise from Hillary that she would submit a prepared text of her remarks and stick to it.

Hillary's speech was somewhat of a group effort, stitched together in three days. Came the commencement ceremonies, Senator Brooke began his speech by praising the student movement for stimulating "healthy self-criticism" and emphasized the need for Americans to work together to serve our "acute social problems." But he had the temerity to condemn "coercive protest" and singled out SDS, suggesting that its antiwar actions might be guilty of giving aid and comfort to Ho Chi Minh.

When her turn came to take the podium, Hillary departed from her approved text and addressed Senator Brooke directly, delivering what *The Washington Post* would characterize as a "mild rebuke," though her words struck many older alumnae as extremely rude. Exactly what Hillary said is a matter of debate, since the text of her remarks available today does not reflect some things that people who were present recall hearing. But the gist of her comments was that Brooke had delivered a defense of Nixon larded with boilerplate "upward and onward" rhetoric that was almost an insult, given the critical state of the country: "We've had assassinations. We're in the midst of a war everybody was confused about and didn't like."

Brooke, it seems, thought of the protest movement in political terms. What he just didn't grasp was that for Hillary's generation, protest was part of the search for self-definition. As Hillary put it,

"Every protest, every dissent, whether it's an individual's academic paper, [or] Founder's parking lot demonstration, is unabashedly an attempt to forge an identity in this particular age. That attempt at forging for many of us over the past four years has meant coming to terms with our humanness."

Hillary went on, scolding the senator by name for expressing his "empathy" with the students' grievances:

> Part of the problem with empathy with professed goals is that empathy doesn't do anything. We've had lots of empathy; we've had lots of sympathy, but we feel that too long our leaders have used politics as the art of the possible. And the challenge now is to practice making what appears to be impossible, possible. What does it mean to hear that 13.3 per cent of this country is below the poverty line? That's a percentage. We're not interested in social reconstruction; it's human reconstruction. . . . We are all of us exploring a world that no one understands and attempting to create within that uncertainty. But there are some things we feel, feelings that our prevailing acquisitive and competitive corporate way of life, including tragically the universities, is not the way of life for us. We're searching for more immediate, ecstatic, and penetrating modes of living. . . .

The immediate reaction to Hillary's impassioned plea was divided. Her fellow students and younger graduates gave her a standing ovation. Many parents and older alumnae were offended, or just perplexed. Hillary herself was a little nervous, fearing that she had gone too far in violating her promise to President Adams. That evening she bolstered her spirits by indulging in another act of symbolic rebellion, heading off for a solitary dip in Lake Waban, a campus landmark where swimming was strictly forbidden. While she was in the water, a campus guard came by and confiscated her street clothes and glasses, leaving the nearsighted Hillary in her dripping-wet bathing suit to feel her way back to the security of her dorm.

In retrospect, Hillary's speech had been an honest attempt to con-

vey the turmoil that so many of her fellow students felt. Her words
vibrated with an idealism that was only a few degrees north of pet-
ulance. The overheated emotionalism, combined with an almost
childish reproof of an older generation who had failed to hand over
a perfect world, probably should have provoked some second
thoughts among educators who were encouraging students to develop
"identities" based on the flux of current events. Instead, it was taken
as evidence of the clear-sightedness of the younger generation. Hillary
was one of three college commencement speakers featured in an article in *Life* magazine as examples of the best and the brightest of the
student activists. (Another of the three was Ira Magaziner, who had
led a movement that replaced Brown University's liberal arts curriculum with a smorgasbord of electives.)

After seeing Hillary's picture in *Life,* a young attorney named Peter
Edelman invited her to attend a summer conference sponsored by the
League of Women Voters that brought together a group of "distin-
guished young and old leaders" to discuss future trends in politics.
Hillary made useful contacts at the conference, including Vernon Jor-
dan, an attorney a decade her senior who was running the NAACP's
Voter Education Project.

After the conference ended, Hillary was off on a student ramble
that took her to Alaska. She worked briefly in a fish-processing plant,
but was fired when she confronted the manager over a batch of fish
that appeared to be spoiling. The desire for "more immediate, ecstatic and penetrating modes of living" that she expressed on commencement day had led her to consider other adventures—travel in
India, or perhaps a job with radical organizer Saul Alinsky, whom
she had interviewed for her poli-sci thesis. In the autumn, however,
she was off to law school, as were a startling number of her Wellesley friends.

The legal profession was just beginning to open its doors to females,
though in some cases grudgingly; Hillary was discouraged from going
to Harvard Law after an encounter with a distinguished gray-haired
professor who told her snappishly, "We don't need any more
women." The remark persuaded her to head for Yale instead, where

she would be one of thirty women in a freshman class of 140—and, of course, the only one who had already been anointed by *Life* magazine as a voice of her generation.

The revolution had a different look at Yale, gritty and frayed around the edges. In May 1969, while Hillary Rodham was negotiating the program for the commencement ceremony at Wellesley, the mutilated body of one Alex Rackley was found floating in the Coginchaug River, twenty-five miles north of New Haven. Rackley, a member of the Black Panther Party, had been shot twice, stabbed with an ice pick, burned with cigarettes and scalding water and repeatedly clubbed. Hours after his corpse was discovered by fishermen, the New Haven police learned Rackley had been put on "trial" by the local Black Panther organization, charged with being a police informant. Tapes of the interrogation were seized and eight Panthers arrested. Later, a new witness surfaced who identified Bobby Seale, the Party's co-founder, who happened to be in New Haven at the time, as the one who gave the order: "Do away with him."

The reaction on the Yale campus was outrage—directed not toward the killers of Rackley but toward the police. There was some legitimate doubt about the testimony of George Sams, the key witness against Bobby Seale. More to the point, the Black Panthers had been romanticized by the New Left as urban Robin Hoods. Despite acting like thugs, and at times boasting of their thuggishness, they were given the benefit of the doubt and more, and the Panthers' argument that they were innocent victims of a police conspiracy was widely accepted.

Bobby Seale was indicted in August 1969, and the attention of student activists across the country focused on New Haven as the prosecution built its case by winning guilty pleas from two of the accused even while wrangling with the state of California over Seale's extradition. Literature tables and protesters' tents sprouted up as the Yale campus became a magnet for freelance radicals like Tom Doustou, a nonstudent who had formerly been the leader of a populist group whose symbol was the Confederate flag but who now emerged as the organizer of the New Haven Panther Defense Committee.

The "real world," as students like to call it, had intruded on campus life, and the effects were jarring. Representatives of the Black Panthers hung around the law school quad selling subscriptions to the party's newspaper, and some students weren't sure if they were being shaken down or not. Strange characters roamed the halls after hours. Professors gave up working late in their offices, and took important papers home for safekeeping. Professor Robert Bork, a strict constructionist on the Constitution, had long tried to challenge the assumptions of his mostly liberal students, but in 1969 all that changed. Students were so tightly wound that they reacted with tears and curses to his attempts to challenge them intellectually.

Yale had long emphasized the social context of the law, and it tended to attract students who viewed a law degree as a credential for a career of social activism. This remains true even today. "You should read the applications we get," Professor Emeritus Ron Brown said in 1992. "They all want to go out and save the world."[1] Such students tended to be more interested in broad social theory than in absorbing volumes of dry case law. Indeed, there was a school of thought that the only reason to study case law at all was to deconstruct it in order to reveal evidence of underlying classist assumptions. Dissatisfied with the more traditional *Yale Law Journal,* three senior students, James F. Blumstein, Jack Petranker and Stanley Herr, set out to found a radical alternative, the *Yale Review of Law and Social Action.* Hillary joined the staff, performing duties that consisted mainly of detail work—copy-editing and production chores.

For the *Yale Review'*s inaugural edition, Blumstein and another student, James Phelan, wrote an article advocating that radicals pick a state and move there "for the purpose of affecting [*sic*] the peaceful political takeover of that state through the electoral process. . . . Experimentation with drugs, sex, individual lifestyles or radical rhetoric and action within the larger society is an insufficient alternative. Total experimentation is necessary." Hillary read the article in typescript and didn't much like it. "She thought it was not practical, too pie in the sky," recalls Blumstein.

But of course, this was just the point. It was a bit of a wacky idea, filled with "a lot of extreme rhetoric which was characteristic of Yale Law School at the time," Blumstein acknowledges. Sometimes, however, there is truth in hyperbole. Two years later, a freelance writer came across the article and used it as the basis for a piece that appeared in *Playboy* called "Taking Over Vermont," crediting Blumstein and Phelan with inspiring the hippie migration to the Green Mountain State. Native Vermonters were none too pleased. Blumstein, whose politics are very different now, observes ruefully that his flight of fancy is still mentioned periodically by writers seeking to explain the transformation of Vermont into a left-wing enclave.

Blumstein dated Hillary a few times—"nothing serious"—and still holds her in high regard personally. Politically, however, he says that "Hillary is stuck in a bit of a time warp. She has not learned as I think I have the power of the market."

Hillary also volunteered for a project organized by one of her professors, Thomas Emerson, affectionately known as "Tommie the commie." Students were assigned to monitor civil rights abuses by the prosecutors of the New Haven Panthers, which were assumed to be plentiful. Ever efficient, Hillary was allotted the task of setting up a schedule so that there would always be a student present in court. The reports were being compiled for the ACLU, but the students worked closely with the Panthers' lead attorney, Charles Garry, sharing their observations, which were delivered to Panther headquarters on a regular basis. The law students noted with indignation that their visits were being observed by a surveillance team in a parked car across the street, presumably FBI agents.

Garry's defense strategy anticipated the overtried cases that have become common today. He raised a multitude of issues about the allegedly unfair treatment of his clients, which ranged from the trivial to the bizarre. The court's refusal to grant bail became a major issue, even though bail is rarely granted in capital cases. Garry also complained bitterly when he was refused permission to have Rackley's body exhumed. The most explosive issue was the judge's decision to issue a contempt citation against Black Panther David Hilliard for his part in a courtroom scuffle that involved a bailiff and another spectator, the French avant-garde playwright Jean Genet. Released on bond, Hilliard appeared at a campus rally where he expressed the view that "there ain't nothing wrong with taking the life of a motherfuck-ing pig." This was too much for some students, who began to boo.

"Boo! Boo! Boo! Boo Ho Chi Minh! Boo the Koreans! Boo the Latin Americans! Boo the Africans! Boo all the suffering blacks in this country!" Hilliard shouted back. "You're a goddam fool if you think I'm going to stand up here and let a bunch of so-called pacifists, you violent motherfuckers, boo me without getting violent with you!" Hilliard then proceeded to stomp a disoriented foreign exchange stu-dent who had approached the podium hoping for a chance to speak.

Afforded a glimpse of the thuggishness underneath the Panthers' veneer of revolutionary charisma, many students simply caved in. "There was a feeling of relief, or release," wrote John Taft, a student at the time. "In their sense of guilt, the whites regarded [Hilliard's threats] as well-deserved chastisement."[2]

Nor was there any leadership to speak of from the older generation. President Kingman Brewster of Yale expressed doubt that it was pos-sible for the Panthers to receive a fair trial in an American courtroom. William Sloane Coffin, Yale's chaplain and an activist much admired by left-leaning students, opined that the "white oppressors" had no right to judge the Panthers' actions: "It might be legally right but morally wrong for this trial to go forward." As a general rule it seemed that liberal ministers of the Gospel, the very people a good Methodist like Hillary Rodham would have looked to for guidance, were the first to lose their bearings. One Protestant chaplain began an affair with a member of the Panther Defense Committee and eloped to Australia with her.

The argument that all protest was legitimate was playing out to its absurdist conclusion. If the rebel was morally superior to the nonrebel, then the loudest, most violent rebels were the most superior of all. With the legal preliminaries winding up, the trial of the New Haven Nine was scheduled to begin in early May, and student protesters from around the country were expected to converge on Yale. Tom Doustou predicted that half a million militants would show up; more realistic prognosticators thought the number would be about 50,000. The Progressive Labor faction of SDS and other radical groups on campus were calling for a student strike. "Shut it down or burn it down!" and "If Bobby dies, Yale fries!" were the slogans of the day.

On April 23, the faculty convened in an atmosphere that one professor would describe as "1789." Student protesters had massed outside the hall demanding to be heard, but the spokesman they offered, a young black activist named Kurt Schmoke, surprised the assembled professors by pleading, "There are a great many students on campus who are confused and many who are frightened. . . . You are our teachers. You are the people we respect. We call on you for guidance and moral leadership." In response, and urged on by President Brewster, the faculty voted to suspend normal operations and welcome outside protesters to the campus. This was either a brilliant strategy for avoiding violence or a capitulation, depending on one's point of view. A compromise resolution offered by liberal psychology professor Dr. Kenneth Keniston, which would have suspended operations only temporarily, not indefinitely, was rejected as inadequate. Many professors left the hall with a sinking feeling that they had betrayed students by voting away the principle of academic freedom. Alexander Bickel, the lawyer who later argued the Pentagon Papers case before the Supreme Court, confessed to feeling "ashamed."

Undoubtedly most students did not want to close the university, but to publicly suggest going on with classes—"business as usual"—was to risk being branded insensitive to the sufferings of the Panthers, black Americans in general and dying Vietnamese peasants. Meetings and rallies, which were many, tended to consist of long, rambling monologues and overheated but pointless shouting matches. Hillary emerged as one of the few students with enough poise to keep order. Well known around campus because of the *Life* article, she presided over a tumultuous meeting on the strike vote. Seated on the edge of the lecture hall stage, she deftly kept the proceedings from degenerating into chaos.

"Hillary was not a radical," according to Michael Medved, now a conservative film critic but then one of the leaders of the antiwar

Moratorium Committee and a friend of Hillary's. She *was* passionately antiwar, antibusiness and suspicious of the police and was beginning to think about women's and environmental issues. What set her apart from the true radicals was that she had little faith in direct action. She wanted to keep lines of communication to the administration open and work through existing institutions.

Of course, in the context of these extraordinary events, the usual political labels had become skewed. Conservatives, moderates and the apolitical, who just wanted to keep on studying and let the courts take care of Alex Rackley's murderers, were for the most part too intimidated to speak up. On the other hand, Hillary and other "moderate" leaders spent a great deal of energy in efforts to conciliate the leftists. One sometime opponent of Hillary's was Kris Olson, later Kris Olson Rogers, a former Wellesley classmate who was also part of Tommie Emerson's ACLU monitoring project.

Olson was "far more to the left" than Hillary. "I remember trying to push her in that direction," Olson notes. "She was the source of some of my frustrations."[3] But the differences that separated them at the time would come to seem minor over the years. Olson later married Jeff Rogers, a nephew of Nixon's secretary of state. As a deputy U.S. attorney in Oregon, she was ordered to obtain an indictment of a former Black Panther on a weapons charge but led the grand jury to a no vote instead, an action that led U.S. Attorney Charles Turner to complain that she was insubordinate. She later lost her job after chairing a committee that advocated decriminalizing street prostitution. Under the Clinton administration, however, Rogers took over Turner's former post, becoming the top federal law enforcement officer in Oregon. She celebrated her swearing-in by having the Gay Men's Chorus sing a Grateful Dead song.

During the week after the faculty resolution, several unexplained fires broke out on and near the campus. The worst of these was early in the morning of April 27, when a blaze occurred in the International Law Library. The fire was a sobering event. Rhetoric was one thing, but as Kris Rogers put it, burning books was "where we drew the line.'"⁴ Several witnesses recall seeing Hillary outside the library later that day, doing her best to calm nervous students. Later, activists met to organize a campus patrol to guard against further incidents.

As May Day grew closer, activists from across the country showed up in New Haven. Saul Alinsky recruited volunteers to go out into the community in an effort to disarm the townies, who were almost universally hostile. Jon Froines, Dave Dellinger and Tom Hayden, all former co-defendants of Bobby Seale in the Chicago Seven trial, were on hand. Many undergraduates had gone home for the duration. Those who stayed were frightened—some of terrorist acts by Panther sympathizers; more, it seemed, of violence provoked by the "pigs." Some students volunteered as marshals to keep order among dem-onstrators; others organized mass feedings of granola and steamed rice drenched in soy sauce.

But the event was an anticlimax. Only about 15,000 outsiders showed up, and many of them appeared to have come in the spirit of an excursion. There were scattered incidents of vandalism and vi-olence, including a fire in the jury box of the moot court hall at the law school, and some demonstrators were tear-gassed. But compared with what the campus had been expecting, the day passed without major incidents.

In a day of many excited speeches, perhaps the most memorable was made by Abbie Hoffman's partner in Yippiedom, Jerry Rubin:

> We know what work is—a dirty four letter word. Things should be free! . . . My brother is a Chinese peasant and my enemy is Rich-ard Nixon. . . . We're destroyed in the first eight years of our life, and what we're trying to do is regain those years. . . . Fuck ration-ality. We're irrational and irresponsible. I haven't taken a bath in six months. . . . If Bobby Seale had been in that car with Mary Jo, he'd be hanging from a tree. . . . Arresting us for smoking dope is like arresting Jews for eating matzohs. . . . We're going to take acid with our kids. We're going to turn the Pentagon into an experi-mental LSD farm and the White House into a crash pad—and paint it black!

Yale is a place where they separate the rich from the poor. . . .
Under that definition Yale is criminal. . . . What kind of sickness is
that when you feel good because you get better grades than someone else? . . . Number one on the program is to kill your parents,
who got us into this mess in the first place. . . .

This, of course, was the kind of wild talk that appalled earnest
students like Hillary Rodham. But Rubin understood better than they
did that a lot of the energy in the movement amounted to a mass
temper tantrum, which he was all too pleased to personify.

Kenneth Keniston, one of the more astute observers of the student
movement, would write an essay called "You Have to Grow Up in
Scarsdale," explaining why the most militant rebels seemed to come
from upper-middle-class bedroom communities. Affluence had allowed young people from well-to-do families to enjoy a period of
postadolescent self-discovery unknown to earlier generations. Keniston predicted that, for most, the passion for political change would
soon burn itself out. They would turn to the pursuit of "meaningful
human relationships" and psychic self-improvement. Some few, however, would make it their mission in life to bring the extraordinary
freedom they enjoyed to the rest of the world—"extending to others
the 'rights' they have always taken for granted."[5]

For Hillary, as for so many others at Yale, the spring of 1970 was
a time of emotional upheaval. In a 1992 speech to law school
alumni, she would tell of being consoled by a professor who observed that "bad things happen in every society." His suggestion
that the apocalypse might not be imminent after all seemed like a
revelation to her.

Nevertheless, in the midst of the excitement, Hillary had met a
woman who would become her most important mentor. Marian
Wright Edelman, the wife of Peter Edelman, had come to Yale to
speak on the subject of "Children and the Law." Hillary had met
Edelman and her husband at the League of Women Voters conference
the previous summer, and she was bowled over by her inspirational
speech.

Born in South Carolina, Edelman was the daughter of a Baptist minister who had graduated from Union Theological Seminary. Her mother reared five children, worked and still found time to be a church organist, a pianist and a community leader. Marian, named after Marian Anderson, attended Spelman College in Atlanta, where she was a student of historian Howard Zinn. She won fellowships to study in Paris and the Soviet Union, took part in lunch counter sit-ins in Atlanta and went on to Yale Law, where she lived for a year with the family of the Reverend William Sloane Coffin. Graduating in 1964, she opened an office of the NAACP Legal Defense and Education Fund in Jackson, Mississippi, and helped to organize the state's first Head Start programs.

A dedicated activist with impeccable Movement credentials, Edelman was also very much a member of the Washington establishment. The winner of numberless awards and fellowships, she had been the subject of admiring profiles in *Mademoiselle* and *Ebony* while still in her twenties. When she married Peter Edelman, who had been Robert Kennedy's legislative assistant, the wedding was held at Adam Walinsky's house, with William Sloane Coffin performing the ceremony. The couple then departed on a trip around the world, courtesy of the Ford Foundation, which was awarding grants to former aides of Bobby Kennedy, to help ease their transition into other careers.

The Edelmans had been young firebrands in the civil rights movement, but they came to maturity during the years of Lyndon Johnson's Great Society. Despite the ferment on the campuses, there was nothing populist or radical about Johnson's program; it was strictly an elitist crusade. More than fourteen presidential commissions made up of public policy experts labored in secret to produce Great Society legislation which promised to wipe out poverty in a decade. As historian John A. Andrew III notes, "The experts were to create the programs, the politicians were to pass them, the bureaucrats were to administer them. The people were to be the beneficiaries of all this missionary zeal."[6]

Great Society legislation did create a foundation of security for the very poorest families and the elderly, but the experts became victims of their own arrogance. They had sold their policies to the voters on the promise that they could wipe out poverty within a decade at no appreciable cost to the middle class. This, of course, was a fantasy. Poorly thought out programs sucked up taxpayer money, and in some cases, especially in the inner cities, they created more problems than they solved. When the public began to lose faith in the experiment, some of the Great Society technocrats concluded that the voters just didn't want to pay for programs anymore because they were selfish and racist, or just plain misled by right-wing politicians. They settled down to become what P. J. O'Rourke calls "the perpetually indignant Left."

While Peter Edelman went on to hold a series of public policy–related jobs in academia and government, Marian was taken up by the Carnegie Corporation. "The deaths of Whitney Young and Martin Luther King left a void in black leadership in America, and focused attention, as perhaps never before, on the rising young leaders who will take over the movement in the coming decade," noted an internal Carnegie Corporation memo in April 1971. The corporation identified Marian Edelman as one of those rising young leaders and supported her work to the tune of $3.2 million between 1970 and 1982.

Edelman saw that Americans were not enthusiastic about giving welfare to adults, especially African-American adults, but found it hard to resist appeals on behalf of children. In a 1971 grant application to the Carnegie Corporation, she observed that "children are the un-recognized, neglected, and mistreated minority in America, much as the poor were prior to their 'discovery' by Michael Harrington. . . . In the 1970's children's rights may well provide the most promising vehicle for addressing the broader problems of poverty and race in this country."[7]

Edelman in effect was arguing that the social welfare state is a gift we owe to our children. The only way to make sure that there are no poor children, at least without removing them from their homes, is to transfer wealth to their parents. This is by no means a new idea.

The observation that some children are born rich while others, through no fault of their own, are born poor has led generations of sensitive young people to espouse socialism. The reason they don't remain socialists is that they come to see that the system doesn't necessarily lead to a family-friendly utopia. Edelman admired the family-support programs of France, Sweden and other Western European nations—though, as it happens, these alleged paradises for children are beset by falling birthrates and find it increasingly difficult to persuade couples to have children at all.

Edelman has usually argued for her agenda on moral grounds. Her writings have a strong element of moral uplift, but she can be quick to demonize anyone who suggests that there might be better alternatives for helping children as "evil" or "spiritually impoverished" or responsible for "threatening the lives of children." Even near allies who dared to differ with Edelman on the details of proposed legislation have found themselves accused of plotting to "rob millions of children." Indeed, her basic argument is that Americans would long ago have voted in the welfare state if it weren't for racism. As recently as 1992, arguing that the call for welfare reform stems from the misconception that the poor are all inner-city blacks, she warned, "We need to stop punishing children because we don't like their parents."[8]

Another of Marian Edelman's favorite themes is that the United States is uniquely callous in its treatment of children. As she put it in a 1975 interview with *Psychology Today:*

> A friend told me a story that points out the disparity between how we treat children and how other countries do. A few years ago, she was in Iran during the earthquake. The Shah sent out a group to gather up the children who had lost their families and their homes. They scoured the villages and came back empty-handed. The homeless children had already been absorbed into the broader community. I doubt this would have happened in America.[9]

Needless to say, it is precisely because we do have child welfare agencies in the United States that children orphaned by natural disasters aren't just "absorbed" into any family willing to have them.

Edelman's fervent commitment to children certainly offered a bracing alternative to the muddy politics of Yale, where it could be difficult to sustain enthusiasm for the cause of Bobby Seale and his co-defendants. Impressed by Edelman's speech, Hillary immediately introduced herself and offered to come to work that summer for the Washington Research Project, the forerunner of the Children's Defense Fund. Edelman was discouraging; she had no money in her budget for another summer intern. Hillary, however, was already skilled at raising funds, and she managed to promote a grant from the Law Students Civil Rights Research Council to cover her expenses. In Washington, she was assigned to work with the staff of Senator Walter Mondale researching the living conditions of migrant farm laborers and their children.

By the time she returned to Yale in the fall, Hillary had found a new direction for her studies, and she signed up for course work at the Yale Child Study Center. As one of a small group of students whose program combined legal studies with child psychology, she assisted the late Dr. Sally Provence, who was observing infants at a local hospital and counseling new mothers on how to communicate better with their babies. She also became a research assistant to Drs. Joseph Goldstein and Albert Solnit, who were about to publish *Beyond the Best Interests of the Child,* an influential if somewhat controversial book that attempted to set up guidelines for child custody disputes.

Dr. Provence suggested that Hillary learn more about custody issues by working with Penn Rhodeen, a young attorney who was handling cases in that area for the New Haven Legal Services Association. At that time, a typical case for Rhodeen involved a very young child in a temporary placement that had lasted for months, even for years. The state welfare authorities would come up with a new custody plan, never considering the traumatic effect of removing the child from the only caregiver he or she had ever known. "That's the child's whole universe, after two years," Rhodeen argues. "Even if on paper you can come up with a superior placement, what you've done is devastate the child." Hillary and Rhodeen often talked about the unfairness of a system that placed such a low priority on family law. If a shipment of fresh fruit were tied up in a civil dispute, the courts would practically convene on Saturday to make sure the issue was settled before the fruit spoiled, but a child could suffer for years in a bad placement and *that* didn't count as an emergency. Hillary's interest in these matters, Rhodeen recalls, was "very deep and thoughtful. It's for real."

Although Hillary cared passionately about children's issues, she didn't see herself becoming a legal aid lawyer. She was already being courted by influential Washington law firms. "I think it was kind of a given that she would go into a big firm," says Rhodeen. "She always anticipated doing something mainstream, powerful legal things, and then she would pursue these interests, perhaps at a board level. . . . She was a big star at Yale and big stars at Yale have lots of options."[10]

ONE day a few months into the new academic year, Hillary was walking through the student lounge at the law school on her way to the Coke machine when she heard a student with a slight Southern drawl, saying, "And not only that, we grow the biggest watermelons in the world. . . ."

"He was good looking. And I thought—you know, I kind of filed that thought away."

First-year law student William Jefferson Clinton wasn't seen much around campus that fall. He spent a lot of time working on the Senate campaign of Joe Duffey, an antiwar liberal with blue-collar roots. When Clinton was on campus, he could often be found in the lounge holding forth on the charms of his home state of Arkansas and the brilliant political future that awaited him there. To the average Yalie, Arkansas meant only one thing: it was the place where in 1957 black schoolgirls in crisp shirtwaist dresses, their skirts held stiff by layers of crinolines, had been forced to walk a gauntlet of screaming white racists to reach the doors of Little Rock's Central High School. Bill Clinton had come to Yale from Oxford, where he was a Rhodes scholar, but he had the sense to see that no one in New Haven would be impressed by this. On the other hand, there was entertainment value in stories about his home state, the land of the Hope Watermelon Festival, the Toad Suck Daze Fair and the McCrory Mosquitofest and Shriners' Parade.

Hillary knew Robert Reich from his undergraduate days at Dartmouth, where she had attended a conference in which Reich took part. Reich, who had been at Oxford with Bill Clinton, introduced him to Hillary one day in the cafeteria line, but somehow they failed to strike up a conversation. For several weeks afterward they were very much aware of each other, but neither made the first move. Then came a day when they both found themselves in the law school library. Jeff Glekel, an editor of the *Yale Law Journal,* was earnestly trying to recruit Bill Clinton to the staff, but Bill was paying more attention to Hillary, who was studying at a table at the other end of the long, narrow room.

Eventually, she got up and walked the length of the library to where Glekel and Clinton were sitting. "Look, if you're going to keep staring at me and I'm going to keep staring back, we might as well introduce ourselves," she told Clinton. He would repeat this story often over the years. While some men found Hillary's direct manner a little off-putting, Bill Clinton was impressed, even flattered.

On their first date, Bill took Hillary to an art exhibit. The museum happened to be closed, but Bill sought out the janitor and talked him into letting them inside after they helped him pick up some trash. Bill had the gift of improvisation and could strike up a conversation with anyone, qualities missing in Hillary, whose direct, slightly cutting manner effectively masked her innate shyness.

Intent on making the most of her education, Hillary had avoided getting seriously involved with the young men she dated. She broke her own rule for Bill Clinton. Whatever else can be said about the Rodham-Clinton relationship, and a lot can be, there seems no real doubt that on Hillary's part, it was love at first sight. And little wonder. Tall and stocky, Clinton, when not working on a campaign, hid his boyish features behind a beard, which gave him the look of an oversize teddy bear. And, of course, he was ambitious, almost as much so as Hillary.

Young women who were stars wanted to marry men who were stars, but the male stars weren't necessarily after the women with the highest grade-point averages. Aside from the obvious, up-and-coming young men tend to want someone who will take care of them, and Clinton needed a fair amount of care. His culinary skills were limited to slapping together peanut butter sandwiches. Clinton had trouble getting places on time and could be flummoxed by simple errands. He missed a lot of classes doing political work for Duffey and stayed up half the night reading books that had nothing to do with his schoolwork—Schopenhauer, a Thomas Wolfe novel, even murder mysteries—then raced around at the last minute looking for someone who had class notes. Raised by his mother and grandmother, both feisty personalities, Clinton was attracted to strong women, but unlike some men who don't mind being told what to do, he was never boring. Clinton had that slow drawl and the Southerner's sense that sometimes the good things in life can come easily—a revelation to Hillary, who had internalized the Rodham belief in hard work and struggle.

In contrast to Hillary, who hadn't been seriously involved before, Clinton had had many girlfriends. "Bill Clinton is not just the first president my age, but the first one that people I know have slept with," says a woman who attended Yale during the early 1970s. Some of his affairs were casual, but not all of them, and he was not shy about declaring himself in love. Cliff Jackson, an Arkansan who was at Oxford with Clinton and later became his avowed enemy, would recall Clinton introducing a young Englishwoman as his future wife. Clinton went all out to impress Hillary, whose excellent Washington connections and leadership during the strike placed her well above him in the social pecking order, even warning his roommates to raise the level of conversation when she was around.

Intellectual sparring played a big role in the relationship from the beginning. Clinton loved to spin stories about his home state and when he got carried away Hillary would chide him, saying, "Oh, Bill, come off it!" When the couple argued about politics, their exchanges could be heated, and Hillary, who could steamroller so many opponents with ease, found her new boyfriend a worthy opponent. Hillary had a razorlike mind and could process a lot of information and extract the essentials. But she had trouble with abstract subjects like advanced math and economics, and her thinking was compartmentalized. Ellen Brantley, who attended Wellesley with Hillary and later knew both her and Bill in Little Rock, put it very well when she observed, "She is clearly highly intelligent, and has succeeded, in part, through the practical application of her intelligence. She is very articulate, very good at *communicating* her intelligence. I don't think she is smarter than Bill though some people will say that. If they were in the same class, she would attend all the classes, read all the assignments, outline her notes, study hard for the exam. Bill would stop by some of the classes, read some of the assignments while also reading other, related things, and then write an exam that brought in some ideas that had been introduced in class, some outside—linking them in an original way." Brantley diplomatically adds, "They'd both get an A." But as the description makes clear, Bill was the original thinker, as Hillary was not.[11]

Hillary was so focused that she had a tendency to neglect her private life. David Brock, in *The Seduction of Hillary Rodham Clinton,* quotes a former Yale Law student describing her as belonging to the "look like shit school of feminism." More accurately, she harked back to the older bluestocking tradition, familiar on certain New England campuses. Her long hair had that studiously neglected look, and her favorite accessory was the black armband, which she wore for the students shot at Kent State as well as after the assassination of Martin Luther King, Jr. Seriously nearsighted, she peered out at the world from behind thick, oversize eyeglasses. In a photo taken at a League of Women Voters banquet in Washington, she looks washed out and very small, like the dormouse from *Alice in Wonderland.* This was probably a very effective presentation. It saved her from having to fend off the approaches of young men she wasn't interested in anyway and increased her chances of being taken seriously. Michael Medved has said that Hillary was not considered "date bait" at Yale—but, as the phrase suggests, women who were "date bait" weren't necessarily seen as smart. Hillary was "more the buddy type, fun to be around and interesting to talk to, perhaps even a bit of a mother hen." Penn Rhodeen had a similar recollection: "She was a good sport."

Dating Bill Clinton brought out the fun-loving side of Hillary. He was living in a four-bedroom beach house on Long Island Sound with three roommates. The house attracted lots of overnight guests, and every weekend was a party. Rhodeen recalls seeing Hillary bopping around New Haven in Clinton's purple Gremlin, wearing an embroidered leather Afghani coat with a sheepskin lining, the height of hippie chic, and Gloria Steinem–style aviator glasses. "She was in love with her fellow," he notes, and when Rhodeen acquired a new suit, she questioned him closely about it, thinking that she might pick up one like it for Bill.

Hillary was generally well liked by people who worked closely with her on projects. Not so Clinton. "[Some] people have always hated Bill," says Nancy Bekavac, a law student who was friendly with both Bill and Hillary. "They hated him at Yale. I've just never really understood it. Maybe it was because he was doing grown-up things, and they weren't." Bekavac explains that most students she knew weren't intellectually stimulated by their classes and were impatient to get out into the world, whereas Clinton was already working on the Duffey campaign and making important contacts. A different explanation, suggested by other Yalies, is that Clinton was too ready to brag about the brilliant political future that awaited him in Arkansas, bad form on a campus where it was assumed that everyone had the potential to become a leader.

During her second year in law school, Hillary was promoted to the editorial board of the *Yale Review* and was serving in that capacity when the *Review* published its fall 1970 issue, which included a long piece in defense of Black Panther Lonnie McLucas, one of the New Haven Nine. The article was illustrated with drawings depicting policemen as pigs. One pig in uniform is labeled "fascist." Another has been decapitated; this drawing was captioned "Seize the Time," a Panther slogan.

When summer came, Hillary went off to Oakland, California, to do an internship with Robert Treuhaft, who, like Charles Garry, defended the Panthers in some of their many brushes with the law. Treuhaft, long known as Oakland's Red Lawyer, is a soft-spoken, courtly man, and his wife, the late Jessica Mitford, was a great wit and the author of such muckraking classics as *The American Way of Death*. The couple had also long been stalwarts of the Bay Area Communist Party. Treuhaft had defended Harry Bridges, the Australian head of the longshoremen's union, enabling him to avoid deportation even though, as is now thoroughly documented, he was a member of the Central Committee of the Communist Party USA.

Treuhaft and Mitford left the Party in 1958, but only because their chapter had lost so many members that it was "ineffectual." Their views remained fixed, and they protected their son, Ben Treuhaft, from meeting people they considered "right-wingers." Today, Ben devotes himself to raising money from American liberals for pianos that are smuggled into Cuba. Analyzing the Cuban situation in an interview with the London *Daily Telegraph,* he said, "They say this is a dictatorship, but it's so nice here. I mean, they lock a few people up in jail but like 600, out of a population of eleven million. . . . Fidel is not Papa Doc. He's a world famous diplomat, and he's famous for loving children."[12]

According to Drucilla Ramey, a 1972 graduate of Yale Law School, Robert Treuhaft began hiring female interns long before most firms would give them interviews, and Yale women tended to pass the word along. Ramey says that Hillary probably learned about the position from her predecessor, just as Hillary recommended it to her. It is possible that Hillary worked on cases involving the Panthers during her time with Treuhaft, but most of her work would have been on routine matters involving indigent defendants. Treuhaft himself says that Hillary was closer to his late wife, who visited the Clintons in Little Rock in 1980, but he remembers Hillary only as a rather drab young woman with stringy, unwashed hair—"not that I was in any way put off by that," he adds.

Jessica Mitford got a great deal of mileage out of portraying her husband and herself as passionate idealists whose devotion to Stalinism was just a lovable eccentricity. Occasionally, however, she let the mask slip a bit, as when she explained in her memoir, *A Fine Old Conflict,* that she and Bob didn't join their friends in quitting the Party after the 1956 Hungarian uprising because the whole thing was instigated by the CIA. In her version of this episode, the Hungarian Freedom Fighters who stood up against Soviet tanks were "grasping neo-Fascist types."

While it is easy to see why a young, idealistic law student might have gone to work for Treuhaft in 1972, one would also hope that she would eventually come to see that communism was not just a colorful variant of leftish idealism. Unfortunately, Hillary has never said or written a single word to indicate that she has. Her second encounter with Mitford, in 1980, would involve Mitford's attempt to influence Bill Clinton, by then governor of Arkansas, to abandon his support for the death penalty. Mitford had taken up the cause of one James Dean Walker, convicted of shooting a North Little Rock policeman after a barroom brawl, who walked away from a prison furlough in 1975. For reasons never quite clear, Mitford argued that if Walker was extradited from California, he had a good chance of becoming one of a long line of "mysterious deaths" in the brutal Arkansas prison system. Mitford was getting nowhere with her pleas to California governor Jerry Brown when she received a long letter from Hillary, assuring her that Arkansas prisons were improving and that she hoped the stigma of unconstitutionality (the death penalty) would be lifted by 1981. Encouraged, Mitford flew to Little Rock only to get short shrift from Bill Clinton, who, realistically, could hardly be expected to let a convicted cop killer continue to live as a free man in Lake Tahoe. The visit ended on a sour note, with Mitford depicting Clinton in a magazine article as an "ultra liberal," too preoccupied with his own ambitions to care about prison reform.

WITH her ability to sift through complex material and focus on a few essential points, Hillary had the makings of an excellent litigator. Penn Rhodeen attended a moot court session where Hillary argued persuasively and was declared the victor by the judge, who happened to be Supreme Court Justice Potter Stewart. (In another instance, in 1972, she teamed with Bill Clinton and lost.) Her attention, however, was increasingly focused elsewhere. Hillary had moved in with Bill Clinton at the beginning of her third year, sharing a cozy one-bedroom apartment, and when summer came they were both off to Texas to work for the election of George McGovern. Clinton was named co-coordinator of the Texas campaign, and Hillary was assigned to a voter registration project. Feminism was the hot topic of the day, and many women volunteers were just beginning to ask why they ended up in the boiler room and the secretarial pool while men found their way into strategy jobs. Hillary was obviously a step ahead of her peers. She came to work in corduroy jeans, wore no makeup, had read Simone de Beauvoir and Germaine Greer and took it for granted that her ideas deserved a hearing. As a result, she provoked varying reactions. Some in the Austin campaign headquarters found her aloof, but there were women who looked up to her as a big sister and role model.

Hillary's job involved traveling around the state, often with Onie Elizabeth Wright, a native Texan known as "Betsey." Only a few years older than Hillary, Wright was already a veteran campaigner. Unlike some political operatives, who become narrowly focused on the mechanics of vote getting and vote counting, Wright was well read and had strong opinions on the issues, invariably staunchly liberal. One of the interesting questions for feminists was whether the entrance of more women into politics would have an impact on the issues. Would women candidates be more "moral" than their male counterparts? "I was less interested in Bill's political future than in Hillary's," Wright told David Maraniss in 1993. "I was obsessed with how far Hillary might go, with her mixture of brilliance, ambition and self-assuredness."[13]

To Wright and others in the Austin campaign headquarters, it was by no means obvious that Bill and Hillary were destined to marry. According to Maraniss, the couple were dating, but not "seeing each other exclusively," and when they were together they often argued.

In the fall, Hillary went back to New Haven alone while Bill stayed on to guide the McGovern campaign through election day. It was a tribute to Yale's emphasis on a broad, conceptual view of the law that he was able to absorb enough in a few weeks to pass his semester's courses.

Hillary, meanwhile, was devoting more of her time to the Child Study Center and children's rights activities. Through Marian Edelman, she was hired as a research assistant by the Carnegie Council on Children, a blue-ribbon panel of eleven experts assembled by the Carnegie Corporation and led by Kenneth Keniston. Keniston preferred not to call the group a "commission," a word he found redolent of "Eisenhower, Kerner, Scranton, etc.," and the panel was the first Carnegie-funded group to be assembled with racial diversity in mind. Its mandate, in part, was to respond to the concerns of sociologist Uri Bronfenbrenner, who had compared child rearing in the Soviet Union and the United States, to the disadvantage of the latter. The council's book-length report, published in 1977 under the title *All Our Children,* is must reading for anyone who seeks to understand Hillary Rodham's vision of the future of American families.

The Carnegie panelists started with the assumption that the triumph of the "universal entitlement state" was inevitable, and the best thing Americans could do for their children was to hasten its arrival. Just as families in an earlier era turned over their children's education to the public schools, the report argued, so in the future would government assume responsibility for many other areas of children's lives. This being so, there was no reason to feel guilty about the rising rate of divorce: "There is nothing to be gained by blaming ourselves and other individuals for family changes." The decline of the nuclear family need not be worrisome, because "schools and doctors and counselors and social workers provide their support whether the family is intact or not. One loses less by divorce today than in earlier times, because marriage provides fewer kinds of sustenance and satisfaction."[14]

All Our Children advocated universal health care, expanded family leave, fully deductible child care, childhood disability insurance, generous income supports for poor families—*and* full employment, through government-created jobs if necessary. "We wanted to help children and the way to do that was to give them more money," Dr. Keniston recalls. The possible economic consequences of these policies were dismissed in a few sentences. Supposedly, the costs of universal entitlements would be offset by social benefits, so that federal, state and local governments would save money in the long run. Inflation might be a problem, but its impact on people living on fixed incomes could be mitigated by cost-of-living adjustments. "Economists have formed additional feasible plans to give the same kind of guarantee to savings and other assets in order to keep investment high and lending rates down in an inflationary economy," the authors wrote—reassurance that appeared somewhat hollow by the time the book appeared, during a period of crippling inflation.

More significantly, *All Our Children* offers a blueprint for undermining the authority of parents whose values the authors consider outmoded. The chapter entitled "Protection of Children's Rights," the section on which Hillary worked, observes that "it has become necessary for society to make some piecemeal accommodations to prevent parents from denying children certain privileges that society wants them to have." The report goes on to advocate laws allowing children to consult doctors on matters involving drug use and pregnancy without parental notification, and preventing schools from "unilaterally" suspending or expelling disruptive students.

But this is just the beginning.

The Carnegie panel further calls for developing a new class of "public advocates" who will speak for children's interests on a whole range of issues, from the environment to race relations:

In a simpler world, parents were the only advocates for children. This is no longer true. In a complex society where invisible decision makers affect children's lives profoundly, both children and parents need canny advocates. What if all parents made relatively small financial contributions to such a cause? It would provide a politically insulated fund for lawyers, ombudsmen, agency monitors, and even attempts at legislative reform.

The assertion that life used to be simple—presumably this was during the Great Depression and World War II—is always a tip-off that one is about to read something very foolish. The report fulfills this promise by suggesting that "child ombudsmen" be placed in public institutions and some form of insurance be introduced to enable individual children to hire "decently paid" private attorneys to represent their interests. The possibilities for child advocacy would seem to be endless. For example, the report suggests, attorneys could bring class-action lawsuits to hold corporations liable for *future* damages their businesses might cause to today's children. Other suits might require product labeling that would "conform to the same standards that a reasonable parent would be held to toward his or her own children." Public advocates would also consider "how to make the rewards in our society more just and how to limit the risks of technology. . . . [They must] ask about tax reform, about reorganizing health care, about racism, about sexism, about energy—all for the sake of our children":

> The critical point is to expand our concept of children's protective services. Child protection should go far beyond the traditional model of social workers looking for neglected or poorly fed children to embrace a federal children's consumer and environmental watchdog agency that screens the practices of private industry and government alike for their effects on children. . . . In the long run nuclear power, disruption of the ozone layer, chemical additives, prescription and over-the-counter drugs, and industrial pollution may well represent more pressing legal problems for whole generations of children than a relatively small number of neglectful or abusive parents.

It is hard to find in this much evidence of great concern for actual, living children, including the "relatively small number" who are victims of neglect and abuse. Rather, this is the voice of people who think they know all the answers and want to use children as a tool to impose their will on others. As the reviewer for *Library Journal* observed at the time of *All Our Children*'s publication, this part of the council's program would "cause even the most receptive to visualize the opening of Pandora's box."[15]

Inevitably, in any foundation study, much of the actual work is done by the paid staff. This time, however, there was some concern at Carnegie Corporation's headquarters that the staff was off on a tangent of its own. At various times, staff members asked for more recognition and complained that it was hypocritical for the panelists to be meeting in comfortable hotels and elegant restaurants while American children lived in poverty.

Faustina Solis, then a professor of community medicine and deputy director of the Public Health Division of the California state health department, recalls that the panelists were slightly in awe of the passion and strongly held opinions of the researchers—"These were very bright students, you know." She remembers Hillary in particular as having a "broad perspective" on the question of children's rights, especially with respect to environmental issues.[16]

During the early 1970s it was easy to imagine that the agenda of *All Our Children* might be fulfilled in the near future. It was, after all, the era of liberation movements—black liberation, women's liberation, La Raza, AIM, even animal liberation. Marian Edelman's prediction that children's liberation would be the next big political movement, the organizing principle for social change in the 1970s, hardly seemed far-fetched. In 1971, a coalition led by Edelman succeeded in getting Congress to pass a comprehensive child development bill that included federal funding for preschool programs. The bill by no means fulfilled Edelman's wish list, but it did succeed in establishing the CDF's goal that "the government would finally take responsibility for the nation's children." When President Nixon vetoed the bill as antifamily, Edelman was "shattered."[17]

It may be necessary to remind ourselves that Richard Nixon was far more accepting of entitlements than even moderate Democrats are today. Nevertheless, he would not go far enough to please McGovern Democrats, and they in turn were under heavy pressure from special-interest groups to make the 1972 Democratic platform even more of a liberation movement wish list. In May 1972, Marian Edelman and Senator Birch Bayh, both members of the McGovern platform committee, held an open meeting in Boston, where they took testimony from fifty or so activists whose demands included homosexual rights and "massive assistance" for blacks. Hillary Rodham testified in favor of a platform that would extend civil and political rights to children.

Undoubtedly, Hillary's appearance was arranged by her mentor, Edelman, but her position on children's rights went beyond the agendas of the Children's Defense Fund or the Carnegie Council. In an article published in November 1973 in the *Harvard Educational Review,* she advocated liberating our "child citizens" from the "empire of the father." This was good feminist reasoning for which the rationale can be found in the writings of Simone de Beauvoir and Jean-Paul Sartre. ("There is no good father, that's the rule," Sartre said. "Don't lay the blame on men but on the bond of paternity, which is rotten.")

In Hillary's own words:

> The basic rationale for depriving people of their rights in a dependency relationship is that certain individuals are incapable or undeserving of the right to take care of themselves and consequently need social institutions specifically designed to safeguard their position. . . . Along with the family, past and present examples of such arrangements include marriage, slavery and the Indian reservation system.

A key court case for children's rights activists was *Wisconsin v. Yoder,* which overturned a Wisconsin law that forced Amish parents to send their children to high school, regardless of their religious objections. One could argue that society has an interest in seeing that all children receive an education, but Justice William O. Douglas, in a dissenting opinion, took another tack. According to Douglas, the court should have taken into account the child's right to maximize his individual potential to become, say, a classical pianist or an astronaut, which could not happen unless he was liberated from the deadening hand of his parents' way of life.

With hindsight, the most interesting thing about Hillary's views is her assumption that all dependency relationships are bad—except the dependence of the individual on the state. Through her volunteer work with Penn Rhodeen, Hillary knew very well that family courts were overburdened, underfunded and generally inefficient. How could she imagine that judges incapable of making timely custody decisions in the most extreme cases would ever have the time, staff support and wisdom to intervene in complicated family disputes?

In fact, Hillary didn't trust judges to make the right decisions, since they tended to uphold "middle-class values." Conceding that "young children in particular are probably not capable of organizing themselves into political groups," she suggests that children's rights cases be handed over to "boards composed of citizens representing identifiable constituencies—racial, religious, ethnic, geographical"—a plan for child rearing by proportional representation.

Like many of her generation, Hillary may have written things in the early seventies that she does not necessarily believe today. However, she continued to espouse similar views as late as 1978, after which she wrote no more on the subject. Since these articles were still being cited in the nineties as the basis of her claim to be an expert on children's rights, one assumes she has not repudiated them.

AS the end of her law school career approached, Hillary's laserlike ambition began to waver. In contrast to so many other Yalies of the era, who could hardly wait to leave academia for the "real world," she had voluntarily taken an eighth year of higher education, supplementing her legal studies with course work at the Yale Child Study Center. Her academic program gave her an excuse to remain in New Haven with Bill Clinton for another year. Nevertheless, the question of their future plans remained unresolved.

Early in 1973, Clinton took a few days off to return to Arkansas to take the bar exam. Hillary went along and sat for the exam with him. Supposedly, she was taking the test "just in case" she should find herself practicing in Arkansas, but a preparation book recently in the possession of L. D. Brown, formerly a member of Clinton's security detail, contains notes in Hillary's handwriting indicating that she studied for the exam.

Clinton had never wavered in his plan to run for office in Arkansas. It was all he talked about at Yale and practically the first thing that Hillary Rodham knew about him. But the Natural State, whatever its attractions, offered no career opportunities that could compare with becoming an associate at a prestigious D.C. law firm. Complicating matters, the last thing Clinton needed on his return home was a girl-friend with whom he had been living out of wedlock for the past three years. The concept of "living in sin" might be laughable to Yale students, but plenty of voters took it seriously, and not just Ozarks rustics, either.

Unable to follow Clinton to Fayetteville, where he had lined up a job teaching at the University of Arkansas Law School, Hillary did not go to Washington, D.C., either. Instead, she went from Yale to a staff position at the Children's Defense Fund headquarters in Cambridge, Massachusetts. Given what she had told Penn Rhodeen and others about her ambitions, this can only have been an interim job, a way of doing good and gaining valuable experience while waiting for her situation to sort itself out.

In January 1974, history intervened. Hillary received a call from one of her Yale law professors, Burke Marshall, offering her a place on the staff of John Doar, the special counsel for the impeachment inquiry being undertaken by the House Judiciary Committee. From advocating children's rights to working on the first presidential impeachment of the twentieth century was a huge leap, but Hillary had the right connections. Doar had been hired by the Judiciary Committee in part because of his bipartisan credentials. He joined the Justice Department during the Eisenhower administration and had the support of Melvin Laird, Nixon's secretary of defense. But Doar had also worked with Burke Marshall and Peter Edelman under Attorney

General Robert Kennedy, and in 1968 he left government service to become the director of the Bedford–Stuyvesant Restoration Project, a model urban renewal effort that had been Robert Kennedy's pride and joy.

Nixon's presidency had been imploding almost since the day he won his second term in a landslide victory over George McGovern. The sentencing of the Watergate Hotel burglars by Judge John Sirica in March 1973 was followed by an investigation into Republican campaign finance irregularities led by Ted Kennedy's Senate subcommittee. Kennedy's probe was succeeded by the Senate Select Committee probe under Sam Ervin, and Ervin's investigators, following a tip from *Washington Post* reporter Bob Woodward, had learned that Nixon had been tape-recording conversations in the Oval Office. Nixon's determination to keep control of the tapes led him to fire Watergate special prosecutor Archibald Cox in the so-called Saturday Night Massacre on October 20, 1973. But he had overplayed his hand. Reaction on Capitol Hill and in the press to the Cox firing was so negative that Nixon agreed to turn over nine subpoenaed tapes to Judge Sirica. When one of these proved to have been partially erased, Nixon's credibility sank to a new low. Meanwhile, Nixon's former attorney general, Richard Kleindienst, had pled guilty to a criminal indictment arising from his refusal to answer questions under oath, and Vice President Spiro Agnew had resigned over unrelated bribery charges.

As these events unfolded, the country was divided between a minority of Nixon haters, who rejoiced in every new revelation, and the majority, who were thoroughly sick of "wallowing in Watergate," as Nixon put it. Nixon appeared to have lost all credibility. Still, as the focus shifted to impeachment proceedings in the House, no one could foresee how the drama would play out.

Given Nixon's reputation as a fierce infighter, many Democrats assumed he would take the offensive, arguing that recent Democratic presidents were guilty of far worse offenses. Lyndon Johnson had both bugged the Goldwater campaign and used the FBI to investigate Nixon and Agnew. Kennedy plotted with the Mafia to assassinate Fidel Castro. Most alarming, perhaps, Nixon was said to have a file on JFK's personal life, detailing his affairs with Marilyn Monroe, Judith Exner and many other women.

Flush with a $1 million budget, Doar assembled a staff that included forty-three young attorneys, a number that by the standards of the day seemed wildly excessive. They were promptly nicknamed "the faceless forty." On January 31, Doar told the Judiciary Committee that all applicants for staff positions were "questioned whether or not they had taken a position on impeachment, and if they had, other than that there should be an inquiry, they were not considered for the job." This was a bit of verbal slicing and dicing. All the attorneys Doar hired may have believed that "there should be an inquiry," but this did not mean that they had no position on the eventual fate of Richard Nixon. Although expressions of partisan feeling were severely discouraged around the office, Hillary loathed Nixon and thought his approval of raids on Vietcong supply lines in Cambodia should be included in the bill of impeachment.

Winter faded into spring, and Doar's approach to his job began to frustrate and then infuriate members of Congress. His staff wasted months debating procedural issues, trying to come up with an abstract definition of "high crimes and misdemeanors" and arguing over the precise language of congressional subpoenas. Doar appointed Bernie Nussbaum, another former member of the RFK Justice Department, to head the independent investigation into Nixon's conduct; however, there was no sign that any investigating was getting done. Instead, members of the legal staff were kept busy churning out statements of fact, summarizing information that was already on the record.

But it soon developed that some of the "faceless forty" were less faceless than others. As a student of Burke Marshall, who was serving as a behind-the-scenes adviser to Doar, Hillary Rodham played a prominent role from the beginning. While other staff attorneys were scarcely ever seen by members of Congress, Hillary usually accompanied Nussbaum and Doar to meetings with committee members, and she was given a number of important responsibilities.

One of Hillary's projects was supervising the compilation of a book-length tome called *Responses of the Presidents to Charges of Mis-*

conduct, edited by Yale historian and Arkansas native C. Vann Woodward. Woodward tried to answer the question of whether Nixon's misdeeds were different in kind from those of previous presidents. (His answers were somewhat beside the point, however, since he did not consider the question of the extralegal uses of the CIA and the FBI. This was a story that was just beginning to come out, and many Watergate insiders believe to this day that Nixon's presidency was a casualty of a power struggle between him and the CIA.) Nevertheless, Woodward made a few interesting generalizations. One of them was that Nixon was unique in that he was acting not to save his administration or its policies but his own skin; he was the "chief personal beneficiary" of his own misconduct and the effort to cover it up. Woodward's manuscript had been completed with taxpayer money and might have been useful to the committee and, potentially, to Nixon's defense, but the Doar staff kept it under wraps, denying its very existence until after Nixon's resignation.

But Hillary's most important assignment was drafting procedural rules for the formal presentation of evidence to the House. Supported by Bernie Nussbaum and John Doar, she came up with a number of dubious suggestions.

One of her ideas was that the President should not be allowed to have an attorney to represent him in impeachment proceedings. Hillary wrote a paper defending this position, which drew on old precedents, ignoring situations that occurred after the Supreme Court established a right to legal counsel.

Rodham, Nussbaum and Doar also favored imposing gag orders on members of the Judiciary Committee to prevent them from disclosing evidence to the public. Other rules they proposed would have kept committee members from cross-examining witnesses or drafting their own articles of impeachment. These rules would have reduced impeachment hearings to a kind of secret show trial run by staff lawyers.

Most strangely, Hillary argued in favor of a bizarre theory that Doar put forward in private conversations. According to Doar, the constitutional definition of impeachable offenses, which specifically mentions bribery and treason, was obsolete. Supposedly, personal and

campaign finances had become so complex and interwoven that bribery was now all but impossible to prove, and foreign relations so
intricate that the term *treason* had lost all meaning.

Jerry Zeifman, who was chief counsel to the Judiciary Committee
proper, had several run-ins with Hillary and felt that on numerous
occasions she lied or withheld information about her work. He came
to regard her, Doar and Nussbaum as "somewhat less than honorable
lawyers." On May 4, 1974, Zeifman wrote in his diary: "I have been
mulling over the events of the past week, and I am incensed with
Doar and some of his top assistants such as Joe Woods and Hillary
Rodham. It seems to me that [Nixon aides] Haldeman and Ehrlich-
man are crude amateurs at arrogance in comparison to the more pol-
ished and sophisticated arrogance and deceit of some of Doar's
assistants."[18]

Zeifman concluded that Hillary Rodham, Doar and some other
staffers were part of an effort led by Burke Marshall to hamstring the
Judiciary Committee. The goal, he believed, was to protect Ted Kennedy's political future by making sure that information about his
brother's misdeeds did not surface during the debate. Zeifman wondered if Doar wanted to keep Nixon in office as long as possible, in
order to help Democratic candidates who were benefiting from
Nixon's loss of credibility.

Zeifman's suspicions are backed up by a 1976 *Atlantic Monthly* article written by Doar's close confidante Renata Adler. Adler revealed in
this article that the effort had indeed been a "charade." Despite his
promise to the Judiciary Committee that he had not formed an opinion, Doar had made up his mind months earlier that Nixon had to be
removed from office. However, he wanted to do it "without any of
the bitterly partisan recriminations which might have divided Congress and the country for many years." To that end, Doar used the
"faceless forty" as a front to gain time. As he explained to Adler, "You
work them very hard and you don't tell them anything." With the
staff turning out pages of legal verbiage, the important decisions were
being made by a small group Adler called the "ad hoc irregulars," old
friends of Doar's who for the most part were not on the staff.

No wonder Zeifman thought Doar was arrogant! Adler never acknowledges that Doar was representing anyone's interest in particular, but she takes it as a given that the elected representatives of the people could not be trusted to do their constitutional duty.

Adler does concede that, in retrospect, she was no longer sure that Nixon's after-the-fact knowledge of the Watergate burglary justified his impeachment—"his departure seems random, arbitrary, and incomplete." Still, she continued to believe that Nixon was guilty of doing something uniquely "monstrous," even though it was difficult to say exactly what that might have been.

Adler's suspicion? Nixon had taken *Asian money*.

With no actual evidence to back up her claim, Adler proposed that the government of South Vietnam had funneled cash to Nixon's re-election campaign. Ironically, considering that she subscribed to the notion that the President could not commit treason, the possibility that Nixon might have accepted cash from a foreign government still shocked her.[19]

In 1996, after the publication of Jerry Zeifman's book *Without Honor: The Impeachment of President Nixon and the Crimes of Camelot*, William Dixon, another former member of the Judiciary Committee staff and later Gary Hart's campaign manager, spoke up about Hillary's role. "I have known the First Lady's husband, have spent time with them in Nantucket, have supported him in all his quests," said Dixon, and yet he agreed with Zeifman that in her work for Doar, Hillary "paid no attention to the way the Constitution works in this country, the way politics works, the way Congress works, the way legal safeguards are set up."

Dixon recalled seeing a draft of a paper on impeachment procedures that Hillary had brought to Representative Bob Kastenmeier for his comments. He considered it "virtually worthless." The question that Zeifman and others on the committee staff found themselves asking was, "Is she incompetent or is she trying to subvert these proceedings?"

"I vote for incompetent," said Dixon.[20]

Representative Jack Brooks of Texas joined with Zeifman to convince his fellow committee members to vote down Hillary's proposals, and the committee went on to do its work, voting three articles of impeachment during the final week of July. Faced with a July 24 Supreme Court ruling against his claim of executive privilege, Nixon did not conduct the street fighter's defense that some had expected. He turned over the so-called smoking-gun tape recording that showed that he had known about the Watergate break-in three days after it occurred and ordered his aide, H. R. Haldeman, to have the CIA initiate a cover-up. Three days after the transcript was made public, Nixon made the decision to resign.

Hillary had been renting a bedroom from Sara Ehrman, a friend from the McGovern campaign in Texas. For six months, she saw nothing of Washington. She worked from early in the morning until late at night, including Sundays, in a cramped, sour-smelling office in the old Congressional Hotel, with a view of a back alley. Doar was a vigilant boss, checking to see who left work to run errands and who had let papers pile up on his or her desk. The young attorneys would sneak out for quick dinners at inexpensive ethnic restaurants on Capitol Hill, but there was little socializing. One staffer compared the atmosphere to a political campaign without the sex.

One of just three women among the attorneys, Hillary was a hard worker and not temperamental, the type to notice other people's sulks and try to coax them back to cheerfulness. She could be counted on to rise to the bait of any antifeminist remark. Once when Republican counsel Albert Jenner made a crack about there being no great women trial lawyers, she told him that perhaps it was because a great trial lawyer needed a wife to press his suits and pack his bags for him. And yet she also told coworkers that she had a boyfriend down in Arkansas who was going to be President of the United States one day, which they thought a strangely frivolous boast coming from someone who seemed so serious.

But despite Hillary's focus on her work, her emotions followed the ups and downs in her relationship with Bill Clinton. Clinton was campaigning for a seat in Congress, hoping that anti-Nixon sentiment would enable him to defeat Republican incumbent John Paul Hammerschmidt. When Clinton called, or on a few occasions made time for a quick trip to Washington, Hillary was exultant. When she learned—apparently from him—that he was seeing someone else, she was crushed. No one who knew Hillary for very long escaped hearing her confess at length her doubts about her future. Up north, she had so many options. She wasn't at all sure she could be happy in Arkansas. And yet, every decision she made pointed her in that direction. She talked her father and brother Tony into driving to Fayetteville, ostensibly to work on Clinton's campaign, though obviously their real mission was to make sure he wasn't getting serious about anyone else. And weeks before her coworkers thought about their effort coming to a close, she took time off to go down to Fayetteville and interview for a position in the law school. She formally accepted the job on July 9, a month before Nixon resigned.

FOUR

The Yankee Girlfriend

IN 1997, a high school classmate of Bill Clinton's, Dolly Kyle Browning, released a self-published novel called *Purposes of the Heart*. The book chronicles the experiences of Kelly, a young woman who conducts a sporadic affair with a self-absorbed Southern politician named Cameron Coulter. Undoubtedly the most memorable scene in this roman à clef is the moment when Kelly shows up at the airport prepared to serve as Cameron's driver for the day and discovers that he is traveling with a frumpy female whom he introduces as his fiancée, Mallory Cheatham. Kelly has heard a lot about her rival and can't quite bring herself to believe that this is the woman in question. Her first reaction is that Cameron must be playing a cruel practical joke:

> The dowdy-looking woman stepped up beside him and Kelly realized that she was of their generation, not middle-aged as her first glance had indicated. Kelly wondered why Cameron would haul such a person in the plane with him in public. She was wearing a misshapen brown dress that must have been intended to hide her lumpy body. The garment was long, but stopped too soon to hide her fat ankles and thick calves which, to Kelly's amazement, were

covered with black hair. Thick brown sandals did nothing to conceal her wide feet and the hair on her toes.

. . . The eyes . . . bulged out of focus behind coke-bottle thick lenses in dark, heavy frames, competing with the dark, thick eyebrow which crossed from one side of her forehead to the other. . . . She noticed that the woman emitted a definite odor of perspiration and greasy hair. Kelly's heart went out to the poor creature.[1]

In a 1997 interview with a London newspaper, Dolly Browning issued a semi-apology for this passage, saying, "It's a reflection on me, a small town Southern girl who had never been exposed to a woman who looked like that. In Arkansas, women shaved their legs." Catty and exaggerated as the description of Mallory may be, it is also a fairly accurate reflection of how Hillary Rodham appeared to many of candidate Clinton's friends, especially the women who were her rivals for the affections of one of the state's most promising native sons. In New Haven and Washington, Hillary's appearance sent a message that she was a young woman with an expensive education and advanced ideas. In Arkansas, only field hands wore heavy leather sandals, and people who could barely pay their electric bills were meticulous about dress and grooming, lest they be labeled white trash.

Browning's view of "Mallory" also explains a great deal about Hillary's relationship to Arkansas and her feelings toward the other women in Bill Clinton's life. Just a few days earlier, she had been part of a small group that was, at least as they saw it, stage-managing the downfall of a president. Now she was reduced to the role of the dowdy Yankee girlfriend, an annoying and totally superfluous accoutrement for an otherwise charming candidate.

Even Clinton's mother considered Hillary too plain for him. Virginia Dwire, as she was then known, would recall the unpleasantly "electrifying" sensation she felt when Hillary and Bill showed up in Hot Springs on an earlier visit, around the time of the McGovern campaign, both dressed in the seventies uniform of well-worn jeans and rumpled T-shirts. Clinton, on that occasion, took his mother and brother aside and sternly warned them that they had better figure out a way to tolerate Hillary: "Look, I want you to know that I've had it up to *here* with beauty queens. I have to have somebody I can talk with."[2]

Virginia, however, was not resigned. When Hillary once again arrived in Hot Springs looking hot and rumpled in a shapeless summer dress and sandals, she did not bother to hide her disapproval. Hot Springs was the Las Vegas of the South, as well known for its flashy women as for its world-class racetrack, and for a young lady to show up at the home of her future mother-in-law without taking the trouble to get a manicure and have her hair styled implied disrespect. Virginia herself spent a full hour and a half at her vanity table every morning "putting her face on." And since her job as a nurse anesthetist involved emergency calls at all hours, she took the precaution of sleeping every night in full makeup. She would rather have died than let anyone see her without her penciled-in Joan Crawford eyebrows and false lashes.

For the daughter of Hugh and Dorothy Rodham, brought up to believe in hard work, thrift and stability, it had to be almost comical that Bill Clinton's family would consider her not good enough. Virginia Cassidy Blythe Clinton Dwire was a woman who still enjoyed nightclubbing, the type of middle-aged reveler who did not hesitate to climb onstage and take over the microphone to belt out a chorus or two of a favorite number. And just as Virginia thought it necessary for a woman to hide her face behind a mask of makeup, she made a habit of forgetting certain unpleasant facts about the past. Clinton's father, Bill Blythe, who died in a car crash before his son was born, had been married at least four times, and two of his wives were sisters. Moreover, gossips in Hot Springs still whispered that Bill Clinton, born seven and a half months after Blythe's return from overseas, was actually the son of a car salesman Virginia had been dating in his absence.

Virginia remembered her second husband, Roger Clinton— "Roger, the Gambler," as she calls him in her autobiography—as a fun-loving man whose "trademark phrase" was " 'God dang it! Let's all get drunk and talk about the chances we had to marry!' He was hilarious." But Roger, who died of oral cancer in 1967, was also an abusive drunk. Virginia's mother had been so upset about this marriage that she talked to a lawyer about suing for custody of her grandson. Virginia divorced Roger in 1962, only to remarry him three months later after he lost thirty-five pounds and took to showing up late at night, bedding down under the front porch "like a derelict."[3]

Roger's son, also named Roger Clinton, grew up to be a surly loafer who decorated his room with psychedelic paraphernalia and spent a lot of time holed up there, lifting weights and practicing his guitar. Roger's own girlfriends conformed to what his mother called "the classic Hot Springs beauty pageant mold," and Roger made no attempt to conceal his opinion that his adored half brother could do a great deal better than Hillary.

At the time of Hillary's 1972 visit, Virginia had been married to Jeff Dwire, a hairdresser whose salon catered to the prostitutes run by a local madam known as Maxine as well as to middle-class matrons. Dwire wore pinkie rings and had done time in federal prison for fraud, but he was the only member of the household who went out of his way to make Hillary feel welcome. Unfortunately, Dwire had suffered a fatal heart attack shortly before Hillary's return to Arkansas, and now there was no one left to act as a restraining influence on Virginia and young Roger. They made no attempt to conceal their distaste for Hillary.

Back in Fayetteville, where Hillary was about to begin her first year of teaching at the law school, matters were hardly any better. She had arrived at the high point of Bill's congressional race, but almost from the day she set foot in Clinton campaign headquarters, the tide began to turn against his candidacy. This was not entirely Hillary's fault, though it would be hard to persuade some Clinton campaign workers of that.

It was a good time to be launching a political career in Arkansas. The state's powerful, long-serving senators, J. William Fulbright and John McClellan, were both nearing the end of their careers, and there was much speculation about who would replace them. Amazingly, considering that his entire work experience consisted of one year teaching at the law school, Bill Clinton was considered one of a half dozen candidates with a reasonable chance to capture one of the coveted Senate seats. Defensive about their state's reputation as an intellectual backwater, Arkansans were impressed when a local boy made good in the big time. To a great extent, Clinton was running for Congress on the strength of his educational résumé—Georgetown, Oxford and Yale, names that offered assurance that he would not embarrass his home state in Washington.

This was exactly the thinking of Paul Fray, a seasoned political operative who first ran into Clinton when the Georgetown student worked as a summer volunteer on the gubernatorial campaign of Frank Holt in 1966. Two years later, when Clinton was back in Little Rock working for Fulbright, Paul and his wife, Mary Lee, had let him stay with them in their apartment in the Quapaw Tower. Now the Frays had uprooted themselves to Fayetteville to work on Clinton's congressional campaign.

In addition to the Frays, Clinton's organization included his campaign manager, Ron Addington, a clean-cut instructor in the University of Arkansas School of Education, and Doug Wallace, who was editor of the student newspaper. Although Paul Fray, the oldest of the group, was just seven years Clinton's senior, they had nicknamed the candidate "the Boy" on account of his perpetual tardiness, temper tantrums and general immaturity. Everyone was willing to put up with his behavior because Clinton was such a hardworking and effective campaigner. Besides the university town of Fayetteville, conservative Fort Smith and the faded resort of Hot Springs, which had never recovered from Governor Winthrop Rockefeller's crackdown on illegal gambling, the Third Congressional District included dozens of remote Arkansas hamlets, reachable only by serpentine country roads. Bill Clinton was tireless when it came to courting votes. He would have his driver take him fifty miles out of their way just so he could sit around the kitchen table of a Democratic poll watcher, savoring her homemade biscuits and plying her with solicitous questions about the state of her husband's gallbladder. Social life in such places revolved around Saturday-night pie suppers, where candidates were expected to entertain the crowd with banter while the ladies' auxiliary auctioned off homemade desserts for the benefit of the local volunteer fire company. Clinton excelled at such events. Even though he knew nothing about farming, he could swap tall tales like a country boy. Paying Clinton the ultimate compliment, Paul Fray said of him, "He never had the trappings of an individual who had been to the East Coast."[4]

Clinton's campaign volunteers were a more upscale segment of the voting public—students, young professionals and homemakers with liberal views by the standards of the district, which was packed with voters who registered Democratic but voted for the most conservative candidate, regardless of party. Clinton's youth, good looks and gift of gab attracted some volunteers who behaved more like fans than political supporters, a situation that at least one reporter who dropped by the Clinton campaign office in Hot Springs found unsettling. "I mingle with the staff for a while and find that I am impressed by the hardihood and charm of the workers," he wrote. "They are young— mostly women—former campus politicians, joiners and gadflies who remind me of the volunteers who beat the bushes for Eugene Mc-Carthy in 1968. These are the kind of people who stand in the rain at high school football games to distribute campaign leaflets. The workers are influenced by the Dexedrine-like effects of campaigning, white-line fever—an inversion that naturally seizes the members of a cult. The volunteers are extremely reluctant to talk about themselves. They constantly mutter the aspirant's name in hushed tones. 'Bill thinks . . . ,' 'Bill feels . . . ,' 'Bill does. . . .' I feel as if I am either in a confession box or am party to the recitation of a first grade primer: There is a monotonous circularity to all the conversation."[5]

By the time Hillary showed up, Clinton had already survived a four-way primary and a runoff election to capture the Democratic nomination. Having begun as an unknown, he was seen as a serious challenger to the popular Republican, John Paul Hammerschmidt. Speaking earnestly about the need to "clean up the mess in Washington," Clinton had won the financial support of the AFL-CIO and managed to finesse the question of his lack of a military record.

At a time when arguments over the situation in Vietnam could easily spark fistfights, Clinton's ability to sail through the campaign without being confronted on the issue of his draft status was nothing short of astounding. One of his primary opponents had unintentionally helped him out by circulating a leaflet identifying him as the "protester in a tree," seen in a widely circulated picture taken during a student demonstration at the university in April 1969. This wasn't so—Clinton had been in England at the time—and the false charge gave him the opportunity to complain about the smear whenever the issue of Vietnam came up.

But, as campaign insiders well knew, Clinton was vulnerable. In 1969 he had indeed received an induction notice; however, Cliff Jackson, another Arkansan studying at Oxford, prevailed on an influential Republican friend at home, Van Rush, who persuaded the head of the state Selective Service Board to cancel Clinton's induction. In the meantime, Clinton's uncle Raymond, a politically connected car dealer in Hot Springs, and Lee Williams, an aide to Senator Fulbright, used their influence to get Clinton a berth in the ROTC program at the University of Arkansas. After all these machinations, Clinton never did enroll at law school. Encouraged by reports that there were no immediate plans to draft graduate students, he returned to Oxford for a second year, where he busied himself organizing antiwar demonstrations.

On December 1, 1969, the Selective Service switched to a new system, a draft lottery based on birthdays. Clinton's date of birth came in number 311 in the newly instated national draft lottery, virtually guaranteeing that he had escaped the draft. Two days later, he mailed a letter to Colonel Eugene Holmes, the officer who had accepted his ROTC application, informing him that he would not be returning to Arkansas to take up the appointment.

In this extraordinary letter, Clinton described his strong antiwar and antidraft sentiments and his activities as a Moratorium organizer in Washington and at Oxford. "I decided to accept the draft in spite of my beliefs," he confessed, "for one reason: to maintain my political viability within the system. For years I have worked to prepare myself for a political life characterized by both practical and political ability and concern for rapid social progress. It is a life I still feel compelled to try to lead." (These "years" of preparation consisted of attending college and working part-time for Senator Fulbright and on a number of political campaigns.) Clinton went on to say that after signing up for ROTC, he began to wonder "if the compromise I had made with myself was not more objectionable than the draft would have been. . . . I hardly slept for weeks and kept going by eating compulsively and reading until exhaustion brought sleep. Finally, on Sept. 12, I stayed up all night writing a letter to the chairman of my local draft board. . . . [I] never mailed the letter, but I did carry it with me every day until I got on the plane to return to England. I didn't mail the letter because I didn't see, in the end, how my going in the army and maybe going to Vietnam would achieve anything except a feeling that I had punished myself and gotten what I deserved."

Clinton's finagling to evade the draft would have been political death in the conservative Third District, and his ability to excuse himself from keeping a solemn promise on the grounds that it would be needless self-punishment should have given even opponents of the Vietnam conflict pause. Whatever led Clinton to write such a self-pitying letter in the first place, by 1974 he realized that it must never be allowed to fall into the hands of the opposition. Someone, apparently a contact at the university, wrote to Colonel Holmes, who had since retired, and persuaded him to get in touch with the ROTC office and ask them to return the original to him. Instead, Holmes was given a copy, which he eventually lost, and the ROTC unit burned the original, in accordance with a policy against keeping records on war resisters. Clinton had no idea that there was another copy of the letter in the files of Lieutenant Colonel Clint Jones, an aide to Colonel Holmes.

So Bill Clinton's first campaign had begun with a cover-up. In reality, there were a number of individuals, including some politically connected Republicans, who knew at least part of the story and were disturbed by it, but none of them felt motivated to pursue the subject. As far as Clinton was concerned the matter was settled, and he was left with an unrealistic confidence in his ability to evade difficult questions about his private conduct.

In any case, Clinton was attracting favorable attention by calling for Nixon's resignation. On August 7, following Nixon's public admission that he had lied about his role in the Watergate cover-up, Clinton told an audience in Flippin, Arkansas, "I think it's plain that the president should resign and spare the country the agony of his impeachment and removal proceeding. I think the country could be spared a lot of agony and the government could worry about inflation and a lot of other problems if he'd go on and resign."[6]

Then Nixon did resign, and Clinton lost his most effective theme.

Hillary, meanwhile, had taken over a desk at campaign headquarters in Little Rock, prepared to shake up the campaign and enforce discipline on the enthusiastic but sometimes unreliable volunteers. Hillary complained that the campaign was spending too freely, and she thought Clinton should get more specific about the issues—not a good idea in a district where the majority of the voters were decidedly to the right of the candidate. She assumed that her status as Clinton's girlfriend gave her the authority to order other people around, and her brusque midwestern manner rankled. Worse yet, she and Clinton seemed to argue constantly. Ostensibly, these clashes were over campaign tactics, but the none-too-subtle subtext of jealousy made others uneasy. On one occasion, the couple was headed for an out-of-town appearance when Hillary got so hot over something Bill said that she ordered the driver to stop the car so she could get out and walk back to headquarters. The tension filtered down to the volunteers, who found they weren't having fun anymore. As Doug Wallace put it in a memo to Clinton, Hillary's intentions were good but she "rubs people the wrong way."[7]

Hillary's presence was especially awkward because Clinton had absentmindedly acquired five or six other girlfriends in various corners of the Third District and in Little Rock. One brunette law student was under the impression that *she* was Bill Clinton's fiancée, and campaign aides were left with the task of trying to keep the two women apart until fiancée number two finally became discouraged and stopped showing up at headquarters.

Dolly Kyle Browning, who had known Clinton since they attended Hot Springs High School together, believed that if it weren't for her unfortunate early marriage, Clinton would have preferred her. Browning's late father had been an associate of Senator Fulbright, and she was distressed by the way Clinton was keeping his distance from his former mentor. The distinguished senator, known for his foreign policy expertise, had lost touch with the voters in his home state and was about to suffer a humiliating defeat at the hands of Dale Bumpers, the popular governor. As for Hillary, Browning had been assured by Billy, as she called him, that Hillary knew all about their occasional assignations, which had been going on since he was at Georgetown. With hindsight, she is no longer so sure. "He's such a liar, who knows what he told her?" she said recently.

Another of Clinton's girlfriends was a gorgeous redhead from Russellville, Arkansas, named Barbara Norris, who later changed her name to Norris Church and married the novelist Norman Mailer. Norris saw Clinton as a man "who had a terrific personality, charisma and possibilities." The trouble was, Clinton never wanted to take her out in public. He always seemed to call at the last minute, proposing a late-night get-together, and Norris eventually lost interest.[8]

Ironically, despite some talk in the district about Clinton's overactive love life, it was Hillary who posed a serious problem. Paul Fray had been listening to Clinton sing her praises for months. "She's the smartest person I ever met in my lifetime," Clinton had told him. When Hillary finally showed up, Fray was taken aback. "Oh, my gracious, this girl's a hippie," he thought.

Hillary actually didn't look like a hippie—rather, she was more of a Birkenstock-style yuppie—but however one chose to characterize her appearance, it didn't go over well in a district where many women still wore housedresses instead of pants. Traditional homemakers couldn't relate to Hillary, and rumors that she was a lesbian began to circulate almost immediately. Someone had to tell Hillary to soften her presentation. Clinton, who didn't have the nerve to broach the subject himself, turned the job over to Mary Lee Fray.

"I don't care how you do it," he said, "just do it."

Mary Lee had become rather fond of Hillary. A Virginia native with liberal views, she agreed with Hillary on many of the issues; the two women would exclaim over how outrageous it was that just anybody who kept a clean house could get licensed to run a home day-care facility. Mary Lee gave Hillary advice on how to accessorize her outfits to make them more feminine, and Hillary took some of it, however reluctantly. Doubtless Hillary had reason to resist. She had no desire to emulate the Southern belles with lacquered hair and little pastel outfits who buzzed around Candidate Clinton, and she only risked further humiliation if she were perceived as trying to compete with them.

Bill Clinton obviously loved Hillary. He talked about her all the time, especially to married women like Mary Lee Fray, who was touched by the way he agonized over asking Hillary to sacrifice her promising legal career by joining him in Arkansas. "She could even be a senator!" he would say. "But if she comes to Arkansas, she'll be on my turf." Clinton had found the courage to stand up to his formidable mother, telling her that she would have to learn to tolerate Hillary because "it's her or no one." At least one of Clinton's aides heard the same declaration, and we can be sure that Hillary had heard it, too, probably many times.

But Clinton was engaging in a classic passive-aggressive maneuver. On the one hand, if Hillary left him she would be guilty of ruining his life, since he could never truly love anyone else. On the other hand, his behavior made it clear that he had no intention of changing his ways. If Hillary stuck by him she would be taking the whole package: Arkansas, his single-minded focus on politics, his hostile family, his chronic lateness, even his inability to resist the charms of other women. Clinton would later confess to Betsey Wright that during this period he had tried to "drive Hillary away."[9]

But Hillary was too proud and stubborn to be discouraged in this way. Hillary had already made a commitment—she had turned her life upside down to come to Arkansas—and if Clinton had changed his mind about their relationship, the least he could do was tell her straight out. Even if Hillary wanted to leave, she couldn't just pack her bag and head for the airport. She had made a professional commitment to teach at the law school and rearranged her family's lives as well as her own. Her father and brother Tony had spent weeks working for the campaign, driving the back roads of the district in Hugh senior's Cadillac, tacking up CLINTON signs on trees and fence posts. Hugh junior arrived at the beginning of the school year, and he and Tony had registered for classes at the University of Arkansas. (Presumably, as relatives of a faculty member they were entitled to substantial discounts.) While her brothers were sharing an off-campus apartment, Hillary had rented an architect-designed three-bedroom home complete with swimming pool.

THE tension at Clinton headquarters came to a head on election day. Paul Fray had cut a deal to take $15,000 in cash from dairy industry interests, money destined for poll workers in the Fort Smith area. Fray explained that this wasn't so much a bribe to skew the vote count as a way of keeping the count reasonably honest, since the other side was presumably spreading money around as well. This was how things had always been done in Sebastian County, he argued.

Hillary was adamantly against the plan. Even a seat in Congress wasn't worth committing election fraud. "If we can't earn it, we can't go," she insisted.[10]

Fray's judgment in such matters was doubtful. He'd seen the election slipping away over the past few weeks and was desperate to win. Mary Lee, however, decided that she wasn't going to risk having her husband go to jail just to elect Bill Clinton. She took possession of the car keys and the cash, refusing to hand them over to her husband.

Later in the evening, when the results began coming in, it became obvious that the money might have pushed Clinton into the victory column. Recriminations started to fly and, as the argument heated up, so did cuss words, papers, staplers, books, even telephones. Paul Fray accused Hillary of treating him from the start like an ignorant redneck. She yelled back that if so, he had proven her right. Mary Lee defended Hillary at first, then became incensed with Bill for standing passively by, doing nothing to stop the fighting. Mary Lee began unloading on Hillary, telling her about all the times she had sneaked Bill's other girlfriend out the back door just as Hillary came in the front. Mary Lee had known Bill Clinton since he was in college and knew that he found it easy to get involved with women but that when it came time to break up, his courage failed him. She began giving Hillary chapter and verse on the messy breakups in his past, going all the way back to high school girlfriends from Hot Springs.

The argument ended Bill and Hillary's friendship with the Frays, but not their relationship. While the details Hillary heard that night may have been new, presumably she hadn't lived with Bill Clinton for three years without understanding his weaknesses.

Ironically, despite the embarrassment that Clinton's womanizing caused, Hillary's place in his life was completely secure. The same soft spot in his character that kept him from telling the truth to the Arkansas belles he collected so thoughtlessly would make it impossible for him to ever break up with her. Hillary, after all, was much more than just a pretty face; she was his companion, his intellectual sparring partner, his most trusted political adviser, the life partner who was willing to take care of his needs and forgive him when he strayed. For a woman who likes to be in control, the sexy bad boy, the man other women want but can't handle, is always an interesting option. Hillary was so ambitious that she would have found it hard to respect a man with modest goals. Bill Clinton might be trying at times, but he brought excitement into her life.

And now that the election was over, Hillary's position had become stronger. Clinton had borrowed money for his campaign, and his rented cottage was cramped and dreary. Hillary was living in a luxurious house, making new contacts and building a reputation as a legal activist. University towns all over the country attract like-minded people, and the informal social life of Fayetteville appealed to her.

Hillary became an avid if mediocre tennis player, saw a lot of movies and enjoyed getting together with other couples to sip wine and debate current events. She was soon fast friends with political science professor Diane Kincaid, a chic brunette whom Clinton had known since 1972 when she was a delegate to the Democratic convention. Kincaid was the driving force behind the Campus Commission on the Status of Women, which was pressing for more females on the faculty. On Valentine's Day, Kincaid was invited to debate Phyllis Schlafly on the Equal Rights Amendment before a joint session of the Arkansas legislature. Hillary and Bill helped her prepare, then shared Kincaid's amusement when the Arkansas *Democrat* covered the event on the women's page, giving a rave review to her "stunning" beige outfit and Schlafly's shrimp-pink dress.

AS law professors, Clinton and Rodham were opposites. His lectures were open-ended, weaving together observations from American history, sociology and current events. If a subject engaged his interest, he didn't worry about its relevance to the syllabus and felt no pressure to move on. His constitutional law class spent three solid weeks on *Roe v. Wade.* Clinton was one of the most popular teachers in the law school, all the more beloved because he was an easy grader. The problem was getting a grade from him at all. He fell weeks behind in grading student papers, and he lost exams, including, once, the blue books from an entire class. As reported by Jeffrey Toobin in *The New Yorker,* Clinton once lost the final exam of a student in his admiralty law class named Susan Webber, later Susan Webber Wright, who as a federal judge would preside over the case of *Jones v. Clinton.* Sometime later, Hillary showed up to explain the problem and offer Webber an automatic B plus. She declined the negotiated grade and insisted on taking the exam again.

Hillary, by contrast, was organized, demanding and opinionated. She expected her students to come to class prepared and brooked no excuses. "People were never indifferent about Hillary," her student Woody Bassett would recall. "They either liked her or they didn't, but they always respected her." Maybe so, though in student evaluations, Hillary was criticized for wearing baggy sweaters. She, in turn, cringed when she heard bad work excused with the complaint "What can you expect from me? I'm a product of the Arkansas school system." Surely, she was correct to have high expectations, yet Hillary also gave the impression that she found Arkansas mores provincial, so it isn't surprising that some students were defensive.

In his spare time, Clinton took on cases for individuals who had encountered him on the campaign trail and showed up asking for his help. One of his clients was a female student on a strict budget who had shoplifted a worm treatment for her beloved Irish setter. She was convicted and fined twenty-five dollars, but Clinton later got her record expunged, earning her gratitude for life. His fee was ten dollars, which the student took months to pay.

Hillary's legal work was high profile. She filed a brief in a death penalty case on behalf of a condemned killer with a sub-par IQ, and she lobbied in favor of a rape shield law. (She was still a bit naive about the criminal justice system, however. While she was visiting the penitentiary to interview a prisoner, one of the guards pointed out the kennel, and Hillary at first thought the dogs were kept as pets for the inmates.)

Hillary's chief accomplishment was founding the university's first legal aid clinic. She lined up funding in short order and created a system for tracking cases from complaint to resolution. The clinic advised 300 clients the first year and took 50 cases to court. In one of these actions, Hillary represented an accused rapist before a judge who had to be persuaded that it would not be improper for a "lady lawyer" to hear the details from the police report.

By now, it was Bill who was eager to get married. He had been living, or at least sleeping over, at Hillary's house. Already during the general election campaign, one Baptist preacher in Fayetteville had criticized him for sharing a house with a woman who was not his wife. If he expected to run for office again, and he did, he had to resolve his personal situation.

Now it was Hillary who was reluctant to commit. In conversations with friends, she continued to gnaw at the same themes: Did she really want to commit herself to living in Arkansas? Could she be married to an elected official and still pursue her own interests? Hillary posed the second question to Ann Henry, a Democratic activist in her own right who happened to be married to a state senator. Hillary said hopefully that Eleanor Roosevelt had done it. Henry was doubtful, pointing out that Eleanor didn't become truly independent until "the marriage was over—she didn't care about the marriage."[11]

The background to Hillary's worries was a change in Bill Clinton's prospects. When she'd come to Fayetteville a year earlier, Clinton had appeared to have a good chance of going to the House of Representatives, with a possible Senate race in the near future. As a congressman's wife, Hillary could have joined a top Washington law firm, taken a government post or worked for a foundation. But that window of opportunity had slammed shut. Without the albatross of Richard Nixon to carry around, Hammerschmidt would be a more formidable candidate in 1976. Dale Bumpers had taken over Fulbright's seat in the Senate and seemed likely to hold on to it. And now the current governor, David Pryor, appeared likely to run for McClellan's seat. Clinton had been so much the underdog in 1974 that his loss at the polls hadn't hurt him, but a second defeat would tarnish his aura of inevitability. Therefore he was now thinking about following a traditional path in Arkansas politics, running for state attorney and then governor.

For Hillary, this meant that there would be no fast track to Washington. Worse news yet, the salary of the attorney general was just $6,000 a year. Back in the 1870s, Arkansans, in their wisdom, had set up a political system that severely discouraged career politicians. It took a constitutional amendment to raise the salaries for statewide offices, and in recent years, inflation had taken its toll. Bill Clinton's attitude toward money had always been "God and my friends will provide." Clinton had borrowed to get through Yale and borrowed again during his run for Congress, and he was already thinking about taking a leave from his $25,000-a-year teaching post so that he could devote 1976 to campaigning. He showed no interest in using his degree to make money in private practice and talked vaguely about becoming a legal services attorney if the unthinkable happened and politics didn't work out for him. Hillary realized if she married Bill, she would be the one responsible for their financial security for the foreseeable future. This seemed to bother her more than his womanizing, or perhaps they were part of the same problem.

Still unable to make up her mind, Hillary left on a late summer vacation, visiting friends in several cities up north. As Bill was driving her to the airport, they happened to pass a little house with a for-sale sign out front, and Hillary commented on how cozy it looked. When she returned from her vacation, still undecided about her future in Arkansas, Bill picked her up and stopped in front of the same house.

Proudly, he announced that he had just become the new owner. "Now you'd better marry me because I can't live there all by myself," he added.

"That's when I finally said yes," Hillary would recall.[12]

Buying the house was a grand romantic gesture and, perhaps, the decisive move Hillary had been waiting for. But there was a catch. Although the house was in a picturesque setting, on California Street, at the end of U of A's fraternity row and not far from a small lake, it needed a lot of work to make it livable.

Hillary would have been content with a small wedding, but Bill wanted a big affair, where he could entertain a few hundred close friends. Many newspaper stories have described how Hillary neglected to buy her wedding dress until the last possible minute, racing into a Fayetteville department store shortly before it closed to purchase a high-waisted Victorian-style dress off the rack. But then, she and Bill had been together a long time, and the wedding was turning out to be mainly a party to launch his campaign for state attorney. Left with the job of getting the house ready, Hillary recruited her brothers to paint. Dorothy Rodham also came down to Fayetteville to help out. Morriss and Ann Henry took over the planning for the champagne reception, which would be held at their home following a small private ceremony at the California Street house. Virginia showed up with a group of friends from Hot Springs on the eve of the ceremony, found the house was still in chaos, paint cans and drop cloths everywhere, and retreated to a Holiday Inn.

On the morning of October 11, 1975, all was ready, if just barely so. A minister whose name was, ironically, Nixon—the Reverend Victor Nixon—performed the simple Methodist ceremony. Virginia Dwire was still reeling from the shock of learning that very morning that Hillary planned to keep her maiden name, a custom she could only imagine was some "new import from Chicago." Roger Clinton, the best man, sulked through the ceremony. At the Henrys' lawn party Bill, not Hillary, was the center of attention.

Dorothy Rodham, discovering that the newlyweds had no plans for a honeymoon, found a package tour to Acapulco advertised in the newspapers. Oddly enough, she booked tickets for the whole Rodham family. But perhaps it took all of them to hijack Bill Clinton into making the trip. Relaxing on the beach was not his idea of a good time, and he spent much of the wedding trip writing thank-you notes.

On his return to Fayetteville, Clinton began to plot his race for the state attorney's job. He would spend most of 1976 crisscrossing the state, displaying a hunger for the $6,000-a-year post that his opponents in the primary simply could not match. The Republicans did not bother to put up an opponent against him in the general election. Clinton devoted the fall campaign to serving as state chairman for Jimmy Carter while Hillary used her connections with Betsey Wright to wangle an assignment as deputy coordinator of Carter's field operations in Indiana. This was a thankless task, because Carter had no chance of carrying the state. Hillary, however, came up with one novel idea. The Carter headquarters happened to be located in the courthouse district of Indianapolis, near the offices of a number of bail bondsmen, and she recruited felons out on bail to work the phone bank.

Bill and Hillary's early support for Carter over Ted Kennedy came as a surprise to some of their friends from Yale, but it was dictated by local politics. Jack Stephens, the younger of the powerful Stephens brothers, had been Jimmy Carter's roommate at the Naval Academy, and there was little point in launching a career in Democratic politics in Arkansas without at least the tacit support of the Stephenses. Witt, the older and more influential brother, was the founder of Arkla, the Arkansas Louisiana Gas Company, and had been the financial angel behind Orval Faubus. Jack was president of Stephens, Inc., the largest bond firm outside Wall Street and a powerhouse in the Southeast. Moreover, from the vantage point of Arkansas, it was possible to see that Jimmy Carter, with his talk of putting morality back in politics, had the best chance to pull together the elements of a Democratic majority—suburbanites, blacks, blue-collar workers leaning toward George Wallace and disaffected liberals. Hillary was not fond of Carter personally—friends hint that she thought he was compromised some-what by his close ties to the Stephenses—but her hard work paid off. In December 1977, Hillary was appointed to the board of the Legal Services Corporation, and Carter let it be known that she was his choice for board chairman, a position she assumed a few months later.

Victory in the attorney general race meant the end of Hillary's teaching career. The California Street house went on the market, and she and Bill prepared for the move to Little Rock. Bill was ecstatic, Hillary a good deal less so. He had never seen teaching as anything more than a way of marking time, but Hillary regretted having to walk away from the legal services clinic and her new circle of friends.

First in Her Firm

Exit the Capital Hotel and turn the corner onto Main Street, and you will find yourself on "Little Rock's Celebrity Walk of Fame." A relic of an earlier era, the "walk" consists of a few slabs of sidewalk divided into squares bearing the names of celebrities who passed through Little Rock during the late seventies and early eighties. Among those who signed their names in wet cement were George Bush, Muhammad Ali and Bob Hope. Among the few famous Arkansans honored are Herb Tarleck, who played Frank Bonner on the TV series *WKRP in Cincinnati,* and Sidney Moncrief, who was a University of Arkansas basketball star before joining the Milwaukee Bucks. The curious feature of this rather forlorn exercise in civic pride is that the celebrities are segregated by sex. Women's signatures are relegated to a separate rectangle of pavement; there aren't many of them, and every one is a beauty pageant winner: Donna Axum, Miss America 1964; Terri Lea Utley, Miss USA 1982; Elizabeth Ward, a Russellville, Arkansas, girl who became Miss America in 1982; and another Miss Arkansas, Lencola Sullivan, who was a Miss America runner-up in 1980.

As the Walk of Fame demonstrates, Little Rock civic culture didn't offer much encouragement to professional women. But it happened that the city did have plenty of potential clients for a young lawyer with a feminist orientation and expertise in the field of family law. Hillary had an opportunity to build a practice in an area of law she cared about passionately and that, with hindsight, raised few if any of the conflict-of-interest questions that would haunt her and her husband in later years. This was an option she never seriously considered. Married to a young politician who was earning just $6,000 a year and already thinking about raising funds to run for governor, Hillary had obvious reasons for preferring the security and steady salary she could earn with an established firm. And when asked if Hillary had ever considered practicing domestic law, her good friend Webb Hubbell dismissed the thought. "She wouldn't have been the first," he said, beaming as if Hillary's need to be first went without saying. "There were already two or three women divorce lawyers in Little Rock. So she wouldn't have been the first."

At the white-shoe law firm of Rose, Nash, Williamson, Carroll, Clay and Giroir, later known simply as the Rose Law Firm, Hillary was the first—not just the first woman, but the first Yankee feminist and the first politician's wife in a firm with a long tradition of steering clear of politics.

Never shy about lobbying on his wife's behalf, Bill Clinton had recommended Hillary to Herb Rule, a bankruptcy specialist at Rose who had supported Clinton's 1974 congressional campaign. But it was another Rose attorney, Vince Foster, who was her strongest advocate. A liberal by Arkansas standards, Foster served on a bar association committee that was trying to expand legal representation for poor people. He had met Hillary on a trip to Fayetteville and was hugely impressed by her work with the legal services clinic at the university. Hubbell, who had met Hillary by chance in 1973 when she came to Fayetteville to take the bar exam, was another of her supporters. Other partners worried that hiring the wife of the state attorney would expose the firm to conflict-of-interest problems. Foster couldn't see how. Rose did very little business with state agencies, he argued, and didn't represent any regulated utilities.

Webb Hubbell's memoir, *Friends in High Places,* stresses the hostility Hillary faced from partners who saw her as a threat to the firm's mas-

culine culture. He tells the story of how Hillary went for her interview with name partner Joe Giroir, Jr., and found him dressed in
sweaty tennis shorts, smoking a smelly cigar and resting his bare feet
on the top of his desk. This does not sound like Giroir, an impeccable
dresser with a taste for French provincial furniture and the perpetually
tanned, professionally groomed look one associates with luxury hotels
and the first-class sections of international flights. Giroir laughingly
dismisses the story as inaccurate. "Now, why would I be in my bare
feet?" he asks. As he recalls the interview, he had just come from
playing squash and while the top buttons of his shirt were open, he
had no idea that Hillary felt uncomfortable until much later when she
told him, "I thought you were testing me."

Like other women of the time who were breaking the gender
barrier in all-male offices, Hillary undoubtedly had a lot to put up
with, but she was more than capable of taking care of herself. Assertive
to a fault, Hillary had a high degree of confidence in her superior
intelligence—Hubbell found her "frighteningly smart"—and no trouble expressing her opinions, even as a first-year associate. Ironically,
though perhaps not surprisingly, most of the overt hostility to the
firm's first "lady lawyer" came from the female secretarial staff. Hillary's refusal to conform to the ideal of Southern femininity marked
her in their eyes as a sad case, and the secretaries made snide remarks
about her unplucked eyebrows and her attempts to diet by munching
on lettuce leaves that she brought from home in a plastic bag. Some
of the firm's male partners also found Hillary abrasive, and perhaps
even slightly intimidating. Although Hillary had been named by President Jimmy Carter to the board of the Legal Services Corporation,
an achievement that for many an attorney would have been the cap-
stone of years of work on bar committees and in the political trenches,
she would occasionally turn up at the office dressed in blue jeans, or
in one instance in a pair of orange pants more suitable for beach wear.

Hillary was assigned to work in the firm's small litigation section,
where she proved to be uncharacteristically nervous in her first court
appearance, defending against a plaintiff who claimed to have found
a mouse's hindquarters in a can of beans. However, she showed a

good understanding of office politics by attaching herself to an influential mentor. Just two years older than Hillary, Vince Foster had graduated first in his class from the University of Arkansas Law School, and he was fast making a reputation in the state as the Clint Eastwood of corporate litigation. Every inch the strong, silent type, he prepared his cases so thoroughly that many an opponent agreed to settle rather than risk a face-off with Foster before a judge and jury.

In a state where it seemed that everyone had some childhood connection with everyone else, it was perhaps not so great a coincidence that Foster had attended Miss Mary's School for Little Folks in Hope, Arkansas, where one of his classmates was Billy Blythe, soon to change his last name to Clinton. Although the boys were near neighbors, they had little in common. Billy was the pudgy boy who worked hard to make other kids like him, sometimes too hard. Vince, nicknamed Pencil, had been the tall, quiet kid whom other boys respected and every girl wanted for her prom date. As adults, Bill Clinton and Vince Foster were friendly, but their personalities didn't exactly mesh. Foster's reserve made it hard to know at times whether he was being serious or ironic. The story is that during his campaign for attorney general, Clinton had tried hard to sell himself to Foster, holding forth at length about all the good things he hoped to do for Arkansas. Foster listened deadpan and when the presentation was finished, commented in a level voice, "Okay. I'm all fired up."[1]

Even in the litigation department, Rose lawyers were expected to be able to attract new clients to the firm. For a young woman with no local connections, this was no easy task, and Hillary's lack of enthusiasm for the business world only made it more difficult. Like many newly minted lawyers of her generation, she was attracted to the power and salary potential of corporate law, but she didn't care much for corporations. In 1977, the same year she joined the Rose firm, Hillary contributed a chapter to a book entitled *Children's Rights: Contemporary Perspectives,* edited by Patricia Vadin and Ilene Brody, in which she once again called for self-appointed children's rights advocates to sue the nuclear power industry, manufacturers of "junk food" and other businesses whose activities might have negative effects on children in the future. (Her essay also proposed that children have a "right" to live in a world at peace, which might be enforced on their behalf through the United Nations.) A year later, she wrote a review of *All Our Children,* the book she had helped research, for the journal *Public Welfare*. Here, Hillary added television, automobiles and drugs (meaning pharmaceuticals) to her list of industries that extract illicit profits through the "extraordinary power they hold over all of us, but particularly over our children." These weren't the sort of views that would have gone over at the Rotary Club even if Hillary, as a woman, had been eligible to join.

Hillary's fierce idealism appealed to Hubbell and Foster, both of whom had reservations about the more aggressive, profit-oriented style of Joe Giroir, who was fast rising to a leadership position in the firm. Her determination to make her own way without sacrificing her identity and her principles took pluck. And the litigation partners, who worked with her on a daily basis, soon learned that there was an earthy sense of humor behind her often frosty exterior. Women's liberation hadn't yet conquered Little Rock, and to men who were used to traditional Southern belles Hillary's bold independence was intriguing. She was daring enough to go out to lunch with Foster and Hubbell at the Lafayette Hotel and later at the Capital, which at that time had a raffish reputation—"one step up from a cat house," recalls one network newsman who occasionally stayed there. On occasion, the trio even visited one of the local restaurants that combined lunch with a lingerie fashion show, a popular form of businessman's entertainment. Hillary could debate the fine points of legal strategy one minute and the next tease her companions for eyeing the shapely models.

Foster helped Hillary polish her litigation skills and protected her from detractors, as when the firm almost lost a major consumer protection case it was trying for General Motors because Hillary was the wife of the attorney general. Hubbell, meanwhile, talked his father-in-law, Seth Ward, into hiring Hillary as counsel for the Little Rock Airport Commission. The position didn't bring much income to the firm, but it gave Hillary an entrée to the local power structure.

In his memoir, the word Hubbell uses to describe his and Foster's reaction to Hillary is "mesmerized." And despite the different directions their lives have taken in recent years, Hubbell's face still lights up when her name is mentioned. She is the onetime protégée who made it good in the big time.

Rose firm partners tended to live in the White Heights, a quiet neighborhood of older homes perched on a cliff above the Arkansas River. They shopped in Dallas, New York and Paris, but at home they mixed with people they had known since high school. Social life was organized around the Country Club of Little Rock, benefits for the Repertory Theater, trips to Fayetteville for Razorback games, private dinner parties and Sunday-afternoon get-togethers beside the backyard swimming pool. The Rodham-Clintons were not immediately recognizable as country club material, and the same qualities that made Hillary fascinating to some of her colleagues engendered suspicions in their wives. At social gatherings, Hillary usually ended up discussing firm business with the men while the women talked about Junior League benefits and their children's schools.

Hillary was used to campus life and the camaraderie of the Watergate committee staff. She would occasionally suggest to Webb and Vince that they go out for a bite to eat or take in a movie after work. When Hubbell told her, "We have to call our wives first," she seemed surprised. Undoubtedly, Hillary was often lonely. She had no close friends in town, social events in Little Rock were invariably organized for couples, and the problem, recalls Hubbell, was that "of course Bill was never around."[2]

Bill Clinton had jumped into his first elective office with both feet and kept his staff working overtime, churning out pro-consumer lawsuits. Hardly a week went by without his office making news in some corner of the state, and from the day he took the oath of office he was laying the foundations of his campaign for governor, showing his face at church suppers in remote hamlets and schmoozing with courthouse pols in Arkadelphia, Fayetteville, West Memphis and the Delta.

Unfortunately, campaigning was never the only activity that kept Bill Clinton away from home. Marriage had done nothing to curb

Clinton's eye for attractive women, and by the spring of 1977 he was involved in an affair with Gennifer Flowers, at the time a zaftig brunette who worked as a reporter for KARK-TV. The married attorney general obviously couldn't be seen in public with another woman, but Flowers, in addition to her day job, was doing a cabaret act at the lounge in the Camelot Hotel, where Clinton occasionally showed up with friends.

According to Flowers, Clinton didn't talk much about his wife. When he did, he sometimes spoke admiringly of her brains and professional accomplishments. At other times, he referred to her as "Hilla the Hun." Observing that Clinton "seemed interested in me as a person," Flowers allowed herself to imagine that his marriage was an experiment that hadn't quite worked out. In December 1977, she says, she discovered that she was pregnant. Clinton was about to announce his candidacy for governor, and when he heard the news, he gave Flowers $200 in cash to pay for an abortion.

Flowers may have deluded herself into thinking that she and Clinton were in love, but there was always a compulsive quality about Clinton's dalliances. Never much of a drinker, Clinton was hyperactive and went through periods when he slept only a few hours a night. Like his mother, who still frequented nightclubs as a middle-aged matron, he was drawn to the world of honky-tonk bars and hotel piano bars. They were good places to pick up gossip—who was having an affair, who was on the take—and there were invariably good-looking women around.

One of the most disturbing stories told about Clinton during these years concerns an incident said to have occurred in a room at the Camelot Hotel during a convention that had brought several hundred nurses into Little Rock. Allegedly, Clinton approached a woman who was involved in the nursing home business and told her he wanted to come up to her room to discuss certain state regulations affecting the industry. Sometime later, in a state of extreme distress, the woman called a friend to her room and confided that Clinton had tried to force himself on her after she refused his advances. The friend applied ice to the woman's swollen lip. The victim of this alleged assault decided not to file charges, but rumors about the incident circulated for years, reemerging in 1998 when Lisa Myers of NBC News reported interviews with four individuals who had knowledge of the event.[3]

Whether—or when—Hillary heard talk of this incident is uncertain, but she knew enough about her husband's other affairs to be deeply hurt. Flowers recalls running into Hillary for the first time at a fund-raiser in Russellville, Arkansas. Flowers had come with a friend, but she was flirting openly with Clinton when she suddenly realized that the woman standing not five feet away from her was his wife. With unapologetic bitchiness, Flowers writes in her memoir that she saw Hillary as a "fat frump" who was behaving oddly, laughing too loud and drawing attention to herself.

Hillary's insecurity when it came to her husband's infidelities was at odds with her assertive demeanor in the office. Friends in whom Hillary confided almost invariably pointed out that infidelity was a professional hazard for politicians. Some suggested that the pursuit of other women was a Southern thing. In Arkansas they had a saying, "The dog was off the porch," and perhaps, indeed, it was just human nature for a male hound to wander off once in a while. Another theory traced Clinton's problem to his youth in Hot Springs, where retired gangsters and their flashy mistresses were local heroes. And of course, the stories told about his hero, JFK, only added to his legend. Clinton himself could always give the excuse that women pursued him, which was sometimes true. "This is fun," he told a woman friend. ". . . All the while I was growing up I was the fat boy in Big Boy jeans." The leap from naive amazement at his good fortune to a sense that women were just another fringe benefit of public office appears to have been made early on.[4]

Unable to control her husband's behavior, Hillary concentrated on taking care of the family finances. As it turned out, during Clinton's term as attorney general, the legislature made a rare move to raise the salaries of public officials to bring them in line with inflation. Bill Clinton was soon earning not $6,000 a year but $22,500, and even in 1977, their first year in Little Rock, he and Hillary declared a joint income of $41,000, a figure that made them comfortably middle-class

by Arkansas standards. Nevertheless, Hillary fretted about their pre-
carious financial situation. Clinton's idea of financial planning was to
remember to bring his wallet when they went out to dinner. If they
were ever going to build a nest egg, it would be up to Hillary to
figure out a way.

Hillary's concerns were well known to Jim McDougal, who was
just six years older than Bill Clinton but fancied himself something of
a mentor. Although still in his thirties, McDougal affected a style that
was a throwback to an earlier era—the country boy made good.
Everything about McDougal proclaimed that he made his living without resort to manual labor. He wore custom-tailored suits, crisp dress
shirts and Bally shoes. A straw hat shielded his face from any hint of
a tan, and he drove flashy foreign cars with an ineptitude that suggested his superiority to mere machinery. McDougal's speech was
salted with quotations from his political idol, FDR, and he could spin
hilarious tales about country characters like Mudd Goad, the "village
Republican" in the hamlet where he grew up. McDougal was not a
simple man, but he did a fair imitation of simplicity. He liked to say
that his father had taught him the importance of dressing up and, as
he put it, "making his manners": "If you watch your appearance and
say 'yes sir, no sir' and are eager and willing, you can go a long way
just on that."[5]

A few years before Clinton came on the scene, McDougal had been
marked as an up-and-comer in Arkansas politics. He was a protégé
of Bob Riley, a much respected World War II veteran, blinded in
combat, who served briefly as governor. He worked in Washington
for both Fulbright and McClellan, and in 1965 he played a role in
the storied "statehouse coup" that toppled the Democratic machine
of Orval Faubus. Alcoholism and recovery through a commitment to
AA put a crimp in McDougal's ambitions, and by the mid-1970s he
was teaching political science at Ouachita Baptist University near Arkadelphia, where Riley was department chairman. There McDougal
fell in love with one of his students, a pretty Latin major named Susan
Henley, and they decided to marry and move back to Little Rock,

where McDougal would devote himself full-time to buying and selling country real estate.

Bill and Hillary attended the wedding, which was held in the backyard of the house in Little Rock that McDougal had renovated for himself and his bride. The neighborhood was "not a bluestocking area," recalled Bob Riley's wife, Claudia, who was also among the guests, but Susan wore a white dress with an elaborate train and the ceremony took place under a floral arbor set up in the backyard while the guests stood around in a circle, discreetly scratching their fresh chigger bites.

The McDougals saw a good deal of Bill and Hillary during Bill's term as attorney general. Jim did not find Hillary especially smart— he considered Susan, for all her lack of sophistication, more intelligent—but he sympathized with Hillary's efforts to make herself agreeable, so often undone by her flat midwestern accent and inability to grasp the rudiments of Southern manners. Susan McDougal, just twenty-one, hardly knew what to make of Hillary. She admired her ability to stand up for herself and for women in general. On the other hand, there were moments when she realized that Hillary did not appreciate how lucky she was to be the wife of the governor. There was, for example, the time when a female Clinton supporter proudly presented Hillary with a pair of handcrafted earrings in the shape of razorback hogs, the U of A mascot. Hillary refused to try them on and, barely waiting until the woman was out of earshot, commented, "This is the kind of shit I have to put up with."[6]

One evening in the summer of 1978, the two couples were having dinner at the Black-Eyed Pea family restaurant, when Jim mentioned that he and Susan had just located a 230-acre parcel up in Marion County in northwest Arkansas. Bill and Hillary had invested in an earlier project called Saltillo Heights in which Senator Fulbright had been Jim's main partner, earning almost 100 percent on their small investment. Now Jim was offering them a half interest with him and Susan on a more desirable property, with frontage on the scenic White River. Best of all, they wouldn't have to put up a penny in cash. The

McDougals had already arranged financing in the form of a mortgage loan for $182,611 from the Citizens Bank of Flippin, Arkansas. The remainder of the purchase price could be covered by a $20,000 personal loan taken out by Bill Clinton and Jim McDougal from the Union National Bank. Jim was confident that he could subdivide the land and start selling off the lots quickly, using the revenue stream to pay off both bank loans. All Bill and Hillary had to do to get the Whitewater Development Company launched was to add their names as co-signers to the mortgage note.

As Jim McDougal would recall, the conversation lasted only a few minutes. Hillary was eager to be involved. Bill gave his assent casually "as though agreeing to buy a potted plant."[7]

In taking McDougal up on his offer, Bill and Hillary had just made the two most common mistakes of naive investors. They took the advice of a friend who appeared to be an expert without doing their due diligence, and they grossly underestimated the risks inherent in the deal. Hillary talked excitedly about Whitewater at the office, telling Webb Hubbell and others that McDougal was setting aside one of the waterfront-view lots for her and Bill to build on. Even so, they were both too busy to drive up to Marion County and inspect the property for themselves. Moreover, when their accountant, Gaines Norton, warned that McDougal was drawing up the partnership papers incorrectly and taking improper tax deductions, Bill told him not to interfere. Since Bill and Hillary weren't putting up any cash, they probably felt that it would be pushy to ask too many questions. Nevertheless, the mortgage note they had co-signed made them personally liable for the entire $182,000 mortgage, an amount more than three times their gross income for the year 1978.

Although Whitewater would take some dizzying twists and turns in later years, the partnership was already fated to end unhappily. Since Jim McDougal appeared to be knowledgeable and the parcel of land was obviously attractive, the Rodham-Clintons never seriously considered the possibility that they might be stuck with the company's debts. When the land deal started to go sour, they chose to blame

McDougal for steering them into a bad investment. But since they were liable for the company's debt, they couldn't just write off their mistake to experience. Rather than put up a substantial amount of their own capital to satisfy the mortgage, they chose to go along passively—and at times, not so passively—with the schemes McDougal cooked up to avert a default.

McDougal may well have persuaded himself that he was doing Bill and Hillary a favor by bringing them in on Whitewater, but of course there was another motive for his generosity. Banks are normally reluctant to give mortgages on undeveloped land, especially to someone like McDougal, who already had liabilities of hundreds of thousands of dollars from other leveraged land deals. The Citizens Bank of Flippin took the risk only because one of the partners was Bill Clinton, who at the time the deal was closed had just won the Democratic primary for governor and was a shoo-in to prevail in the general election.

Clinton's participation was all the more important because the Whitewater land had already been flipped once, and was selling at a premium relative to other acreage in the area. Frank Burge, the loan officer who handled the transaction, assumed that Clinton's friends would be buying lots at above-market value as a discreet way of subsidizing his political future. Since Jim McDougal was not exactly an innocent when it came to such customs, presumably the same thought had occurred to him. But if Bill and Hillary were aware of any such expectations, they did nothing about them. None of their friends bought lots.

The purchasers who did come around, and there weren't many of them, were low-income individuals looking for an inexpensive summer camp or retirement homesite. McDougal made these sales on a contract basis, a form of financing once common in rural areas where bank mortgages can be hard to come by. In a contract sale, the landowner retains the title until the property is paid off, thus protecting himself from the expense and inconvenience of having to foreclose on a deadbeat buyer. McDougal liked to think of this financing method as a boon to working people who otherwise would never have the opportunity to buy rural property, but contract sales by developers have a bad reputation and now are illegal in many states. The buyer in a contract sale has no equity. If he or she misses a payment for any reason, the land is forfeited.

Ironically enough, it soon became apparent that the contract sales were a bad deal for the partners as well as for the purchasers. Essentially, Whitewater was undone by the runaway inflation of the Carter years. Even as interest rates on bank loans were climbing toward 20 percent, Arkansas's usury laws prohibited private lenders from charging more than 10 percent. Slack sales and marginal profits never allowed the partners to make up for the effects of the interest rate squeeze. As a result, McDougal couldn't afford to maintain and develop the property. Some buyers, seeing that "Whitewater Estates" never lived up to the glowing copy of the sales brochure, simply abandoned their contracts, and the vacation enclave perched above the sparkling waters of the White River soon degenerated into a drab encampment of fishermen's trailers and garbage-strewn lots.

As the property's troubles mounted, Bill Clinton and Hillary Rodham would be called on to ante up $22,620 in 1978 and $12,490 in 1979. The following year, the Citizens Bank refused to refinance the mortgage unless Whitewater started to pay down the principal, so McDougal asked Bill and Hillary for an additional $9,000, plus $4,350 in interest. Even so, he and Susan were absorbing more than half the venture's losses. Embarrassed about luring his friends into a disastrous investment, McDougal refrained from telling them how bad the situation actually was.

Bill Clinton was no businessman, but when it came to electioneering, he was ready to try the most up-to-date tactics. Freelance political consultants were still a novelty, but Clinton hired a New Yorker named Dick Morris who used polls to pretest the issues, so that Clinton knew which positions would play well to voters before he committed himself. Clinton was fascinated by Morris's adaptation of opinion sampling techniques originally developed by Hollywood studios to pretest the plot lines of big-budget films. After it became clear that he was the heavy favorite in the primaries, he and Morris turned their attention to plotting against Representative Jim Guy Tucker, who was considered likely to be his chief rival in the future. Governor David Pryor was running against Tucker for the Senate seat formerly held by John McClellan, who had died in office. His own victory all but assured, Clinton talked Pryor into hiring Morris to run some negative campaign ads. Barbara Pryor found Morris's methods so repellent that she banned him from the governor's mansion. Hillary, like Bill, thought Morris was brilliant.

Even Bill Clinton's extramarital escapades took a backseat to the election cycle. Not long after Flowers ran into Hillary at the Russellville fund-raiser, a volunteer at his campaign headquarters received an anonymous call from a woman who threatened, "You'd better tell Bill Clinton we know about his affair with Gennifer Flowers, and it had better stop." Clinton, who happened to be in the room when the call came in, laughed off the threat, joking that he was "flattered" by the accusation even though it wasn't so. Gennifer Flowers wasn't at all sure that the threat was a hollow one. She had recently auditioned as a backup singer for country star Roy Clark, and she was relieved when Clark offered her the job. For the next seven years she would live away from Little Rock, mainly in Tulsa and Fort Worth, seeing Bill Clinton only occasionally.[8]

Once again, however, it was Hillary who became a campaign issue. Clinton opponent Frank Lady questioned the propriety of her employment by a firm that represented the business interests of politically powerful brothers Witt and Jack Stephens. Lady's charges had little impact on the voters, largely because the public was scarcely aware of Hillary. Dick Morris had chosen to feature Clinton's mother and his first-grade teacher in the TV ads he produced, keeping the candidate's Yankee wife out of sight.

ARKANSANS did not hear much about the state's new first lady until after the election, when she was introduced to them in two starkly contrasting newspaper interviews. The first, which ran in the *Arkansas Gazette,* was a conversation between the newly elected governor and

Bill Simmons, who covered Little Rock for the Associated Press. Before the interview got started, Clinton allowed Simmons to listen in while he cooed endearments to Hillary over the phone. He then spoke "with something akin to reverence" about the merits of "my little wife," wrote Simmons. "She's just a hard-working, no-nonsense, no frills, intelligent girl who has done well, who doesn't see any sense to extramarital sex, who doesn't care much for drink, who's witty and sharp without being a stick in the mud," Clinton enthused. The reference to Hillary's disapproval of extramarital sex was just one of the unexplained oddities of this conversation, during which Clinton also confessed to mooning cars as a teenager in Hot Springs.

Hillary greeted the people of Arkansas in her own way a few weeks later, giving a tour of the private quarters of the Governor's Mansion to a woman's page reporter from the rival *Arkansas Democrat*. During the course of the tour, Hillary explained that she did not expect to make her position as first lady a full-time job. Also, she would be keeping her own last name because, she said, "I need to maintain my interests and my commitments. I need my own identity, too." The suggestion that married women who used their husbands' names somehow lacked separate identities was an unfortunate example of Hillary's tendency to make sweeping pronouncements, oblivious to the possibility that she might be insulting people. The *Democrat* headlined the piece "MS. RODHAM: JUST AN OLD-FASHIONED GIRL."[9]

Arkansans were already a bit wary of first ladies with identity problems. In 1975, Barbara Pryor had announced that she was taking "a year off from marriage" to find herself, using the time to go back to school and start a movie production company. Pryor also adopted a frizzy permanent, which became the symbol of her declaration of independence and the talk of the state—after which Hillary, then a law professor in Fayetteville, had permed her own hair in a gesture of sympathy. Barbara Pryor's experiment ended happily, but the public had reason to be wary of the possibility that they were about to become a captive audience for another marital drama in the governor's mansion, and Hillary's decision to call herself Ms. Rodham would become more controversial over time, not less so.

But the name issue was still just a minor annoyance as she and Bill danced together at their "Diamonds and Denim" inaugural ball. Clinton, as the youngest governor in the nation, had already been the toast of the Democratic Party's midterm convention in Memphis, and a few reporters from national publications filed items speculating that he might be named Ted Kennedy's running mate in 1980. (Since Clinton would not be thirty-five until the summer of 1981, they were getting a little ahead of the game.) For the moment, even the normally skeptical *Democrat* allowed itself to hope that Camelot was being reborn on the banks of the Arkansas River.

To understand what Clinton himself expected to accomplish in office, one has only to look back to his letter to Colonel Holmes a decade earlier: "For years I have worked to prepare myself for a political life characterized by both practical and political ability and concern for rapid social progress." Today, in a decade that tends to be more skeptical about the ability of activist government to foster prosperity, one can only wonder at the hubris of this statement. Arkansas, while a poor state by some measures, had always produced its share of talented politicians. Clinton's belief that he could spur the "rapid social progress" that had eluded his native state in the past was founded on the Great Society principle that cleverly designed programs could act as catalysts for growth. And he wasn't alone in thinking this. As late as 1988, Hillary's professor friend Diane Kincaid Blair would publish a study of Arkansas political history in which she argued that the state's economic backwardness was a function of its tradition of limited government. Simply put, Arkansas did not have a low tax base because it was an economically underdeveloped state; rather, Arkansas was poor because taxes were too low.

Clinton's first message to the legislature called for rebuilding the highway system, reorganizing the public schools and creating a network of rural health clinics. And this was just the beginning of a detailed program that consultant Dick Morris would summarize as "admirable but indescribable." By priming the government pump, and drawing in more federal dollars, Clinton planned to work miracles. As he promised in his inaugural address, Arkansas would soon become the "envy of the nation."

But the most noticeable accomplishment of the new administration was to bring to an end a long tradition of frugality in state government. One of Clinton's first acts as governor was to ask the legislature to increase his administrative budget by 15 percent. This $200,000 increase, soon augmented by a $300,000 federal grant, allowed him to hire additional aides. Soon Arkansas, forty-ninth of the fifty states in terms of average income, supported the twenty-third-largest governor's staff in the nation. Clinton doled out appointments to personal friends like Rose Crane, who had lived across the street from him in Hot Springs, and his cousin Sam Tatom, a private detective who became director of the Department of Public Safety. More controversially, he brought in friends from out of state at higher salaries than native Arkansans had been earning in the same jobs.[10]

Instead of one chief of staff, Clinton had three, quickly nicknamed by the legislature "the Three Beards." Rudy Moore, Jr., was a former state representative about the same age as Clinton who had managed his campaign for governor, developing a computerized list of donors so effective that the campaign ended with a surplus. Steve Smith, who had served in the attorney general's office, was an ardent environmentalist who favored banning the clear-cutting of timberland, a combustible issue in a state in which International Paper happened to be one of the largest landowners. The third aide, John Danner, and his wife, Nancy Pietrafesa, were old friends of the Clintons. A management consultant with degrees from Harvard and Berkeley, Danner ran brainstorming sessions for state employees, tacking a length of butcher's paper onto a wall and scribbling down ideas for policy initiatives as fast as the group could call them out. He also used state money to allow public employees to attend a seminar entitled "How to Tell What Turns You On."

Hillary's schedule was suddenly full. Her work for the Legal Services Corporation, Children's Defense Fund activities and business travel kept her on the go, with frequent trips to Washington and New York. Marked as young Democrats with a future, she and Bill were guests of President and Mrs. Carter at a White House dinner. At home in Little Rock, they were together more than they had been at any time since their marriage. At the Governor's Mansion, they played host and hostess to a permanent policy seminar. Almost every evening Clinton aides would gather around the kitchen table, munching chocolate chip cookies baked by the mansion cook, Liza Ashley, while they debated ideas for new programs. Hillary often excused herself and went to bed early, but Bill enjoyed these sessions so much that he would keep the conversation going until one or two A.M., then have to drag himself out of bed the next morning.

Hillary also had the company of Nancy "Peach" Pietrafesa, an outspoken feminist. Pietrafesa and Danner had come to Arkansas a few months before the election to work as volunteers on Clinton's campaign. They stayed with Bill and Hillary to save money, a situation that quickly gave rise to rumors that Pietrafesa was Hillary's lesbian lover, imported from out of state. After the election, Clinton appointed Pietrafesa liaison officer to the Ozarks Regional Commission, making her the highest-ranking woman in his administration.

It was the era of "small is beautiful" economics. A much-discussed report issued by the Club of Rome in 1972 had forecast that the world would soon run out of oil, and Clinton's brain trust was as passionate about saving Arkansas from pollution and conserving energy as it was about attracting jobs. Fairly or not, Danner and Pietrafesa became identified with this point of view, which infuriated a segment of the legislature. Although right-to-work laws gave the state an edge in attracting manufacturing jobs, Danner spoke disdainfully of "chasing smokestacks" and promoted handicrafts, service-oriented businesses, experimental farms and "community development" projects. After fourteen months, when the legislature began introducing resolutions mocking Danner's self-important style, he and his wife were sent packing.

Typical of the "small is beautiful" vision, though it didn't happen to be one of Danner's programs, was the Special Alternative Wood

Energy Resources program, known as SAWER. Funded primarily by a $410,000 grant from CETA, a federal job-training act, SAWER was supposed to train Arkansans to chop wood—roughly the equivalent of teaching Minnesotans to shovel snow. According to one progress report by the *Arkansas Democrat,* six SAWER trainees had managed to cut just three cords of wood at a cost of $62,000. The only person to get much benefit out of SAWER was Monroe Schwarzlose, an elderly turkey farmer and perennial candidate, who promptly announced that he personally had cut three cords of wood at no cost to the government whatsoever. On the basis of this claim, Schwarzlose would capture 31 percent of the Democratic primary vote for governor in 1980.

It is universally agreed that the worst gaffe of Clinton's first term was raising the cost of car license tags. Strangely enough, considering how often this issue comes up in the reminiscences of Clinton supporters, some of them still miss the point. In an essay he contributed to the 1993 book *The Clintons of Arkansas,* Rudy Moore writes, "It was not a matter of dollars—the increases amounted to ten or twelve dollars per car." Webb Hubbell's memoir calls the car tag issue "symbolic." But in the late seventies the average annual income in Arkansas was about $12,000, which means that many a family was getting by on $9,000 or $10,000, or even less. Tag fees, moreover, were assessed by the weight of the vehicle, not the value. In a largely rural state where it was not unusual for a family to have three or four old heaps in the driveway, many a taxpayer showed up at the county revenue office to find that he would have to come up with an extra forty, fifty or even seventy-five dollars just to keep his vehicles operating legally. This was not a symbolic problem.

The Clintonites' inability to understand that these small amounts of money actually mattered to people blinded them to the erosion of Clinton's support until it was too late. When Clinton hit the campaign trail in 1980 he was shocked to find that the same audiences that cheered him two years earlier had turned hostile. "Rudy, they're killing me out there," he reported to Moore, who was amazed to hear it.

Hillary, meanwhile, had launched herself into another high-risk investment. On October 11, just three months after entering into partnership with the McDougals and on the eve of Bill's election to the governorship, she opened a commodities trading account at the Springdale, Arkansas, office of Refco, the Chicago-based broker. In ten months of trading, her account netted an amazing $99,540— nearly a 10,000 percent return.

This time Hillary's mentor was Jim Blair, the fiancé of her best friend, Diane Kincaid. Taller than Bill Clinton, Blair was an imposing physical presence whose arched eyebrows gave him a look of devilish glee. Like Jim McDougal, Blair had worked for Senator Fulbright— almost a rite of passage for ambitious Arkansas Democrats—but he soon became a lawyer in private practice, sometimes acting as a straw buyer for his chief client, chicken king Don Tyson. Blair would purchase land and shares in rival operations from owners who had no idea that they were selling out to the competition. Not the least bit apologetic about being a capitalist, Blair represented a type that Hillary can't have encountered often in her progress through Wellesley and Yale. He liked vintage wines and high-stakes poker and threw lavish parties that can't possibly have been as wild as some of the stories told about them.

Blair had been introduced to commodities trading by one of his clients, Robert L. "Red" Bone, a former Tyson executive who played in high-stakes Vegas card games and competed regularly in the World Series of Poker. Blair had seen Bone through two divorces and represented him when he was accused of manipulating the egg futures market while on the Tyson payroll. By 1978, Bone had moved on to become a trader in the Springdale office of Refco, and Blair became his customer. Not one to do things halfway, Blair kept a primitive Radio Shack computer on his desk for tracking commodities prices. "I was on a streak," Blair told *The New York Times*'s Jeff Gerth in 1994, "on a streak that I thought was very successful, and wanted to share this with my close friends, as I did with my fiancée, as I did with my law firm, as I did with my children, as I did with my fiancée's children, and as I did with the person who was the best person at my marriage, which was Hillary, my tennis partner and friend."[11]

Hillary would indeed stand as "best person" at the Blairs' wedding in September 1979. She did not, however, wear a tuxedo, as reported in several sources. It was Bill Clinton who wore formal dress—not a tuxedo but white tie, tails and a top hat—to perform the ceremony. Hillary was dressed for the occasion in an unremarkable floral print dress, and the bridal couple's four children from previous marriages joined her as attendants. Over the years, the Rodham-Clintons would be frequent guests at the Blair summer house on Beaver Lake, where the entertainment ran to marathon games of hearts, organized by Bill, and charades, during which Bill and Hillary met the challenge of miming such book titles as *Being and Nothingness*.

On Blair's advice, Hillary was about to plunge into the high-risk arena of commodities trading. The commodities market is essentially a socially useful form of gambling, and compared to the stock market it is very much an insider's game. Amateur speculators can theoretically make huge profits, but an estimated 80 to 90 percent of them lose money in the long run. Commodities contracts are also highly leveraged. Under normal circumstances Hillary, who earned $24,250 at the Rose firm in 1979, would not even have been allowed to open an account.

Explaining her venture into commodities during a 1994 press conference, Hillary would claim that she made all her trades herself. Jim Blair had explained to her that cattle futures were entering a bull market of historic proportions, she recalled. And so, after studying *The Wall Street Journal* and other publications, she jumped in and began to trade. "I was lucky," she said of her 10,000 percent gain.

Business writers and financial analysts were quick to point out that if riding a bull market were all it took to score big in commodities, a lot of people in the Farm Belt would be driving Ferraris. William Harvey, who succeeded Hillary as chairman of the board of the Legal Services Corporation, checked his files and discovered that Hillary was involved in LSC functions on the days of three of her more substantial trades. This information, which Harvey provided to the author, shows that on January 2, 1979, the day of her largest single trade, Hillary chaired a daylong legal services conference with only a fifty-minute break for lunch. She could not have been on the phone to Refco making the trades charged to her account.[12]

In any event, far from riding the bull, Hillary began her trading career by selling short, betting on a temporary dip in the upward trend, a highly unusual move for a beginner. From her first day in the market, she traded like a pro. By December 11, just three months into her trading career, she controlled contracts with a value of $2,300,000. And yet, Hillary had only $6,000 on deposit with Refco to meet margin requirements.

Hillary did not have a discretionary account, so it was against the rules for someone else—say, Blair or Red Bone—to trade for her. One can safely assume, however, that this would not be the first time this particular rule had been bent. As for the margin requirement, presumably the brokerage was letting Hillary piggyback on Jim Blair's credit. But it takes more than these irregularities to account for Hillary's astounding success. A careful analysis by Caroline Baum and Victor Neiderhoffer, who examined Hillary's trading records for the *National Review,* found that a high percentage of successful trades in her account were made at or very close to the moment when the market had reached its high (or in the case of shorts, its low) for the day. Such near-miraculous timing is often an indication of a practice known as straddling: Over the course of a given day, a broker buys and sells large blocks of contracts for his own account. At the close of business, he allocates profitable trades to certain favored clients, dumping losing transactions off on the less favored.

Red Bone has denied resorting to straddling, but lawsuits filed against Refco tell another story. For example, one Refco client named Stanley Greenwood complained that trades of over a million dollars were assigned to his account without his knowledge. He then received a margin call that wiped out his equity, leaving him with a loss of $100,000. Baum and Neiderhoffer also point out that the original records of two of Hillary's most profitable trades are missing, while others show certain anomalies—missed keystrokes and the like—suggesting that they may have been drawn up after the fact.[13]

If Hillary had really been trading without a net, she would have needed nerves of steel. On July 12, 1979, her equity was minus $20,000 and, had she been an ordinary client, she would have faced a margin call of $117,500. Seemingly, she was on the brink of financial ruin. Nevertheless, the next activity in her account was a doubling of her short position by another fifty contracts. Four days later her balance was back on the plus side, and in another two weeks she was able to close out her account, withdrawing $60,000 in profits.

When a politician's wife starts making big money on the eve of an election, a degree of suspicion is always in order. And in fact, about two months before Hillary opened her Refco account, Don Tyson had brokered a deal with Clinton to raise the weight limit on trucks, bringing Arkansas in line with neighboring states. In a 1994 interview with journalist Michael Kelly, Tyson confirmed that Clinton had promised to push the weight limit change through: "Damn right he did. He promised me, personally, in my car driving to the airport three or four months before the election. He said that if he was elected, he'd do it. He said he'd take care of it. Now, there's a bunch of chicken folks in this state, and we all had a big interest in this weight thing, so we supported Clinton."[14]

There are several problems with the theory that Hillary's commodities score was a disguised bribe, however. Even for Tyson, who did things in a big way, $99,000 would be a huge payoff for an untried politician. Clinton, moreover, failed to keep his promise on raising the weight limit, caving in when he ran into opposition from the Highway Commission. Clinton's defeat on this issue squelched a plan to get truckers to pay higher license tag fees and led directly to his ill-fated increase of car tag fees for motorists.

On the other hand, Hillary's timing in closing her account in July 1979 was amazingly lucky. Jim Blair would not fare so well. One day in October, Refco chairman Thomas Dittmer, Red Bone's chief source of market information, indicated in his morning conference call that he was going long on cattle. Instead, he started liquidating contracts, forcing prices down. Within two days, Bone was wiped out, and so was his favorite client. Blair personally lost $2.5 million. Refco eventually faced federal charges of market manipulation in connection with the sudden end of the bull market in cattle. The firm was found not guilty but paid $250,000 in fines levied by the Chicago Board of Trade. Jim Blair sued, settling out of court for an undisclosed sum. He later went to work for Tyson Foods as general counsel.

Hillary's explanation for her decision to quit trading on July 23 was that she had finally become pregnant after years of trying and couldn't take the nervous strain of worrying about her account. It would seem, however, that she could have been only one or two weeks pregnant on the day she closed the account. At any rate, in October, a few weeks after learning about Jim Blair's disastrous losses, she opened another commodities account, this time with Bill Smith, a Stephens, Inc., broker. Smith traded on Hillary's behalf, nearly doubling her initial $5,000 stake in six months. Since Hillary could sleep well with her money invested by Smith, she must have understood that her account in Springdale had exposed her to a much higher order of risk. Indeed, had she stayed in the market a few more months she would have been bankrupt, with devastating consequences to Bill Clinton's future in Arkansas politics.

With hindsight, the most remarkable aspect of Hillary's adventure in the commodities market was her faith in her own incorruptibility. The experience appeared to occupy its own compartment in her consciousness, which had nothing to do with her views on other people's business ethics. *New Yorker* profiler Connie Bruck relates the story of a certain Refco insider who happened to be present at a party in early 1979 where Hillary and a friend were making "rather pious pronouncements" about the commodities market: "They saw it as too laissez-faire. They were leaning toward more government regulation, rules." Moreover, they disdained the whole principle of market-making, condemning it as "not socially acceptable, dealing in such large sums of money, such greed." The hypocrisy of this infuriated the insider, who found himself shouting, "Give the money back—I'll give it to a charity! Give the money back!"[5]

Earning a quick $99,000 did not put an end to Hillary's anxieties. When tax time came around, she looked into the idea of investing in a gold mine as a tax shelter but was talked out of it by one of her law partners. Next, she approached Jim McDougal, asking if she and Bill could take deductions for interest he and Susan had paid on the Whitewater loans. McDougal thought not. He also tried to tell Hillary about straddling, and how losses on her commodities account had doubtless been charged to some other client. It infuriated him that she pretended to be unable to follow his explanation.

Bill and Hillary eventually deducted only their own contributions to Whitewater. Even so, some of the money they listed as interest actually represented payments for the principal of the loan and operating expenses. On their 1980 return they also neglected to list Hillary's profits from her account at Stephens, Inc., an omission that was rectified only in 1994, after the existence of the account was reported in news stories.

When it came to keeping personal and public business separate, both Bill and Hillary exhibited a certain carelessness that may have been a holdover from the mentality of sixties activism, which celebrated improvisation and disdained petty formalities. After the rise in car tag fees became an issue, it was revealed that Hillary had failed to renew her tags for the Fiat she drove in 1977 and 1978, and Clinton gave the excuse that the aide in his office who was supposed to take care of the matter had forgotten.

The state of Arkansas's handling of the governor's compensation didn't help matters. Bill Clinton's salary would remain fixed at $35,000 throughout his career as governor. Since it would have taken another constitutional amendment to give him a raise, the legislature compensated by offering extras, which increased in value over the years. In addition to receiving the usual fringe benefits, such as health insurance and a pension, the governor and his family lived in the official residence, a redbrick Georgian constructed in 1950, and had a staff of servants, including a cook, maids and the mansion administrator. A twelve-man detail of state troopers provided round-the-clock security and chauffeured Clinton in his official Lincoln automobile. Even household furnishings like bed linens somehow wound up being charged to the state. The governor also received a food budget, which by 1992 amounted to $51,000, and a $15,000 "publicity fund." In all, the cost of maintaining the governor and his family exceeded $750,000 a year, but Clinton paid taxes only on his base salary.

To her credit, Hillary made an effort to avoid fuss and formality, which she genuinely disliked. She drove her own car and did without an official trooper escort. But Hillary was always in a rush—juggling her career, her duties as first lady, her activist projects, her personal life—and the habit of ignoring boundaries became ingrained. During Clinton's first term one aide, Russ Perrymore, clocked so many miles on his state-issued car running personal errands for the governor and the first lady that legislators threatened to order an investigation. Hillary took state business into the Rose firm and asked state troopers to deliver personal items to her at work, including Tampax—a chore they saw as an attempt to humiliate them. By the late 1980s, she would be criticized for dispatching the convicts who did yard work for the Governor's Mansion to take care of the outside chores at a Little Rock condo owned by her parents. It turned out that good intentions were inadequate protection from these petty temptations.

SIX

Survival Plan

EARLY in the evening on February 27, Bill Clinton arrived home from a trip to Washington just in time to drive his wife to Little Rock's Baptist Hospital. Hillary had been having a stressful time, trying to wrap up a particularly emotional child custody dispute, when she began to experience labor pains. At the hospital the obstetrician became concerned about the position of the baby in the womb, and shortly before midnight Hillary was wheeled to the delivery room, where she gave birth by cesarean section to a girl. The baby would be named Chelsea, after the Joni Mitchell song.

Grandmother Virginia—now Virginia Kelley, because of her fourth marriage—cooled her heels in the waiting room for hours. At last, she asked a nurse what was taking so long. "They're bonding," the nurse replied. At the Yale Child Study Center, Hillary had learned that in order to develop properly, babies need to establish an emotional attachment to their parents in their first hours of life.

Hillary herself would tell the story about her first attempt to breast-feed. When bubbles began coming out of Chelsea's nose, she became frightened and called out for the nurse. "Well, if you hold her head up a little higher that won't happen," the nurse said. The point of this story, as Hillary told it, was that all new mothers need expert advice. But Hillary, in this instance, *was* the expert, having studied infant development under Dr. Sally Provence, so the moral becomes blurred.

Hillary had always believed that no woman could be truly fulfilled unless she had children. A mother at last at thirty-two, she was determined to do everything by the book. As a result, Chelsea Clinton would be subjected to the safest environment that expert advice could devise. Diane Blair recalled Hillary and Bill arriving for a weekend visit with Chelsea riding in a backseat infant bed, "more thoroughly protected against any possible automobile injury than any infant I'd ever seen. She was thrilled to be freed from her restraints."[1]

According to Hillary, she took four months off from practicing law to stay home with the baby, followed by two months of working part time, mostly from the Governor's Mansion. Webb Hubbell would remember Hillary's leave as being closer to six weeks. Hillary had just made partner the previous year, and some of her colleagues at the Rose firm were skeptical of her ability to keep her billing up while taking care of a baby and serving as the state's first lady besides. But life in the Governor's Mansion gave Hillary an advantage over most working mothers. Chelsea's nanny, Dessie Saunders, who began work the week after Chelsea's birth, was put on the state payroll as a security guard at a weekly salary of $75.

Clinton, whose own father had died before he was born, was delighted to be a parent. But doting fatherhood did not reverse the negative trends in his personal life. In a 1993 reminiscence, Clinton aide Rudy Moore, Jr., wrote, "As I look back, it is more evident that Bill Clinton was not the same person psychologically in 1980 that he had been before or that he has been since. It must have been something personal, perhaps in his relationship with Hillary, but he was ambivalent and preoccupied."[2]

Wives tend to expect that the arrival of a child will inspire their husband to settle down and take more responsibility. The opposite seemed to be the case with Clinton. Women who had occasion to meet him at official functions joked about the "Clinton handshake"— a little too warm and too prolonged to be anything but a wordless come-on. At times, he flirted outrageously with women in front of

Hillary, and even in front of the women's husbands. Moreover, now that he was governor, Clinton had an escort of state troopers wherever he went, and he would regale them with lewd comments about the attributes of any attractive woman who happened to cross his path and occasionally ask the troopers to get the phone numbers of good-looking women he spotted at political rallies. Gary Gentry, not one of the troopers who went public so controversially in 1993, told Little Rock journalist Meredith Oakley that he attributed much of this behavior to a misplaced desire on Clinton's part to be "one of the boys." There may have been at least some truth to this. Clinton always wanted to be admired, and sex was one of the few interests that he and his trooper escorts shared. At any rate, with Hillary preoccupied with the baby and her job, Clinton was feeling sorry for himself, and he compensated by seeking out other women.[3]

The heady days of the endless policy seminar were over, and the dismissal of John Danner and Nancy Pietrafesa, a month after Chelsea's birth, removed Hillary's best friend in the administration. On the political scene, there was nothing but trouble. Jimmy Carter had opened a detention center for Cuban boat people, the Marielistas, at Fort Chaffee, near Fort Smith, Arkansas. Clinton, as a staunch Carter supporter, at first welcomed the refugees, accepting the administration's assurances that the camp would be temporary and limited in size. Before long, however, some 1,900 refugees, mainly male and some with criminal histories, were packed into the facility, living under squalid conditions. On the day before the Democratic primary, several hundred of the roughest detainees stormed the fences. Their military guards, under orders not to turn their weapons on camp inmates, simply let them go. Homeowners in the area lived in terror until the Cubans were rounded up, and the voters reacted by giving 30 percent of the vote to the chicken farmer Monroe Schwarzlose, previously considered to be in the race mainly for his entertainment value.

On June 1, there was another breakout from the center, and once again soldiers stood by as hundreds of rioters marched down the highway carrying clubs and broken bottles toward a confrontation with outnumbered state police units. Clinton activated the National Guard and toured the area, speaking to agitated local residents from the back of a pickup truck. He thought he had Carter's promise not to send any more refugees to Fort Chaffee, but that summer Carter decided to close camps in three other states and dump them all in Arkansas. Clinton was furious, but he was also scheduled to give a prime-time speech at the Democratic convention. He avoided an outright break and kept his speaking date, leaving many Arkansans with the feeling that he had chosen ambition over protecting their interests.

Clinton reacted to his woes by becoming strangely disengaged. He had installed a pinball game in the basement recreation room of the Governor's Mansion and spent hours hunched over the machine, his eyes glazed over. Arrested for speeding on his way to a campaign appearance, he joked that perhaps he and Hillary should have called their daughter Hot-Rodham. In Little Rock, the governor was a regular at popular nightspots like Busters. One longtime Clinton watcher would recall that during this period Clinton became a "whirling dervish on the bar circuit. . . . That big red nose of his was unmistakable during the Christmas season. He was always with women, with an entourage, and you never knew whether this was his 'action' or just staff members, but he was always with a pack. His wife was never there."[4]

Rudy Moore fired one Clinton aide whom he blamed for encouraging Clinton's nightclub excursions, but nothing changed after the aide's departure. The governor, however, was subject to bad influences closer to home. His half brother, Roger, had developed a heavy cocaine habit and was running with a rough crowd. As Roger recalls in his autobiography, he had fallen into the habit of using the guest house on the grounds of the Governor's Mansion as his private party shack. He says he would show up late at night after Bill and Hillary were in bed, "hammered" and "stoned out of my head," and "sometimes a friend or two was with me." Others recall that Roger's late-night visits were far from discreet. Some of them involved piercingly loud music and raids on the mansion kitchen for munchies.

According to Roger, Hillary put up with his behavior for quite some time before politely but firmly telling him that he could "at least call first." Somehow this doesn't sound like Hillary. Nor would many working women with a young child at home be so tolerant. Rumors and allegations abound that Clinton partied with his brother on occasion—perhaps when Hillary was out of town on business— and had developed a drug habit of his own. The eyewitnesses who claim to have seen Clinton actually doing drugs tend to be people who do drugs themselves, so the case must be regarded as unproven. As Meredith Oakley put it, "Just because there had been gossip and snide remarks in some quarters about Bill Clinton's chronic runny nose, this did not mean that he was a cokehead, too." But cocaine use, which after all was the reason a lot of people became inordinately fond of the club scene in 1980, is one obvious explanation for Clinton's uncharacteristic lack of interest in getting reelected.

By late fall, Clinton was being spotted around town in the company of Lencola Sullivan, the reigning Miss Arkansas, who had won the swimsuit competition in Atlantic City and come home to work as a reporter for KARK-TV. Whether their relationship was anything more than a flirtation is uncertain, but Little Rock wasn't ready for a married governor who appeared to be dating an African-American beauty queen. The rumor mill ran hot for a time until Sullivan left the state to pursue a show business career in New York and L.A.

And yet it was Hillary's conduct, not Bill's, that had become a campaign issue. Clinton's Republican challenger, Frank White, was being advised by a Republican consultant named Barbara Pardue, who began running ads that targeted Hillary's use of her birth name. Hillary had legitimate reasons for wanting to be known as Ms. Rodham; she had taught law, published and made useful professional contacts under that name. But the custom of husbands and wives using different names was all but unknown in Arkansas, and Hillary didn't help matters by telling an interviewer that keeping her name made her feel more like a "real person." Did this mean that she thought women who used their husbands' names were not real people? Dick Herget,

Clinton's campaign manager, begged Hillary to consider becoming a Clinton, to no avail.

As often happens, the symbolic issue of Hillary's name reflected a vague unease about the Clintons' marriage that was entirely justified. While the average voter knew next to nothing about Hillary Rodham as a person, those who did have occasion to see her around Little Rock or at campaign functions sized her up as an unhappy person whose perpetual scowl seemed to reflect her distaste for Arkansas and everything connected with it. Dolly Kyle Browning, who saw Hillary at social events, would observe, "It took her a long time to learn to pretend to like us."

The state troopers, who were stationed at a guardhouse on the grounds of the Governor's Mansion, drove the governor and first lady to official appointments and often traveled with them, observed that the couple's relationship had undergone a dramatic change. Where before they were affectionate and fun-loving in private, there was now a distinct chill. And if Hillary was cool to Bill, she was openly hostile to the troopers themselves. To some degree this was understandable, since some of the troopers had become her husband's accomplices in woman-chasing. But to a greater degree, the problem was a clash of cultures, exacerbated by Hillary's bad manners. Trooper Larry Gleghorn itemized their complaints: "She was a bitch day in and day out. She always screamed we were taking the wrong route when we drove her to an event. She was a die-hard antismoker and most of us smoked. She hated it when we wore cowboy boots with our uniforms. It made her furious. On official trips, you'd go to her room and say to her, 'We have to leave in fifteen minutes,' and she'd snap, 'We know it.' Then they'd be late and they'd give us a good cussing."[5]

It is tempting to think that Hillary was fed up and had decided to let the campaign take its course. This was not quite the case, however. A mere eight days before the election, Hillary took a look at some new poll numbers and realized that Frank White might actually win. She placed a panicky call to Dick Morris, who left the Paula Hawkins campaign in Florida to come to Little Rock, but it was already too late.

★   ★   ★

ON election night, Clinton held a lead through the early vote count, and NBC News officially declared him the winner. The next morning, Little Rock awoke to hear a popular local DJ breezily announcing that Governor Clinton and Hillary Rodham would soon be seen around town collecting cardboard boxes. At the Governor's Mansion, the phone was ringing off the hook. Dick Herget, one of the first to pay a consolation call, found himself sitting in the kitchen while Bill, still barefoot, took consolation calls from Ted Kennedy and Walter Mondale. The calls were a revelation for Herget, who'd had no idea that leaders of the national party were taking such a personal interest in Bill Clinton's future.

All morning long, friends and supporters beat a path to the mansion, congregating on the lawn. Some were teary-eyed, hiding their red eyes behind sunglasses and consoling each other with hugs. Roger Clinton was beside himself, pumping his arm up and down in a gesture of defiance. The display of emotion was excessive, more like the grieving of relatives after a plane crash than a gathering of friends to console a thirty-four-year-old professional politician who had lost an election.

Clinton had never been shy about predicting a brilliant future for himself. In defeat, he was overcome by the feeling that he had let everyone down. In the words of one Arkansas friend, "He was so low he could play handball against the curb."

After some reflection, many of Clinton's friends saw his defeat as an opportunity. Ever since his high school years, when classmates tagged him as "Billy Vote Clinton," he had been almost obsessed with campaigning. Perhaps it wouldn't hurt for him to take a break. As America's youngest former governor, he would be an attractive commodity on the job market. Sympathetic friends from around the country were soon calling in with suggestions, some more serious than others. There was talk of Clinton's being offered the chairmanship of the Democratic National Committee, and Jerry Brown proposed that the Rodham-Clintons might want to move to California, a state more congenial to their liberal politics. More practical possibilities included a university presidency in Kentucky, a partnership in a Washington, D.C., law firm and the chairmanship of People for the American Way, a liberal lobbying group.

In a 1992 interview, Clinton said that he never for a moment considered these options. Even as he was conceding the election to Frank White, he was already thinking about his comeback. Undoubtedly this was so, though for months after the election Clinton was in such a deep funk that he was in no shape to organize his own political resurrection. Hillary stepped into the breach. In the early morning hours after the vote count, she was on the phone to Jim McDougal, who had left Little Rock and was living with Susan in northwestern Arkansas. Frantic, she delivered the news of Bill's defeat and in the next breath talked about how she and Bill would be needing financial support from their friends.[6]

"You need to send us money," she said.

It was only a year since Hillary's big commodities score, so the Rodham-Clintons were hardly facing a personal financial crisis. Rather, they needed money because Bill Clinton's next race for office was beginning immediately.

In retrospect, surely the friends who urged Clinton to take a few years off were correct. A vacation from politics would have given him and Hillary a chance to live a normal life out of the public eye and work on their problems, whatever they were. Hillary's instant decision to engineer a comeback effort suggests that even at this early stage she couldn't see her marriage apart from her and Bill's shared political ambitions.

Within days, Hillary had called Dick Morris and negotiated an agreement for his services. He would soon begin making regular jaunts to Little Rock for consultations.

The Clintons also reached out to Betsey Wright, the political operative from Texas whom they had met while working for the 1972 McGovern campaign. Wright was closer to Hillary than to Bill, and the cause dearest to her heart was getting more women elected to public office; nevertheless, she left behind a good job in Washington working for the National Women's Education Fund and moved to Arkansas to organize Bill Clinton's political rehabilitation. At first there was no money to pay Wright's salary. She moved into the guest house and began sifting through the files that were stored in the basement rec room, which doubled as Clinton's personal office. Wright set out to reorganize the card file containing the names and addresses of 10,000 or so contacts Clinton had made since his undergraduate days and sent out brief notes, thanking the governor's loyal supporters and promising, "We'll have another day."

Clinton's campaign had ended up about $50,000 in the red. In a similar situation, after his failed Senate race, Jim Guy Tucker retired from politics temporarily in order to earn money to pay off his debts. This was an option that never seemed to occur to Clinton. On the contrary, he was still spending money. Finance chairman Skip Rutherford was trying to close out the books on the 1980 race, but bills kept coming in. Clinton stalwart Mack McLarty was also alarmed by the mounting debts: "Bill and Hillary wanted to keep on having these 'thank you' parties. . . . I mean, it was understandable that they wanted to do it, but we didn't have any money to spend." McLarty eventually covered the $50,000 debt himself by tapping fifty friends for $1,000 donations.[7]

MEANWHILE, Clinton was looking for new financial angels. A day or two after the election, former state senator George "Butch" Locke got a call from Clinton, asking him to set up a meeting with Dan Lasater, his partner in the bond brokerage house Collins Locke & Lasater. "I mean he didn't take a day off," Locke mused. "I figure[d] Bill would be somewhere vacationing. He was running. I was shocked he wanted to have this meeting but he wanted to know what he did wrong, and taking advice more than anything else, and trying to gather support."[8]

The get-together took place in Lasater's apartment, number 12B of the Quapaw Tower, not far from the Governor's Mansion. Even though Clinton was, in Locke's words, "a crippled duck," he was revved up about his prospects and talked for two hours or more about his vision of the state's future economic development. Every fifteen minutes or so, he would pause in mid-monologue to race to the phone and call Hillary, promising her that he would be home soon. Clinton was furious that the Stephens brothers had supported Frank White against him, and he was hoping that Lasater and Locke would want to back a candidate friendly to them.

At the time, Dan Lasater was Arkansas's golden boy, better known, in some circles at least, than Bill Clinton. Just three years older than the governor, he had been born Danny Ray Lasater in a remote corner of White County. The Lasaters moved to Kokomo, Indiana, when Danny was eight, and he got his first job mopping the floor of the local McDonald's when he was a teenager. By nineteen, he was Mickey D's manager of the year, and he borrowed money to start his own fast-food restaurant, Scotty's, which sold burgers for two cents less than the national chains. After building eleven Scotty's restaurants, Lasater sold out to found the Ponderosa steak houses, and then sold Ponderosa for a reported $15 million.

At twenty-eight, Lasater returned to Arkansas, where he raced Thoroughbreds at Oaklawn Park in Hot Springs. His stable was Oaklawn's top money winner for seven years in a row, and for three years straight in the mid-1970s, he won Eclipse awards as American racing's top owner/breeder. Lasater hadn't been in the bond trading business very long. In fact, his brokerage house had yet to make any money. With interest rates spiking to over 20 percent, it was hard to do a profitable business in bonds, but when rates began their downward trend he would be well positioned.

Lasater was a man who boasted that he had never read a book from cover to cover, yet he had owned horses with George Steinbrenner and was a personal friend of Kentucky governor John Y. Brown. Clinton hadn't actually met Lasater before, but doubtless he had heard a lot about him from his mother and half-brother. Virginia Kelley was an avid racing fan who attended Lasater's famous Derby Day party, an annual event celebrating the end of the Oaklawn racing season. For a year or so, Lasater had also been friendly with Roger

Clinton. "Roger was around the races and was involved with cocaine and I was involved with cocaine and that kind of pulled us together," Lasater said later.[9]

Lasater's Quapaw Tower apartment was the scene of wild parties where dishes of cocaine were set out on the coffee tables like hors d'oeuvres and the women were young, in some cases high school girls seduced into partying with an older crowd by the availability of free drugs. It is difficult to imagine that Roger Clinton, who took few pains to hide his own drug use, had failed to mention this aspect of Lasater's character to his brother.

The friendship of Bill Clinton and Dan Lasater has been subject to historical revisionism. Lasater, now a born-again Christian who says his main interest is the Fellowship Bible Church, has insisted that he and Clinton were never more than "distant distant friends." He told the Senate Whitewater Committee that he had met Clinton perhaps eight times in his entire life, a number that rose to eleven as committee lawyers refreshed his memory about certain public encounters. Other witnesses recalled it differently. Michael Drake, who handled underwriting for the brokerage, said, "There was no secret that Dan and Bill were friends. There just—there was none."

At any rate, Lasater and his companies would contribute $16,000 to Clinton's 1982 campaign, with substantial additional contributions coming from Lasater's partners and bond salesman. Around the same time, Clinton began cultivating other Little Rock brokers who competed with Stephens, Inc., including Dabbs "Doobie" Sullivan. In fact, Michael Drake recalled that Clinton for a time kept a desk in Sullivan's offices.

During his first term, Bill Clinton had espoused a populist, pro-environment agenda, counting on tax increases to fund his programs. The voters had rebelled, and in his farewell speech to the legislature he would scold both the lawmakers and the public for being too selfish to pay for progress. Now, literally overnight, he was signaling a new political strategy. Clinton was beginning to reach out to the financial community. He would raise a record $600,000 for his 1982 race, even while denouncing White as a sellout who took money from "people who wanted decisions from the governor's office and paid for them." Clinton would also begin to see that bond issues could be used as a way of paying for programs that would never get funded through direct taxation. It is difficult to attribute these insights to Clinton, who knew so little about financial matters that he had trouble balancing a checkbook. Hillary, a partner in a firm that structured bond deals for Stephens, Inc., is more likely to have foreseen the coming bond bonanza.[10]

FOR Hillary, the 1980 election had been a losing battle on two fronts. Even as the governorship was slipping away from Bill Clinton, she watched the storm clouds gather around the Legal Services Corporation, the organization whose board she had chaired since 1978. As an active board chairman, Hillary had worked hard to build up the LSC during 1979 and 1980, only to see the agency's future threatened by the victory of Ronald Reagan, an avowed enemy of the program.

The legal services movement is often confused with the Legal Aid Societies, a much older group dedicated to providing representation for people unable to afford attorney's fees. By contrast, legal services was based on the theory that the American system of justice is inherently corrupt, designed to perpetuate the power of the ruling classes. Therefore, in this view, the most effective way to get at the root causes of poverty was to turn the justice system against itself, using court actions to challenge the power of established institutions. As longtime legal services executives John A. Dooley and Alan W. Houseman noted in an unpublished history, legal services was a movement whose mission was to "redress historic inadequacies in the legal rights of poor people." To do this, it was necessary to go beyond what was known in legal services jargon as "uninformed demand"— in other words, the services poor people actually wanted, such as help in avoiding eviction, collecting child support, divorcing abusive husbands and filing for bankruptcy.

Legal services began in the 1960s with a few community-based programs funded primarily by the Ford Foundation. One of the first of these was the New Haven Legal Assistance Association, set up by Jean Cahn, a recent Yale Law graduate, and her husband, Edgar. In 1964, the Cahns persuaded Sargent Shriver to include funding for a legal services agency under the aegis of the Office of Economic Opportunity. Eager to avoid the bad publicity that the AMA had garnered through its opposition to Medicare, the American Bar Association lent its support to the new program, despite the qualms of some of its members.

In garnering support for the agency, legal services supporters invariably stressed the more traditional side of its role in funding groups that would provide direct services to poor people. It was assumed that the agency would also fund some class-action suits, to enforce existing civil rights laws and provide a check on abusive bureaucrats. In practice, some of the funded organizations went much further. Radical attorneys discovered new "rights" that no one had previously imagined and used taxpayer money to thwart the will of the voters. A contentious, at times even violent, campaign by California Rural Legal Assistance frustrated Governor Ronald Reagan's attempt to change the state's welfare and Medicaid policies. In Missouri, legal services lawyers supported rent strikes by tenants in public housing. In New Orleans, they defended a black militant group that engaged in a shootout with police. In Camden, New Jersey, legal services filed suit to halt an urban redevelopment project. These kinds of actions led to complaints that radical lawyers, accountable to no one, were spending public funds to frustrate the work of legally elected and appointed officials.

In 1971, in the same veto that killed the bill for a comprehensive child-care program advocated by the Children's Defense Fund, Richard Nixon refused to continue the mandate of the Office of Legal Services, citing "abuses which have cost one anti-poverty program after another its public support." The bitter three-year battle that followed saw the firing of OLS director Terry Lenzner and his replacement by conservative activist Howard Phillips. After much debate, Congress passed legislation restructuring legal services as a nonprofit corporation, comparable to the Corporation for Public Broadcasting.

Under Carter, the Legal Services Corporation fared better. Hillary Rodham took over as LSC board chairman during a period when the corporation was expanding its reach. Once confined mainly to urban areas and a few programs serving migrant workers, the network of LSC grantees was extended into virtually every county of the fifty states, as well as the U.S. Territories. The corporation's budget grew apace, rising from $71.5 million in 1976 to $321 million by 1981. Thanks to nonconfrontational leadership and closer monitoring of grant programs, the LSC had enjoyed a honeymoon from controversy.

Hillary's appointment began on a note of controversy when she refused to pledge that the Rose Law Firm would abstain from doing business with organizations funded by legal services grants. As a first-year partner whose position as the governor's wife was already creating conflicts for the Rose firm, she would have been hard put to extract such a promise from her colleagues. The Senate confirmed her nomination despite some members' reservations about her unusual interpretation of conflict-of-interest principles.

The tone of Hillary's term as chairman was set at a two-day retreat at Airlie House in Virginia, where the board affirmed its goal to "empower the poor." The LSC's idea of empowerment was to use legal action to expand the welfare rolls and get as many people as possible on food stamps and Social Security disability. The reasoning behind this somewhat cockeyed notion of progress was that the LSC needed to expand its client base to include the working poor, working through the courts to bring about a redistribution of income. This goal would be expressed most clearly by Gary Bellow, executive director of the Legal Services Institute, in a March 1982 speech entitled "Strategy and Substance for the 80's," which he delivered to a group of legal services attorneys and community activists. Bellow stressed that one effective way to ensure continuing support for legal services was to "increase people's sense of grievance and entitlement"—a strange goal for a government-funded agency.[11]

The board's first major task under Hillary's leadership was choosing a new president for the corporation. Their choice fell on New York attorney Frederick A. O. Schwarz, a friend of Vice President Walter

Mondale, who combined very liberal political views with a reputation for being a stickler for procedural correctness. Schwarz, now with the firm of Cravath, Swaine and Moore, recalls his interview with Hillary, who struck him as being at once "extremely impressive" and also "very young." He eventually declined the job because he didn't want to move his family to Washington.

Some months later, the post went to Dan Bradley, a former deputy of Terry Lenzner who had served for a time as head of the Florida Gaming Commission. Bradley was an able executive, well liked by politicians of both parties, but he also epitomized a trend within legal services toward emphasis on identity politics. An orphan who grew up in a Baptist-run group home, Bradley had his first homosexual experience at the age of twenty-eight, an event that plunged him into a suicidal depression. There followed a decade of therapy, more depressive episodes and a long period when he was "neither straight nor gay." A much discussed 1982 magazine article by Taylor Branch told how Bradley had lived a double life, concocting a sham love affair with a platonic friend named Barbara for the benefit of his Washington colleagues and FBI background investigators, while in private he made what Branch called the "heartening discovery" that he could express his sexuality as part of the gay bath scene.[12]

"What sustained me all those years and as I struggled with myself," Bradley said in a tearful 1982 interview with *The New York Times,* "was Legal Services. In a way, it was my ministry. What Legal Services represented to me was the embodiment of everything I believed in politically, morally, professionally and otherwise. All those years when I was struggling with who I was, when I had no social life or meaningful personal relationships, when I was literally hibernating, Legal Services was my life."

In this interview, Bradley told of suffering emotional anguish as he sat in the House gallery listening to the debate over an amendment that would have restricted the LSC from spending grant money on homosexual rights cases. In fact this measure, proposed by Representative Larry McDonald of Georgia, never became law and was regarded as extreme even by many legal services critics. Nevertheless, in its clumsy way, it did highlight the difference of opinion over the LSC's role. For Bradley, defending homosexual rights was obviously right, part of the LSC's mission to fight for embattled minorities and redress historic injustices. But for many conservative and moderate critics, such cases amounted to diverting tax money meant for emergency aid to the poor into the pursuit of a special-interest agenda.[13]

The LSC's sense of mission was fragmenting. Where once young radicals had talked of mobilizing "the poor," there were now competing constituencies, each preoccupied with its own goals. For the first time in the history of legal services, the board Hillary presided over included individuals participating as representatives of client groups, Hispanics and Native Americans. Among grantees, there were bitter arguments over funds that never seemed adequate for the growing number of interest groups. There were bitter complaints that Native Americans living on reservations and migrant workers received more than their share of grant money. Tenants' rights organizers, often recruited from church groups, were not necessarily sympathetic to spending money on homosexual rights issues or on "disability" cases, such as a suit in New York that attempted to prevent the transit authority from denying employment in the subway system to men and women with a history of heroin abuse. Meanwhile, elderly residents of nursing homes—a group whose problems held little appeal for young attorneys—were chronically neglected.

The same radical analysis that led legal services theoreticians to attack the courts and Congress was now being applied by activist groups to the authority of the LSC. Using the skills she had developed as a student activist at Wellesley and Yale, Hillary proved adept at presiding over fractious meetings and building support for the LSC bureaucracy among local organizers. But relative harmony came at a price. Efforts to monitor how grant recipients spent their money were relaxed. No one really knew which legal services programs were most efficient or how much grant money was being spent on political organizing. News stories favorable to the LSC invariably stressed that only about 1 percent of the cases handled by legal services attorneys were class-action suits. But this statistic was meaningless, since the money and resources spent on, say, helping a woman divorce an abusive husband could hardly be compared to the cost of a major class-action lawsuit.

By the late 1970s, LSC training programs were teaching harassment techniques, such as nailing a dead rat to an opponent's front door. Training manuals in use during this period also advocated "muckraking"—digging up dirt on opponents' personal lives by interviewing their ex-wives or unearthing "juicy tidbits" from court records. Another approved tactic, often used against farmers and small business owners, was filing many small harassment suits in order to break the opponent financially. One manual used in training community group activists advised, "If the group starts its selection by looking at the law, the result can be narrow, unimaginative, or restrictive tactics." A manual entitled "Strategic and Tactical Research" explained how to target corporations, using as an example a fictitious company called Chemkill, "whose NASDAQ symbol is KILL." Literature handed out at a 1982 conference in New York City explained that plant closings presented an "exciting" opportunity to move toward "rebuilding a broad-based American socialist movement."

Hillary backed Dan Bradley in opposing a movement called Judicare, which would have allowed local bar associations to use LSC funds to compensate private practice attorneys who took on cases for indigent defendants. Judicare would have dealt efficiently with the kinds of problems that brought poor people to legal services clinics—landlord/tenant disputes, domestic cases—but it was opposed by legal services staff because it steered money away from groups with a larger political perspective. The LSC position was that only specialists in poverty law were qualified to represent poor people's interests.

Congressman James F. Sensenbrenner represented two rural counties in Wisconsin where everyone, including the local welfare rights groups, favored Judicare. Hillary and Dan Bradley paid a call on Sensenbrenner's office and informed him in no uncertain terms that Judicare was unacceptable because they wanted the LSC bureaucracy to have a presence in every county in America. "Back then I reached the conclusion that she is a control freak," Sensenbrenner would recall. "She wants to utilize the power of the federal government to make sure she, her people, and her philosophy have control."[14]

In practice, poverty law "specialists" were often young attorneys fresh out of law school, recruited through publications like *Guild Notes,* the newsletter of the National Lawyers Guild, a longtime Communist Party front. (Since LSC money was used to pay the professional dues of legal services attorneys, the American taxpayer indirectly became a supporter of such National Lawyers Guild causes as the Palestine Liberation Front and the Baader-Meinhof Gang.) The Guild became such a force within the LSC that it ran its own employees' association, the National Guild of Legal Services Workers.

Ambitious, politically active young attorneys were not interested in handling their three-hundredth child support case—in some cases, they viewed errant fathers as society's victims who should not have to pay child support at all. Their primary focus was on finding novel ways to attack the system. Even Dooley and Houseman acknowledge that orientation got out of control during the late 1970s. Grant applicants who proposed specific goals, such as reducing utility rates in their area by 20 percent, were more likely to receive funding, even though the goals were never actually going to be achieved: "Priority setting became synonymous with a grand event, a big open retreat meeting attended by program staff, clients, board members, and others (social agency staff, activists, etc.) where priorities—goals, objectives— were determined by a series of votes after open brainstorming. . . . The real 'client' of the program was all the people who were present at the retreat where the goals and objectives were set."[15]

This approach resulted in a series of court decisions that expanded entitlement programs at a cost to taxpayers of many millions of dollars. A major case in Texas established the right of the children of illegal aliens to free public education. Supplemental Social Security benefits were extended to alcoholics and drug addicts, on the grounds that addiction was a disability. In Connecticut, Hartford Neighborhood Legal Services sued to require Medicaid to pay for a sex change operation on the grounds that the client's gender confusion was causing him "frustration, depression and anxiety." The effort failed only because the court ruled that the treatment was still experimental. A 1979 suit filed by California Rural Legal Assistance against the University of California asked the courts to bar the university from doing research on laborsaving agricultural machinery.

The case of *Ross v. Saltmarsh* offered an especially pointed demonstration of how litigation to protect the rights of a few individuals could lead to injustice for many. In Newburgh, New York, Mid-Hudson Legal Services and the Children's Defense Fund joined to sue the school board after several African-American students were suspended for taking part in protest demonstrations. Originally filed in 1974, the case carried over into Hillary's tenure as LSC chairman and led to a 1980 consent decree that required the school district "to eliminate all racial disparities in suspension rates." The result, not only in Newburgh but in many districts across the country, was that school boards who feared financially ruinous litigation adopted a de facto policy of disciplining one white student for every nonwhite, or else avoiding suspensions altogether. Disruptive behavior came to be tolerated in the classroom to the detriment of students, including minority students, who needed a solid education to have any hope of escaping poverty.

In another children's rights case in Michigan, *Martin Luther King, Jr., Elementary School v. Michigan Board of Education,* an LSC grantee representing fifteen elementary school pupils filed suit demanding that the state Board of Education recognize "Black English" as a distinct language. Judge Charles Joiner ruled that teachers must familiarize themselves with Black English and take into account the students' "first language" in their approach to teaching reading.

Today, Alan Houseman defends the LSC even while arguing that Hillary Rodham cannot be held personally responsible for its more dubious cases. These, after all, were undertaken by independent organizations that just happened to be receiving LSC grant moneys, and were never directly approved by the board. Indeed, the genius of legal services as an engine for cultural change was that absolutely no one was responsible for the social impact of the legal battles being fought with taxpayer dollars. In setting up the LSC as a nonprofit corporation, Congress had enacted restrictions that were supposed to guarantee accountability. But many of these restrictions were overturned during the Carter administration, and the LSC charter itself insulated the corporation from outside control. LSC staffers were not subject to the same laws and regulations as federal employees, nor could Congress exercise oversight on spending, as it could with federal agencies.

Under Hillary's leadership, the LSC plunged into partisan political activity. In 1980, LSC funds were used to pay for an extensive lobbying campaign organized by Hillary's friend and fellow board member Mickey Kantor against Proposition Nine, a California ballot tax limitation initiative. The "No on 9" campaign would turn out to be just the beginning of the LSC's political ventures. Despite a November 1980 ruling by the comptroller general that the use of federal appropriations and staff hours for political lobbying was illegal, LSC grantees joined with union and activist groups to oppose the election of Ronald Reagan, a longtime critic of the legal services movement who had proposed doing away with the LSC and distributing legal aid funds to the states in the form of block grants.

Within days of Reagan's election, the LSC and its satellite organizations went into emergency mode. In many regional centers, service to poor clients virtually came to a halt. Funds earmarked for helping the poor were diverted to a massive lobbying campaign, and women who showed up at legal services clinics looking for help were put to work writing letters to congressmen and the media. Staffers were told that if a client did not wish to write such a letter, they should write one for her.

In order to pay for this survival campaign, Houseman and two other highly placed LSC executives came up with a strategy they called "saving the rubies"—an allusion to the rescue of the Romanov crown jewels even as the Bolsheviks were storming the palace. Another name for the program was the Government in Exile. Money allocated for training and other administrative purposes was allowed to build up in interest-bearing accounts and was ultimately transferred to state-level agencies, where it could be used to fund even more lobbying. Some transferred funds were laundered through a series of accounts in Puerto Rican banks and so-called mirror corporations.

Reagan did not get around to dealing with the LSC issue until the end of 1981, when he named seven new board members as recess appointments. The news touched off the final, hysterical phase of the Government in Exile campaign. During the first two weeks of December, Dan Bradley sat down and wrote $260 million in checks, disbursing the LSC's entire 1982 budget to prevent the new board from exercising any control over the money. There were reports of wholesale document shredding at LSC headquarters.

"When we arrived at the first of the year the files were gone," says William F. Harvey, a former dean of the University of Indiana Law School, who would become president of the new board of directors. "The corporate files were just taken." As for the board's dealings with the permanent staff, "those meetings were sort of like the first wave at Iwo Jima, that's about the level of cordiality."[16]

A few weeks later, Harvey was in Dan Bradley's office when Bradley took a call from Little Rock. It was Hillary, informing Bradley that she and the other members of the Carter-era board were planning to file for an injunction to prevent Reagan's people from taking office on the grounds that the LSC charter did not give the President the power to make recess appointments. Bradley, recalls Harvey, was "pretty adamant" against the idea. For one thing, Hillary and some of the other Carter directors had been recess appointments themselves. Bradley argued that the suit had no chance of winning and would only antagonize Congress.

Harvey speculates that Bradley feared the lawsuit, if it ever got to the discovery stage, would uncover the extent of the financial shenanigans involved in the Government in Exile scheme. Hillary, however, would not listen to reason. When Bradley tried to tell her that the suit would be ill advised, she became testy, snapping, "I want you to know that I have private counsel here in the Rose Law Firm, and that's Vince Foster."

Hillary, at this point, was no longer board chairman, having been replaced by William McAlpin, who had a more middle-of-the-road political profile. That she would go to the trouble of consulting her own attorney was evidence of a high degree of personal involvement in the fight against the Reagan board. The lawsuit, *McAlpin v. Dana,* was duly filed and got nowhere, undoubtedly to Bradley's relief. However, a smear campaign against the Reagan appointees, charging them with everything from racism to padding their expense accounts, did succeed in keeping them from being confirmed by the Senate. (A subsequent audit showed that they had actually undercharged the corporation.) Control of the LSC board fell to moderates who decided that it would be better to reform the corporation than to dismember it, though reform proved difficult to achieve when dealing with entrenched staff whose expertise lay in slicing and dicing the law.

Days after his phone conversation with Hillary, Dan Bradley resolved his personal problem by declaring his homosexuality in an interview with *The New York Times* and resigning from the LSC to become a gay rights advocate. In 1983, Bradley was called back to Washington to testify before the Senate Committee on Labor and Human Relations, which was looking into malfeasance in the Government in Exile campaign. Bradley acknowledged that the survival plan was "stupid" and very likely illegal; however, he had kept Hillary Rodham and the other board members apprised of what was going on and they did not interfere.

The Senate hearings led to a General Accounting Office review of the LSC's finances as well as an investigation by the Criminal Division of the Justice Department. Because of technical language in the LSC charter, which Alan Houseman had helped write, LSC personnel escaped indictment on federal fraud and theft charges; however, the Justice Department's final report concluded that "the activities of the Corporation and many of the people associated with it are uniquely reprehensible and beyond the scope of the LSC's original mission. . . ."

Bradley, who died of AIDS in 1988, was unusual if not unique among LSC insiders in expressing regrets about the way the corporation had managed its affairs. The press and liberal supporters of the program appeared ready to accept the LSC insiders' view that since their mission was helping the poor, they should not be held to the same standards of fiscal responsibility that applied to others in public service. While it lasted, the survival campaign had been exhilarating, uniting squabbling interest groups in battle against their common enemy, Ronald Reagan. Nor, in the short run, did legal services appear to have paid much of a price. The corporation's budget, even after a 25 percent cut, was still more than it had been when Hillary took over the board. As for Hillary, she had escaped indictment and even serious embarrassment, to emerge as something of a heroine to the activist community.

IN January 1981, the Rodham-Clintons had moved to Midland Street in the Hillcrest section of Pulaski Heights, a pleasant neighborhood of older homes and tree-lined streets that Webb Hubbell would describe as somewhat "more populist" than the prestigious White Heights. They had used a good portion of Hillary's commodities profits as a down payment on a handsome older brick house with a swing on the front porch, hardwood floors and built-in bookshelves. Chelsea was still cared for during the day by Dessie Saunders, the former Governor's Mansion "security guard," who was now on Bill and Hillary's personal payroll. Completing the household was Zeke, a neurotic cocker spaniel, who barked all night and wandered into the neighbors' backyards, rummaging through their garbage cans.

The ex-governor now spent his days in a cramped cubicle in the offices of Wright, Lindsey & Jennings. Clinton was listed on the firm's directory as "of counsel" and he worked fitfully on a few cases, but it was understood that his real work was getting reelected. Bruce Lindsey, a friend since he and Clinton had met while working for Senator Fulbright and a son of one of the founding partners, had invited him to make the firm his base while he plotted his comeback.

Although he had been governor for just two years, Clinton complained that he was having a hard time adjusting to life as a private citizen. With no staff to take care of Chelsea when neither the nanny nor Hillary was around, Clinton was often called on to take over baby-sitting chores. (Gennifer Flowers would recall talking on the phone with Clinton one evening when he interrupted their conversation to rescue Chelsea, who had fallen off the bed.) Clinton was even more morose about having to handle tasks like doing the laundry and grocery shopping. Still trying to cope with the voters' rejection, he would go up to strangers in the supermarket, apologize profusely for the failures of his first term and beg their forgiveness. Hillary would refer to this time in their lives as the period of Bill's "supermarket confessionals," when her husband "got crazy in the incessant quest for understanding what he did wrong, which was masochistic."[17]

Clinton's state of mind may have been self-punishing, but he was doing some serious thinking about the meaning of the 1980 election. His defeat had been part of a larger trend that saw working people, especially men, whose families had been Democrats since FDR, desert the party to vote Republican. As Clinton told David Maraniss in a 1992 interview, the Democratic Party couldn't expect to win elections as long as they presented themselves as "the party of blame."

Not surprisingly, given her immersion in the legal services survival campaign, Hillary took a more partisan view of the situation. Writing to her high school mentor, Don Jones, she expressed dismay that in political matters "people become so irrational"—in other words, the voters just weren't thinking straight when they chose Frank White over Bill Clinton and Ronald Reagan over Jimmy Carter.

In a talk at the University of Arkansas that struck some in the audience as ungracious to say the least, Hillary blamed her husband's defeat on Frank White's use of negative advertising. One White campaign ad had incorporated footage of the Fort Chaffee riot, images Hillary dismissed as an appeal to racism. Of course, for many voters the issue was not the race of the rioters but why Clinton had allowed the Carter administration to dump its refugee problem on the state of Arkansas. Many suspected that Clinton had been reluctant to take a firm position that would hurt his standing with the national Democratic Party.

Well into the spring of 1981, there was a real question of whether

Clinton was psychologically fit for another campaign. While Hillary was going out of her way to cultivate *Arkansas Gazette* editor John Robert Starr and Betsey Wright tried to patch up Clinton's differences with Don Tyson, Clinton himself was in a funk. Normally eager to schmooze, he avoided talking to local reporters for several months. In April, when he was asked to be the guest of honor at a biannual press event called the Farkleberry Follies—the sort of event he would normally have loved—he declined the invitation. It took heavy persuasion from Hillary to get him to change his mind. In the end, he played himself in a skit about a team of Cuban skyjackers, a spoof on the way the Fort Chaffee disturbances had hijacked his reelection bid.

As Dick Morris saw it, "When Clinton lost the governorship in 1980, Hillary felt she had to step in and provide a firm managerial hand. She began to see Bill as well-intentioned and highly intelligent, but deeply impractical and hopelessly naive."[18]

Hillary's plan to manage her husband back into the political spotlight relied heavily on Morris, who was now advising on matters well beyond campaign tactics, including what office Clinton should pursue. Clinton had dreamed of serving in the Senate and Betsey Wright thought he might have the right profile for the office, but a Morris poll in October 1980 showed that Clinton would face an uphill battle against either Dale Bumpers or David Pryor. In his race for governor, his chief challenge would be getting past Jim Guy Tucker in the primary to face off against the vulnerable Frank White.

Hillary and Morris were united in their opinion that the amiable but unsophisticated White was a "cretin," and they were equally disdainful of his wife, Gay, a devout Christian who expressed few political opinions but went around the state speaking to Bible study groups on "The Power of Prayer." And it was true that White was politically inept. He got into trouble when he signed a bill that mandated the teaching of creation science, apparently without realizing its implications. Before long, the *Arkansas Gazette*'s political cartoons were portraying the new governor as a monkey eating a banana. Unfortunately for Hillary, the voters continued to find Gay White just about perfect, while Hillary persistently scored high negatives in Morris's polls, especially with older voters.

A makeover was clearly in order. Hillary got rid of her frizzy perm and acquired a more conservative wardrobe, exchanging her floral prints for a Talbots catalog look, and her thick eyeglasses for contacts. Except for learning to wear contacts, which was difficult for her, these changes conceded little to local standards. The seventies were over, and even in New York and Washington feminists weren't wearing long hair and granny dresses anymore.

But when it came to her use of the name Rodham, Hillary was stubborn. Everyone, including Morris, agreed that the name issue—known in Arkansas as "the Hillary question"—had cost Clinton votes, but Hillary didn't want to accept this. She continually sought out the opinions of Arkansas friends, who almost invariably told her the name she used didn't matter, undoubtedly out of politeness. In later years, Clinton would often insist that he never put pressure on his wife, saying, "I'm the only man in Arkansas who never asked Hillary to change her name." In fact, Clinton desperately wanted Hillary to take his name but he lacked the nerve to confront her. Instead, he broached the subject with Webb Hubbell one day during a round of golf at the country club, begging Hubbell to make Hillary see reason. Hubbell was reluctant to take on such a personal errand—"It was her *name!*"—but Clinton had a way of wheedling people into doing such things for him.

Approached by Hubbell, Hillary seemed resigned, but she told him that the change would be hard. So many of her friends already thought she had sold out when she got married and moved to Arkansas. Now she would be disappointing them again. The sensible thing at this point would have been to acquire new friends. But this was a step Hillary never felt prepared to take. At thirty-five, she still had one foot in Arkansas and the other firmly planted in the world of politically correct leftism.

Even the Clintons' practice of religion was a subject for image adjustment. Hillary had quietly been doing pro bono work for the

Methodist church since her arrival in Arkansas, but her religion tended to express itself in social activism rather than in regular attendance at services. In a churchgoing state, some voters complained that Hillary Rodham "had no religion" and that neither she nor the governor was seen regularly in church. Clinton now appeared every Sunday, singing in the choir of the Immanuel Baptist Church, while Hillary attended services at Little Rock's First Methodist Church. Clinton's sudden onset of religious fervor may not have been as calculated as it appeared. He and Hillary had recently traveled to Israel with a church group led by Immanuel's pastor, the Reverend W. O. Vaught, and Clinton had come to see Vaught as something of a father figure. Still, cynics took note that Immanuel's services were televised, allowing home worshipers a good view of Clinton as he raised his voice with the choir on Sunday morning.

From another perspective, Bill and Hillary's inability to agree on a common religion was indication of a failure to meld their lives on any front save his electoral ambitions. Since Hillary had moved to Arkansas and changed her name one might think it was Bill's turn to compromise, but nothing of the kind occurred. On the other hand, Hillary's concessions tended to be grudging at best. Although she changed her name on her business stationery in early 1981, she would continue to file legal papers, pay her income taxes and vote as Hillary Rodham. More disturbing than any of the specific decisions the couple made was their need to negotiate such issues through third parties like Webb Hubbell and Dick Morris.

Each in a different way, Bill and Hillary were exceptionally inflexible individuals. Dick Morris had insisted that Clinton should kick off his campaign with a TV ad in which he apologized to the voters for mishandling such issues as the car tag tax increase and Fort Chaffee refugee situation. Despite his "supermarket confessionals," Clinton had a difficult time wrapping his tongue around the word "apology" in front of the cameras, and he argued with Morris that the ad copy suggested his decisions had been wrong, which he didn't believe. In the end, as David Maraniss would note, Clinton "managed to say he was sorry without saying he was sorry." The commercial, which be-gan running on February 8, 1982, provoked a negative reaction at first, but as the campaign wore on Clinton's gesture of contrition made it all but impossible for his opponents to get any mileage out of attacking his record. Clinton's show of humility didn't stop him from running one of the most negative electoral campaigns in Arkan-sas history, however, as well as the most expensive. He attacked Jim Guy Tucker as a bleeding-heart liberal, then savaged Frank White in the general election as a sellout to big-money interests. Clinton's op-ponents tried to respond in kind, but his supporters accused them of being mean-spirited for harping on errors Clinton had already ac-knowledged.

ON February 27, Chelsea's second birthday, Clinton formally an-nounced that he was running for a second term as governor. At the same press conference, Hillary was publicly identified for the first time as Mrs. Bill Clinton, and she revealed that she would be taking a leave of absence from her law firm to campaign at her husband's side. This was twice the time Hillary had allowed herself to care for her newborn daughter, so the decision was hardly likely to convince anyone who really cared that she was a second Gay White.

Far from the young man with big-city ideas and more than a touch of big-city arrogance who had run in 1978, Clinton was so timid in this campaign that he refused even to come out in favor of teaching evolution in the schools, saying only that he probably would have vetoed the legislature's creation science bill "in the form that was passed." Hillary barnstormed the state at her husband's side, often with a visibly tired and cranky Chelsea in tow. Her presence helped to dispel gossip about Clinton's fondness for after-hours clubs, and it certainly discouraged anyone who might have been tempted to ques-tion him about the subject. Behind the scenes, Hillary was more the stern personal manager, ordering Clinton away from in-flight card games with reporters and keeping an eye open to make sure he did not collect any women's phone numbers. She had even gone so far as to hire a private detective to track down background on his numerous affairs.

Ivan Duda's name had been in the news in connection with the case of Mary Lee Orsini, the so-called Lady Macbeth of North Little Rock. Orsini's saga began with the shooting death of her husband in March 1981. North Little Rock police considered her the prime suspect, but she spun a complex tale about her husband's involvement in cocaine trafficking and a feud between two Little Rock nightspots, the Kowboy Kountry Klub and BJ's Star-Studded Honky Tonk. To bolster her credibility, Orsini staged numerous attempts on her own life and seduced a man named Yankee Hall, who obligingly murdered Alice McArthur, the wife of Orsini's defense attorney and co-owner of BJ's. After Alice McArthur's murder in July 1982, her husband, Charles, became the media's favorite suspect. Mary Lee Orsini's story got sympathetic play in the *Arkansas Gazette,* and Pulaski County sheriff Tommy Robinson believed that thanks to her information he was on the brink of breaking a major cocaine distribution ring. There was no cocaine ring, and Orsini was eventually convicted of arranging Alice McArthur's death to further her fantastic alibi. In the meantime, however, Duda, one of several private investigators hired by Orsini, was mentioned in the papers as having knowledge of Little Rock nightlife and the drug scene.

Around this time, Duda received a call from Hillary, who hired him, sight unseen, to compile a list of women her husband had been seeing. Citing confidentiality, Duda declines to mention names, but he says that he identified more than half a dozen women who were involved with Clinton, including one who worked on the support staff of the Rose Law Firm. Hillary showed some emotion when he mentioned the Rose employee, Duda recalls, but she never expressed any anger toward her husband or indicated that she was thinking of leaving him. He assumed she wanted to be prepared for any charges that might come up in the course of the campaign.

Duda admits that he can't document his dealings with Hillary. She paid him in cash and they communicated by phone, "except for the one meeting, and that one backfired." While he and Hillary were conferring in his car in a restaurant parking lot, they were spotted by two state troopers and, he says, "the next thing I knew they were going after my license."

What happened was that sometime later, apparently after Clinton's reelection, Duda was charged with carrying a badge, a violation of state regulations. He considered fighting the case, but, he says, "my lawyer told me, 'Look, the deck is stacked against you,' and I was thinking of getting out of the business anyway."

In another of those untimely deaths that so intrigue the devotees of Arkansas conspiracy theories, Sergeant Bill Eddins, the head of the state police unit that took Duda's license, later died in a one-car accident, crashing his state police cruiser, allegedly while under the influence of alcohol. Duda says he received a call from a relative of Eddins's who suspected foul play, but by this time he was retired for good and he declined to get involved.[19]

SEVEN

A Woman of Many Projects

By 1983, when Bill and Hillary moved back into the Governor's Mansion, Arkansas was changing in ways unimagined by the eager policy wonks who had sat around the living room brainstorming ideas for new programs just four years earlier. Wal-Mart, a business built on the old-fashioned concept of thrift, was making millionaires out of middle-class couples who had bought the stock early and held on. Don Tyson, working from his famous Oval Office in Springdale, was following a business model just as old, gobbling up competitors right and left. Strangest of all, sleepy Little Rock had morphed into the bond sales capital of the universe.

Stephens, Inc., had long been one of the largest brokers of investment-grade paper to regional banks, but now it seemed that almost every month a new brokerage house was opening in Little Rock. With individuals and deregulated savings and loans scurrying to lock in high rates on fixed-income securities, marginal players and outright con artists were getting into the game and found Little Rock, with lax securities regulations and virtually no law enforcement, an ideal place to do business. The Pulaski County prosecutor's office had no trained bank fraud specialists and an annual travel expense budget of just $700 for white-collar crime investigations. In one of the rare fraud cases to come to trial, a former Razorback football star, convicted of selling $485,000 in worthless bonds to a paralyzed man and his wife, was sentenced to 100 hours of community service. Little wonder that *Forbes* identified Little Rock as a "hot spot" for boiler-room operations selling "tax exempt trash."

Newcomers to the brokerage business like Dan Lasater were living high. Lasater kept a sixty-foot houseboat on Lake Hamilton and his personal plane was a Learjet. His company plane was a German Hansa. Like a few other wealthy Arkansans, Lasater at times made his planes available for the use of politicians, and in 1983 he flew Bill and Hillary to the Kentucky Derby as his guests. Even some of Lasater's salesmen, kids a few years out of the University of Arkansas, were driving Ferraris. It was hard to see how the firm could be generating enough in profits to support this kind of spending.

At the high end of the spectrum, Rose firm attorneys were earning huge fees structuring bond sales for Stephens, Inc. and New York underwriters. Joe Giroir, the firm's rainmaker, was bringing in most of the business, and under the firm's unusual partnership contract, he and the bond specialists who worked under him kept most of the profits, an arrangement much resented by Vince Foster, Webb Hubbell and Hillary Clinton. Giroir was using his profits to buy banks, and by 1984 he would become a major investor along with Jack Stephens in Little Rock's largest bank, the Worthen Banking Corporation. Another major investor that entered the picture in 1984 was Lippo Holdings, owned by Indonesian billionaire Mochtar Riady. James Riady, Mochtar's son, became Worthen's president, and the bank opened an international division. "International banking is not basically about money," wrote James Riady in his annual letter to shareholders, "it is a matter of buying and selling trust."[1]

According to Webb Hubbell, he, Vince and Hillary did not become really close until after Bill Clinton's loss to Frank White. Some Rose partners, worried that Hillary's presence at the firm would cause White to cut off the flow of state business, had urged Hubbell and Foster to pressure Hillary into offering her resignation, suggestions they indignantly refused. Hubbell paints a touching picture of the three-way friendship that was cemented during these trying months, calling it "the time of our greatest intimacy. . . . We all had relationships with our spouses that, for different reasons, prevented totally frank discussions. Hillary knew that Bill was more wrapped up in his own career than in her need for a life of her own."

And yet, Hubbell acknowledges, the three friends were not quite on a par. In the same election that cost Bill Clinton the governorship, Hubbell had won a place on the Little Rock City Board, a seven-member panel that controlled patronage appointments to the boards that ran the local airport, the Art Center, and so on. The City Board promptly made Hubbell the mayor of Little Rock. Under the city charter, the administrative work of city government was performed by a professional city manager. The mayoralty was a nonelective but time-consuming position that required Hubbell to spend his evenings doing "civic things" while Hillary and Vince often worked late at the office.

It soon became apparent to Hubbell that Vince and Hillary had another dimension to their friendship. In Little Rock society, intellectuals risked being seen as social misfits. Hillary and Vince had discovered that they could indulge their shared passion for books, political movements and ideas in conversations that excluded Webb Hubbell. Foster was known for his glacial reserve. He refused to discuss cases with his wife and sometimes went for days without speaking to his secretary. But Hillary had ways of breaking through his defenses. Underneath Foster's sober exterior lurked the soul of a Don Quixote—and, perhaps, the spleen of a Holden Caulfield. Hillary found Foster's reserve a challenge, and perhaps she realized that it was also a symptom of deeper problems, something that never seemed to occur to male colleagues who equated silence with strength. She teased Foster, made up nicknames like Vincenzo Fosterini, and once hired a belly dancer to deliver birthday greetings to Foster at the office.

"I confess that I felt jealous that she sometimes seemed closer to him than to me," Hubbell writes. But if the relationship was more than a friendship with romantic overtones, he was too discreet to notice. Indeed, his studied lack of curiosity made him the ideal third musketeer.

Trooper Lawrence Douglas Brown, known as L.D., gives a less idealized view of Hillary and Vince Foster. A holdover on the governor's security detail from the White administration, Brown was an avid reader who often borrowed history and law books from Clinton and soon became his confidant as well as his bodyguard. Brown escorted Clinton on more than a dozen out-of-state trips, including a family vacation to Disney World. He also excelled at trolling for women, collecting the phone numbers of "tens" who caught Clinton's eye or talking them into joining the governor's party at public events. If there were two women together, as often happened, Brown would pair off with the other one, a practice he called taking the governor's "residuals." Hillary, in his opinion, knew that her husband was unfaithful but had no idea of the extent of his one-night stands while on the road.

One woman Hillary did know about was Clinton's favorite at the time, a local attorney whom her husband had appointed to a judgeship, a woman Brown has described as a "beautiful, gleaming kewpie doll with brains." On one occasion, according to Brown, he was present when the Clintons and the Fosters had dinner with the judge and her husband at an Asian restaurant. The room was empty except for their party, and after drinking wine for several hours Clinton and the judge, and then Vince and Hillary, became openly affectionate. Leaving the restaurant, Lisa Foster and the judge's husband walked on ahead, while the other two couples lagged behind. "Hell, it was a twenty minute walk—they are drunk, running tongues down each other's throats," Brown recalled. "It was a diagonal swap."[2]

Brown felt that both the Clintons held working people, law enforcement types and "rednecks" in contempt, though Bill was better at disguising his feelings than Hillary. Nevertheless, Brown eventually began to date Chelsea's nanny, Becky, and became somewhat closer to Hillary, at times confessing to her his guilt over having fooled around while his first marriage was falling apart. Hillary advised him that sometimes you have to go outside your marriage to get what you need, telling him, "L.D., sometimes you just have to make a leap of faith."[3]

L. D. Brown had a falling-out with the Clintons in 1985 that led him and Becky, who eventually became his second wife, to a bitter parting, not only from the Clintons but from Becky's mother, who was by this time serving as the administrator of the Governor's Mansion. On this count, he perhaps can't be considered an objective witness, though other troopers also believed that Hillary and Vince Foster were having an affair, based on a few incidents in which they were observed being affectionate toward each other in public and, quite simply, Vince's tendency to show up at the mansion as soon as the governor went out of town.

If Hillary never gave way to the temptation to cheat on a husband who continually cheated on her, then she was a saint. On the other hand, it is difficult to imagine that sex played a dominant role in her relationship with Vince Foster over the years. What drew her and Vince together was that they were two people who had freely chosen lives that made them unhappy. Hillary loved Bill Clinton and was certainly enamored of the success she saw them sharing one day in the not too distant future. Her life was on track as long as she could avoid spoiling things by allowing herself to be eaten up with bitterness. Foster, on his part, was a prisoner of his own introspective nature. He complained that his wife was too clinging, yet shut her out emotionally; he found his life too constricting. Neither was prepared to risk the kind of affair that would lead to seriously disruptive consequences.

Professionally, however, the shared unhappiness of Hillary, Vince and their friend Webb Hubbell was a potent combination. For different reasons, each was out of sympathy with the era of go-go deal making that was enriching some of their law partners. Hillary saw the law as a vehicle for furthering her goals as a political activist and the wife of Bill Clinton. Foster loved the scholarly side of the law and waxed nostalgic over the firm's more genteel past, which may have existed only in his imagination. Hubbell loved playing the role of the city father and was never happier than when Bill Clinton appointed him to a term as chief justice of the state supreme court. Unfortunately, all three litigation partners also needed money. Hubbell and

Foster were living expensively in the White Heights, with Hubbell, in particular, spending far beyond his means, a situation he blamed on his need to live up to the expectations of his irascible and demanding father-in-law, Seth Ward. Hillary was married to a man with a limited cash income and an insatiable need to feed the kitty for his next campaign.

All three partners were under pressure to increase their billings, none more than Hillary. Hubbell notes that some partners complained about her frequent leaves of absence, whereas a male partner could take nine months off in the wake of a rumor about his affair with a secretary and still expect to be paid his full salary plus partnership share. But this (hypothetical?) partner may have earned the Rose firm a great deal of money in the past, something Hillary had never done.

Hillary stayed on at Rose even though she was having trouble carving out a niche for herself. Recruited by Foster on the basis of her work for the University of Arkansas's legal clinic, she took occasional criminal and child custody cases, at times partnering with William R. Wilson, a criminal defense lawyer from outside the firm. But corporate attorneys often see domestic practice as déclassé, and Joe Giroir was trying to focus the firm on service to institutional clients. As Vince Foster explained in a 1992 interview with *The American Lawyer,* domestic and criminal cases were "not something we were trying to encourage or promote at the firm."

But what were the highlights of Hillary's career at the Rose firm? In the same interview, Foster singled out her work for the data processing software giant Systematics, Inc.: "We did not have any intellectual property specialists when she joined us. She became sort of self-taught in all this. And you don't find a lot of intellectual property subspecialists in Arkansas. Quite frankly, the rest of us thought of it as a foreign language."[4]

Ironically, Alltel, the telecommunications company that now owns Systematics, refuses to confirm that Hillary Clinton ever represented it. After some prodding, a company spokesperson told me that even if the First Lady had done work for Systematics, "that would be proprietary information." In 1995, moreover, Alltel went so far as to hire a San Francisco libel specialist, Charles O. Morgan, to threaten legal action against anyone who claimed that Vince Foster had been an attorney for Systematics.

Alltel has reason to be sensitive. One of the most convoluted and persistent of all Internet conspiracy theories, a variation on the long-running Inslaw scandal, turns on claims that Vince Foster and Hillary Clinton, acting on behalf of U.S. intelligence, were responsible for getting Systematics to write a "back door" into its banking software, making it possible for the government to monitor private banking transactions. Supposedly, Foster—and perhaps Hillary—later became double agents, selling the computer codes to Israeli intelligence.

The notion of Vince Foster as a Mossad agent is of course fantastical, and Jack Steuri, the president of Alltel Information Systems (formerly Systematics), is on record as saying that Systematics software doesn't even support the kind of private money transfers that were supposedly being monitored. Nevertheless, despite the company's denials, Hillary and the Rose firm, if not Vince Foster personally, did do work for Systematics. The relationship began with a 1979 lawsuit that was a forerunner of the BCCI banking scandal. The Worthen Bank and Systematics, then part of the Stephens family's financial empire, were sued in connection with the attempted takeover of a Washington, D.C., bank by the same group of investors later associated with BCCI. Joe Giroir, Webb Hubbell and Hillary, who apparently did much of the research on the case, represented Systematics and succeeded in getting the company dismissed from the suit. In 1983, Hubbell was again hired by Systematics to take on a competing software firm in an intellectual property case. According to Charles R. Smith, who writes on computer encryption issues for *Insight* magazine, Hillary also did work for Systematics in connection with a 1990 contract that a subsidiary of the firm received to build something called a Sensitive Compartmented Information Facility at Fort Gillem, Georgia. While none of this supports the conspiracy theories that have been woven around the Rose-Systematics connection, it does suggest that Hillary may have taken a more active interest than previously suspected in formulating the Clinton administration's often controversial policies on computer encryption issues like the Clipper Chip, which was intended to give the government the capability to monitor everything from wireless phone calls to electronic data transmissions.

On the other hand, Hillary's work on these issues hardly seems to have been extensive enough to qualify her as a specialist, or even a subspecialist, in the arcane field of intellectual property law as it relates to computer software. Throughout the 1980s, Hillary was at best a part-time lawyer. Although *The National Law Journal* twice named Hillary to its list of the country's 100 most influential attorneys, her professional visibility was primarily a testament to her networking skills. A 1992 investigation by *The American Lawyer* found that she tried just five cases in her career. William Kennedy III, then the Rose firm's chief operating officer and later an assistant counsel to President Clinton, told the journal, "We're very proud of Hillary, and if she had ever quit having two lives and concentrated on the law practice, she'd have been a superb lawyer."[5]

One of Hillary's rare court appearances was on behalf of the fragrance company Aromatique, also one of the few clients she personally brought to the Rose firm. An aristocratic woman who dresses for work in riding boots and a houndstooth jacket, Patti Upton started her business in 1982 with the help of $10,000 in financing from her husband. Her first product was a potpourri called The Smell of Christmas, made up of pinecones, gumballs, wood chips and various oil-based scents that she packaged herself at home and sold in a friend's gift store in the resort town of Heber Springs, Arkansas. By 1984, Aromatique had 500 employees and its products were being sold at high-end gift shops around the world, including Harrods of London. Aromatique's success soon inspired imitators, and Upton, a social acquaintance of Hillary Clinton, hired an attorney to take action against a company called Gold Seal, which was selling similar packaged fragrances. Beginning in 1985, the attorney sent out letters threatening legal action against Gold Seal for infringing on its federal and state registered trademarks. In 1988, Aromatique filed suit and Hillary handled the matter personally. The trouble was, Aromatique didn't have trademark protection.

"I had the impression that she just didn't get into court every day," recalled Gold Seal's attorney Hermann Ivester of Hillary's performance at the trial. Ivester had some sympathy for his opposing counsel, since she wasn't the one who had sent the letters. The elements of the disaster were already in place by the time she stepped into the case.[6]

Judge Morris Arnold of the Eighth Circuit Court of Appeals was less indulgent. In a 1994 decision, he ruled that the Rose firm acted "in reckless disregard for the law," adding "there can be no doubt that Aromatique's counsel knew that the claim . . . was false." The court punished Aromatique by stripping it of trademarks it had acquired since the suit was filed.

To add insult to injury, in 1992 the Clinton presidential campaign would distribute an Aromatique knockoff called The Smell of Hope. Patti Upton, however, remains a friend of Hillary's and a contributor to the Clintons' defense fund.

IF practicing law was often unrewarding, Hillary's first major venture into policy making would be deemed a resounding success, winning national attention and greatly enhancing Bill Clinton's attractiveness as a future candidate for national office. Unfortunately, Hillary's work as point person for her husband's campaign to raise the level of Arkansas's public schools also illustrates why educational reform has often been so frustrating for so many.

Hillary's ideas about education were basically conservative. She believed that success came through taking challenging academic courses and working hard at them—after all, she and Bill were living proof that high College Board scores and admission to a competitive college could be a springboard to success. But Arkansas, whose schools were among the worst in the nation, had a long way to go. Poor, mostly rural and geographically isolated, from the mountain hamlets of the northwest to the largely black counties of the southeastern Delta, with their heritage of plantation agriculture and sharecropping, the state was dominated by a mind-set that the historian W. J. Cash called "a terrified truculence toward new ideas from the outside." Students who did well academically often left Arkansas in search of high-paying jobs and a more stimulating cultural environment, a brain-drain effect that made it hard to sell taxpayers on the argument that spending on education was an investment in their community's future.

Beginning with Bill's first term as governor, Hillary held an annual reception at the Governor's Mansion for high school valedictorians from all over the state and sponsored a project known as the Governor's School. At the beginning of Clinton's second term, however, a pending court decision on school funding presented an occasion to push education to the top of the state's agenda. The governor was about to ask the legislature for a tax increase to pay for a miscellaneous assortment of new programs, when the *Gazette*'s John Robert Starr gave him a timely warning. "I told Clinton, 'You'd better not do that, because you're going to need that money,' " recalls Starr.

The Arkansas supreme court was about to rule that the formula for distributing state aid to public schools was discriminatory. Achieving parity between rich and poor school districts—what Bill Clinton would soon refer to as the "historic obligation to equalize funding"— is always controversial. In a small state with no fewer than 371 school districts, all centers of political patronage, it promised to be inflammatory. Clinton hoped to turn the situation in a more positive direction by proposing a slate of reforms and using them to win tax increases, enlarging the pool of state aid and leaving students in all districts better off.

As Clinton himself told Arkansas reporter Roy Reed, he was wondering aloud whom to appoint to chair his education committee when Hillary volunteered herself, saying, "This may be the most important thing you ever do, and you have to do it right."[7]

In June, shortly after the court handed down its long-awaited decision, Hillary took a leave of absence from the Rose Law Firm, her third in four years. She spent the summer conducting hearings, traveling with her committee to each of the state's seventy-five counties and listening to as much as nine hours of testimony per day. Hillary's message, straight from the pages of the recently published study *A*

Nation at Risk, stressed the need to raise expectations of student performance and enforce higher standards through objective testing. Hillary could be a very effective speaker, although, on paper, many of
her remarks have a patronizing tinge. She talked of how her move to
Arkansas had been a kind of "re-entrance to the human condition"
and expressed regret that many of the state's students were not exposed to the kind of "broad education" that children in Illinois had
enjoyed "in the forties and fifties." Still, Arkansans expected to be
lectured on the subject of their failing schools, and Hillary was respected for having the courage to take on the sacred traditions of
football and basketball, even supporting the heresy that reading teachers ought to be considered as important as coaches.

Hillary's hard work won her many admirers. John Robert Starr,
who had relentlessly attacked Clinton as "billion-dollar Bill" during
his first term, recalled, "My wife is a teacher, and she complained to
her that 'you really have to see what it's like in my school, with
twenty-seven kids in a classroom.' She said this at a conference one
night, and the next morning Hillary called her up and said, 'When
can we go?' She was that vigilant."

But the hearings, supposedly an effort to gather input from teachers
and parents around the state, were really a dog-and-pony show whose
aim was to build support for a sales tax increase. The effort was sup-
plemented by a public relations blitz, financed with a $130,000 fund
raised from corporate donations. This exercise in creative financing
would create a dubious precedent. Officially, the fund was philan-
thropic, since its goal was to improve education, though in the long
run the ads also served to enhance Bill Clinton's image.

While Hillary and her committee were mobilizing the grass roots,
the governor was also busy, conferring with educational organizations,
labor unions, business groups and political clubs. If Hillary's role as
chair of the commission never became controversial, it was largely
because of the perception that Bill Clinton was firmly in control of
the policy initiative.

As John Robert Starr acknowledges, the fifteen-person commission's recommendations were essentially complete before the tour be-

gan. Borrowed from proposals by the Arkansas Education Association, with a dash of *A Nation at Risk* ideas thrown in, they included a longer school year, limits on class sizes and a required minimum of 13½ units of academic courses for high school graduation. Districts were mandated to provide one guidance counselor for every 450 elementary students or 600 high school students. Mandatory kindergarten was provided for in a separate bill. School districts would have to spend at least 70 percent of their budgets on salaries, a provision that would guarantee raises as state support increased.

Another part of the commission's plan called for students in the third, sixth and eighth grades to pass a standardized test as a requirement for promotion. If 85 percent of the students in a district repeatedly failed to pass the eighth-grade test, the district would be decertified. This guaranteed that classroom instruction would follow officially approved course content guides. Some teachers and administrators—one suspects, the weaker ones—were enthusiastic about the system. "It tells us exactly what to teach," said one principal. "I think it's great for all students in the state of Arkansas to be studying exactly the same thing." Others were not charmed. History had been replaced with a "global studies" curriculum that skimmed lightly over dates and mandated that Africa must receive as much classroom time as Europe. "Some committee came up with these things that were important in American history; others were left out," complained one teacher. "I don't understand how one person can decide, for example, to leave out the entire War of 1812."

Overall, the standards were designed to enforce conformity at a higher level. Most of the proposals were worthy, though none of them, including smaller classes, was a magic bullet that would fulfill the expectations the governor was raising when he predicted that the reforms would lead to "the most dramatic improvement in our system of education in our lifetimes."

A subtext of the standards was that they would be budget breakers for the state's smaller school districts, some of which had fewer than 350 students. Residents of these small rural communities mostly wanted to keep their schools open, but enlightened opinion equated small schools with backward, hillbilly ways. Communities in other parts of America had already discovered that school consolidation was no panacea. Big regional high schools with a full menu of course offerings bred problems of their own, including student disaffection and gang violence. Nevertheless, while Hillary gave lip service to the principle that "there are advantages to small schools as long as small-ness is accompanied by excellence," she also made it clear that her own high school in Park Ridge, Illinois, which had offered a choice of foreign language studies and a wide range of advanced placement options, remained her ideal. Whether a consolidated school in a poor Arkansas county would turn out to be another Maine South was an untested assumption. Instead of looking for creative ways to improve the schools that communities wanted, the education reform plan erected new and probably insuperable barriers for small districts al-ready slated to lose special state subsidies in the name of equalizing funding.

Hillary's committee submitted its preliminary recommendations in September, and a month later the governor formally delivered the plan to a special session of the legislature. Members of the Arkansas Education Association, which had enthusiastically supported the re-form effort, were stunned when Clinton's televised address included a call for every teacher in the state to take a minimum competency test. The state undoubtedly had a stake in getting illiterate teachers out of classrooms; anecdotal proof that they existed included one par-ent's complaint that his child was being taught about "world war eleven." But the Clintons had chosen to address this critical issue in the most confrontational way possible. Justifiably, the teachers felt they had been blindsided.

Dick Morris's memoir, *Behind the Oval Office,* confirmed what many in Arkansas had long suspected: Teacher testing had been added to the reform package as a sop to voters who were suspicious of the Clintons' cozy relationship with the AEA and thought that some teachers were already being paid more than they deserved. Morris writes that it was Hillary who first suggested adding the teacher testing provision in order to demonstrate that "Clinton's commitment to schools was motivated by a love of children, not an affection for teachers' campaign money." (This was quite a sentiment, considering that the Clinton administration not only loved the AEA's campaign donations, it was positively enamored of the organization's pension fund, as we shall see.)

Morris calls the teacher testing initiative one of Clinton's first attempts to "merge Democratic compassion with the Republican notion of responsibility." But responsibility isn't a "notion," and it has little to do with conditioning segments of the public to respond like laboratory mice to the government's delivery of shocks as well as treats. After a two-year battle that won the Clintons national recognition as advocates of standards, the teachers of Arkansas submitted to the test. Three and a half percent of them failed, mainly African Americans working in the poorest sections of the state. Unfortunately, the reform package offered no plan for training or recruiting better-qualified teachers, so their replacements were likely to be only marginally better prepared.

Some Arkansas teachers were just about ready to strangle Hillary Clinton over the testing issue, and state legislator Pat Flanagin complained that she was too ready to accuse anyone who challenged any aspect of her plan of being "against education." Most legislators, however, were won over by her presentation of the committee's report. "I think we've elected the wrong Clinton," commented one. These compliments came more easily, since Bill Clinton was the one pressing the lawmakers for their votes on the tax increases. The governor tried every tactic he could bring to bear, from late-night sessions and old-fashioned deal making in the corridors to bringing in Razorbacks coach Lou Holtz to plead with the holdouts. In the end, only the most regressive of his tax proposals passed, a one-cent hike in the sales tax, with no exclusions for groceries or utility bills.

As always, Hillary had worked very hard, but the long-term impact of the education reform project was mixed at best. Every school district now had to submit copious paperwork to prove it was in compliance with state regulations. In Fort Smith, superintendent Benny Gooden kept track of the forms he submitted between March and

July 1987, and found that they weighed a total of eleven pounds. A year after the reform passed, eighty-four districts raised real estate taxes in order to meet state requirements and avoid forced consolidation. State spending would continue to rise as well, doubling between 1984 and 1992.

What did the voters get for their money? Classes were smaller, and high schools offered more advanced placement classes for the college-bound. Most districts had no trouble getting an 85 percent pass level on the eighth-grade minimum performance test. Opinion varies as to whether this was because the test was quite easy or because of rampant cheating. In 1986, Arkansas students actually scored worse on the American College Test, or ACT, a standardized test used by college admissions offices. By 1992, the scores had risen slightly, to the same level they attained in 1972. (Only 6 percent of Arkansas high school students took the better-known SAT.) A 1988 study by the Winthrop Rockefeller Foundation found that school districts were weaker financially under the new system, and a whopping 77 percent of teachers surveyed complained of lower morale.[8]

"We used to have it great. Now it's ho-hum," said one teacher in a typical comment. "The standards did not pull us up to meet everybody else but pulled us down to meet everybody else."

Added another: "We're not reaching for the stars; we're shooting for the horizon."

One of the stranger footnotes to the Clintons' impact on education in Arkansas was the Governor's School, a six-week program for 400 gifted high school juniors, begun in 1979. Although Bill Clinton once called the program his "dream come true," the summer school was actually the pet project of the state's first lady. Held on the campus of Hendrix College, a Methodist liberal arts school in Conway, Arkansas, the Governor's School was a reincarnation of the Reverend Don Jones's University of Life.

Much of the Governor's School curriculum consisted of the modernist masterpieces (or in some cases, old chestnuts) familiar to college freshmen everywhere. Students read Beckett's *Waiting for Godot* and Sartre's *The Flies;* studied the theories of sociologist Herbert Gans, who believes that capitalist society deliberately maintains people in poverty in order to have buyers for heroin, as well as "cheap wines and liquors, Pentecostal ministers, faith healers, prostitutes, pawn shops"; and watched such films as *Do the Right Thing, Night and Fog* and *The Times of Harvey Milk*.

But the Governor's School was a long way from the mild-mannered existentialism of Reverend Jones. According to a former Governor's School student named Eddie Madden, who took copious notes during his attendance in 1980, the program that year included a speech by Hillary Rodham, who told students that she would trust big government over big business anytime. Another speaker was a physicist who said that science was the antithesis of religion and no good scientist could be a religious believer. "The Book of Genesis should read, 'In the beginning man created God,' " he told the class. When some students in the class objected to this, the physicist retorted, "After all, doesn't Genesis say, 'In the beginning was the word'?"

"No! No!" several in the audience shouted. "That's the Book of John."

"Well," the physicist shrugged, "it's in the Bible."

A third speaker, feminist theorist Dr. Hope Hartnuss, lectured on the fall of woman from the glorious matriarchal societies of pre-Christian times to her lowly state today, explaining, "Women built the society of communal living; marriage and private property ended it. . . . Through asceticism, Christianity was anti-woman and anti-sexuality. . . . Persecution of the church caused women to fall back. . . . The church had misogyny—a fear and hatred of women. . . ."

A decade later, the program of the Governor's School was more of the same. One of the featured speakers in 1991 was feminist Mary Daly of Boston College, whose topic was "Re-Calling the Courage to Sail: the Voyage of a Radical Feminist Philosopher Pirate." Among the assigned readings—and by no means atypical—was an essay by Emily Culpepper, a self-described "freethinking witch" who dismissed Christianity as "compost" and condemned belief in Christ's divinity as "no longer just implausible [but] offensive."

Chris Yarborough, who attended the 1991 summer session, described a program that made no attempt to present controversial subjects in a balanced way or encourage reasoned dialogue. Rather, the goal seemed to be to deprogram students from their traditional values and then indoctrinate them with postmodernist, multicultural clichés. Course work at the school was divided into three areas: academic courses in the student's chosen area of specialization, "critical thinking" and "feelings." Students were told that the unit on "critical thinking" was designed to "take you apart" so that the "feelings" unit could "put you back together." Discussions in the second area typically consisted of the professor posing challenges like "prove you exist" to students who were invariably ill equipped to outdebate him. After these intense sessions, the students moved on to the "feelings" segment, in which they were encouraged to share their most private thoughts and emotions without judging one another. The hostility inherent in the school's curriculum, which assumes that people with traditional values are small-minded and in need of deprogramming while wiccan academics who call Christianity "compost" are smart, is enough to make one think that Arkansans are justified in feeling a bit "truculent" about outsiders who want to sell them on new ideas.[9]

HILLARY Clinton was a woman of many causes, some more worthy than others. Over the years, she would be amazingly successful at shaping her public image, identifying herself with the more benign and moderate programs while avoiding undue publicity about the others. Thus, one hears very little about the Governor's School, but a great deal about the Arkansas HIPPY program.

Home Instruction for Preschool Youngsters, developed in Israel by Dr. Avima Lombard of the Hebrew University of Jerusalem, is a program that encourages poor mothers to take an active interest in helping their children learn. The American version, HIPPY, whose motto is Learning begins at home, encourages mothers to develop the habit of sitting down each day to go over their children's schoolwork and read to them from children's classics like *Goodnight Moon*, *Corduroy* and *A Snowy Day*. The costs, about $750 per child, are modest.

In 1985, Hillary was attending a National Governors Association function in Miami when she happened to read a newspaper article about one of the first HIPPY programs in the United States. On her return to Little Rock, she handed the clipping to Barbara Gilkey, an aide on the governor's staff. "That's all it was in the beginning," says Gilkey, "just a little yellow Post-it note that said, 'Look into this.'" Gilkey was new on the job and, she says, "a little bit nervous. I was very impressed that she had asked me to work for her."

The state budget didn't include funds for preschool programs, Gilkey recalls, noting, "Our challenge was to see if we could develop a preschool program that didn't cost anything." With Hillary's guidance, Gilkey managed to tap federal resources including Title 1 remediation funds, the Job Training Partnership Act, public housing initiatives and child abuse prevention moneys. The HIPPY name caused problems in some localities during the early years, adds Gilkey, recalling, "Word got around that we were founding Communist youth camps." But the first four community-based programs were so successful that the organizers were overwhelmed with applications.

Arkansans familiar with HIPPY think of it as Barbara Gilkey's program, not Hillary Clinton's. Gilkey herself, however, is an avid admirer of Mrs. Clinton. "This country has missed out a lot on not knowing her the way we know her," she says.[10]

Hillary's considerable skills as an executive and motivator often seemed squandered because of her inability to find a job large enough for her ambitions. Similarly, her political philosophy often appeared unfocused, and perhaps even insincere. In 1986, as chairman of an ad hoc group called the Arkansas Adolescent Pregnancy Child Watch Project, Hillary traveled the state promoting school-based health clinics. Allaying parents' anxieties, she promised that the clinics would provide basic health care services for children whose parents could not afford regular doctor visits. She also spoke of her personal belief that children should be taught "not birth control, but self-control."

But the Clinton-appointed head of the state Department of Health was Dr. Joycelyn Elders, who proudly displayed an "Ozark rubber tree" in her office and boasted of distributing more than one million condoms to Arkansas schoolchildren.

Elders achieved national stature promoting condom distribution as the antidote to ignorance. In a typically feisty 1994 speech to a Methodist church group in Austin, Texas, she challenged ministers to "stop moralizing and get out there and help me save the children." But teenage pregnancy rates, which had declined in Arkansas between 1980 and 1985, actually rose during her tenure as the state's chief health officer. And the rates rose most steeply—an average of 12 percent—in the eleven counties with school-based clinics. The clinics, moreover, have lived up to fears that the quality of medical care and advice would be substandard. Clinic staff had time to hand out condoms and refer even junior high students for Norplant, but not enough time or funding to run gonorrhea tests or to counsel young people that condoms do not offer much protection against some of the most common sexually transmitted diseases.[11]

Most alarmingly, in 1993, FDA inspectors found that four lots of LifeStyle condoms purchased by the Arkansas Health Department had a failure rate of over 5 percent. The FDA attempted to obtain a court order to recall the condoms but was thwarted by the Arkansas Department of Health, which took the position that a recall would undermine the public health goal of getting young people to use condoms regularly.

As for abstinence education, a visitor to the clinic at Little Rock's Central High School in 1992 found a stack of informational comic books in the waiting room promoting the message that "safer sex is having fun without taking risks." Among the "safer sex" activities suggested were oral sex and mutual masturbation. Thus, it would seem that when Hillary reassured parents that school-based clinics were consistent with a pro-abstinence message, she was, at best, ill informed.

In the area of children's services, the contradiction between promises and performance came to light in 1992, when Arkansas Advocates for Children and Families, an organization Hillary helped found, filed suit against the state's Human Services Department. The case of Daniel Toric, a seven-year-old tortured so cruelly by his foster mother that he lost a leg, had focused attention on the failures of child welfare services. With both Clintons, developing new programs and expanding the power of state over local government had taken precedence over keeping existing state functions running smoothly. The number of child welfare caseworkers had dwindled as the Human Services Department grew in other directions, and its managers had unaccountably failed to obtain millions of dollars in federal funds to which the state was entitled. The governor responded to the suit by commissioning a study of the department, but the reorganization never went forward, probably because he was distracted by his campaign for the presidency.

The fall of 1984 should have been a good time for the Clintons. The Quality Education Act was still seen by most Arkansas voters as a big step forward for the state's schools, and Bill Clinton coasted to an easy victory in his bid for a third term. Instead, the Clintons were embarking on one of the more difficult years of their often troubled marriage. Clinton, observed John Robert Starr, "was not very firmly in control of the 1985 legislature." Hillary was not often seen at her husband's side, and when she was her expression telegraphed a desire to be elsewhere. Reporters who covered the political scene were hearing the usual rumors—a love affair, drugs, clinical depression.

But one didn't have to look far to find a reason for the Clintons' unhappiness. In late July 1984, Roger Clinton had been indicted by a federal grand jury in Fort Smith on six counts of drug trafficking. According to David Maraniss, however, a heads-up about the investigation came when a state police official tipped off an aide in Clinton's office. The aide told Betsey Wright, who scoured the city looking for Hillary and found her at a west-side restaurant, having one of her long lunches with Webb Hubbell and Vince Foster. The two women then confronted Clinton, impressing on him that this time it was too late to save Roger and he must do nothing to interfere.

It is characteristic of Clinton that he would manage to turn this near disaster into a touching illustration of his own nobility. In the official Clinton version, he had no knowledge of his brother's involvement with drugs until a few weeks later, when state police chief Tommy Goodwin brought him the news. Saddened but resolute, Clinton ordered Goodwin to handle the matter like any other case.

In fact, Roger Clinton's drug problem was hardly a secret. His band, which played occasional gigs at local clubs, was named Dealer's Choice, and Roger's none-too-discreet use of marijuana had been a cause of concern to Clinton aides as early as the 1974 congressional race. In 1981, after Roger was arrested for failing to pay two speeding tickets, Clinton got him released into the custody of their cousin Sam Tatom, and arranged for Roger to be appointed to the Juvenile Advisory Board of the Arkansas Crime Control Commission, apparently on the theory that getting Roger involved in crime control would help him to mature. In March 1982, Roger was arrested for driving under the influence and possession of drug paraphernalia, but the charges were postponed and eventually dropped.

That Clinton himself used cocaine on occasion was widely suspected, if never proved. In 1990, for example, a federal drug informant named Sharline Wilson told a Saline County grand jury that while working as a bartender at a Little Rock club named Le Bistro, she was often hired to work private parties, where she sold coke to Roger Clinton, who then gave the drugs to his brother. At one party, said Wilson, the governor was so out of it that he slid down the wall, ending up slumped in a trash can.

By 1983, however, Roger was not just a user who occasionally sold a few grams to friends. Through a former girlfriend, he had won the confidence of a New York dealer with Colombian connections. At first he received small shipments by Federal Express. Then he began shuttling back and forth between New York and Little Rock with packages of coke strapped to his body, once making a run while traveling with his brother, the governor. Roger said later that he was confident that his all-American looks would keep him safe. Others who knew him at the time recall a young man whose pasty complexion and perpetual sniffles made the nature of his problem all too obvious.

In the latter part of 1983, Roger got into a jam. Thieves vandalized his convertible and stole a shipment of cocaine that Roger was selling on consignment. His suppliers threatened to kill him, his mother, even the governor, unless he found a way to pay them back. Panicked, Roger approached Dan Lasater, begging for $20,000. Lasater balked but eventually gave him $8,000. Lasater also shipped Roger off to his horse farm in Ocala, Florida, where he remained through the Christmas holidays. According to an FBI investigation of Lasater, Bill Clinton had asked him to stash Roger at the farm to get him out of harm's way. In 1995, Lasater would say that he did not recall who asked him to give Roger a job in Florida.

Even before he left for Florida, Roger's activities had come under surveillance by a joint state-federal task force on narcotics trafficking. After his return, in the spring of 1984, he was caught on tape snorting coke with an informant and negotiating a $30,000 payoff in exchange for getting "Big Brother" to arrange a sewer permit for a time-share condo project in Hot Springs, partly owned by his friend Dan Lasater. "We're closer than any brothers you've ever known," boasted a bleary-voiced Roger. "See, I didn't have a father growing up, and he was like a father to me growing up, all my life, so that's why we've always been close. There isn't anything he wouldn't do for me." At a later meeting, Roger reported a problem: One member of the Control Board responsible for issuing the permit was honest and couldn't be bribed. Roger predicted that Big Brother would find a way around the holdout.

Roger could also be heard bragging about having sex with women in the Governor's Mansion—in the main house as well as the guest house. "Oh, they love it!" he exclaimed.

Another April 1984 recording, made by a Hot Springs police detective, Travis Bunn, captured Roger trying some coke and saying, "I've got to get some for my brother, he's got a nose like a Hoover vacuum cleaner!"

Of course, it is possible that Roger was letting his imagination run wild while under the influence, but his talk of partying at the Governor's Mansion was confirmed by visitors' logs that showed Roger making late-night visits about three dozen times after the spring of 1983, often accompanied by one or two women. Most of these visits seem to have taken place when Hillary was out of town.

Bill Clinton would later recall how painful it was to know that Roger would soon be arrested and to be unable to warn him or interfere in the investigation. In retrospect, it is astounding that Clinton would manage to turn the incident into a story that placed his conduct in a virtuous light, when in fact the tip to Betsey Wright had broken the secrecy of the investigation just when it appeared that the governor himself might be implicated in criminal behavior. Roger was now using coke an average of sixteen times a day, and he was so close to the edge psychologically that when he learned he had been indicted he threatened suicide. Clinton then had to confess that he'd known about the pending indictment for weeks.

Not that Roger was exactly thrown to the wolves. On August 14, 1984, he was represented at his arraignment by William R. Wilson, the prominent criminal defense attorney who had partnered with Hillary on a number of cases. Roger had the good fortune to have his case assigned to Oren Harris, an elderly judge with a reputation for sleeping on the bench. Despite the seriousness of the charges, Roger was released on $5,000 bond and was soon enrolled at Hendrix College in Conway, the same Methodist institution that played host to the Governor's School. After attending classes for a short time, Roger was asked not to return. Virginia Kelley was furious with the college for its "cruel" treatment of her son, jumping to the conclusion that the dean had "caved in" to pressure from parents who didn't want an indicted felon on campus. But the truth was, Roger was still using cocaine and, for that matter, still partying occasionally at the Governor's Mansion.

In November, three days after Bill Clinton was safely elected to a third term, Roger appeared in federal court and entered a plea of guilty to one count of drug trafficking. The whole family, including

Hillary, appeared for his sentencing on January 27, 1985. After hearing testimony from a drug rehabilitation counselor, who explained that Roger had continued to use cocaine even after his arrest, Judge Harris declared that he could not consider probation. Roger was sentenced to two to five years. (He would be released on good behavior after doing half the minimum.) The sentence, Virginia would declare later, was "more than we had been expecting." But considering that Roger had been dealing in quarter pounds, half pounds and whole pounds—not exactly street-corner quantities—the sentence was amazingly lenient. Defense attorney Bill Wilson had certainly earned his fee. (In 1993, Wilson would be appointed by President Clinton to the federal bench.)

In February, Roger would appear as a witness at the trial of Sam Anderson, Jr., a childhood friend from Hot Springs. Anderson was a customer, not a dealer, and thus lower on the prosecutorial food chain, making Roger's deal all the more remarkable. During the trial, however, it came out that Clinton backer Dan Lasater had been one of the targets of the narcotics probe. Roger Clinton made the feds' job easier by testifying about the $8,000 Lasater had given him to pay off his debt to his supplier; however, Lasater's indictment was still more than a year and a half in the future.

From Virginia Kelley's point of view, Hillary was the villain of this sorry episode, resented if not exactly blamed. Virginia believed that family—and her boys in particular—came first. Bill had known about the sting operation and failed to derail it, or even to warn his baby brother of his peril, and this betrayal was all Hillary's doing. Understandably, Hillary didn't see it that way. While out on bail, Roger briefly entered a rehabilitation center, and on his release he was referred to a family therapist, Karen Ballard. Hillary attended a few of these sessions and used the opportunity to deliver some brutally frank observations about the syndrome of denial and irresponsibility in the Clinton clan.

Co-dependency was a hot concept in therapy in the mid-1980s, and it didn't take a Ph.D. to see that Virginia and her sons fit the profile of the family group organized around the alcoholism of Roger

Clinton. Young Roger was eaten up with anger at his dead father, yet became more like him every day. Virginia, with her mask of heavy makeup, was consumed by the need to put "a good face" on the family situation. Virginia had been sued twice for malpractice after young women died on the operating table while she was administering anesthesia. She had failed to keep up with developments in her field and was seen by witnesses reading the racing form in the operating theater. Except for a questionable ruling by the state medical examiner, Dr. Fahmy Malak, she might have faced criminal charges in the second case. Nevertheless, when the Hot Springs hospital terminated its contract with her, she filed a federal antitrust suit, convinced that she was the victim of a conspiracy to replace nurse anesthetists with M.D.'s trained in anesthesiology.

Virginia had babied both her boys. Although the family was not wealthy, Bill had been the neighborhood kid with a well-stocked Coca-Cola freezer in his garage and his own car. Both boys knew they could avoid punishment by being cute, and when Virginia was called to the hospital and couldn't cook their breakfasts, she would set out the ingredients on the counter, even pre-separating the bacon strips. And yet she had remarried Roger Clinton, even though he was so abusive that he once took a shot at her in front of the boys and on other occasions caused her to take her sons and hide out from him at a motel. Moreover, although young Roger's drug problem had been a decade in the making, Virginia had somehow failed to take note of it.

Bill Clinton fit the profile of what therapists call the "Family Hero," the child of an alcoholic parent who overcompensates for the messy situation at home by collecting friends and public honors. Ever the conciliator, he strove to charm everyone around him and avoid open confrontations, even while channeling his unacknowledged anger into behavior that hurt the people closest to him.[12]

Hillary's acid observations on the family structure during counseling sessions sent Virginia fleeing the therapist's office in tears. As Clinton saw it, Hillary was attacking his mother when she was already down. (There was never, it seems, a good time to bring these matters up.)

Clinton himself was miserable, more abject than he had been after his 1980 defeat. Arkansans Charles F. Allen and Jonathan Portis write of this period, "Intimates say the vacuum of a lost childhood finally drained him of his self-respect. He began to see himself as a failure. He hit a low in his mood swing and became self-destructive."[13]

Clinton sought solace from girlfriends, including Gennifer Flowers, who had returned to Little Rock and moved into a second-floor apartment in the Quapaw Tower, a convenient stop on Clinton's morning jogging route. At one point, Flowers was hired to sing at an outdoor party on the grounds of the Governor's Mansion. It was a crisp fall day and she appeared in a bright red dress. Hillary cut Flowers dead, passing by her without ever making eye contact. But the governor, Flowers says, tried to talk her into having a quickie in a first-floor rest room of the mansion.

On another occasion, Flowers and Hillary crossed paths at a fund-raiser held in the Capitol Club. This time, Hillary tried to draw her husband away from a conversation with Flowers, saying, "Bill, I'm going over to the bar for a drink. Will you come with me?" Even this modest power play failed. Clinton brushed her off, saying, "You go ahead," while Flowers crowed silently. "It was a showdown," she later observed, "and I had won."[14]

During the summer of 1985, Hillary confided to friends outside of Arkansas that she'd had enough and was about to file for divorce. She sounded so determined that they assumed the marriage was finished. But the next news that came out of Little Rock was that Bill Clinton, Hillary at his side, had announced his candidacy for a fourth term.

What had happened to change Hillary's mind?

One answer is that the family therapy sessions, which seemed such a disaster at first, were showing some positive results for everyone, including Clinton. He had begun to delve into the literature on addiction and ACOA, the adult child of alcoholics syndrome, and was fascinated and amazed by the insights he gained into his own motivations. Clinton discussed these subjects rather freely with his old friend Carolyn Staley and with Betsey Wright and others.

Apparently Clinton also made a sincere effort to change his ways, though in the long run it would seem that his reform effort was undone by creative rationalizations, such as the now famous theory, first shared with trooper Larry Patterson, that the biblical definition of adultery did not include oral sex. (Biblical guidance on this point was reinforced by the threat of AIDS and sexually transmitted diseases. By 1985 it was no longer possible to ignore the health risks of promiscuous sex.) In the meantime, the psychological point of view offered a new level of self-justification. Clinton's affairs were not acts of disloyalty to his wife or even perquisites of power but just one more expression of his compulsive drive for emotional validation, which allowed him to be easily victimized by predatory females.

With hindsight, the key to Clinton's character was his belief, expressed in his 1969 letter to Colonel Holmes, that preserving his "political viability" was a goal that justified cutting ethical corners. Clinton was surrounded by people who supported this notion to one degree or another, and none more than Hillary. In order to keep her marriage going, Hillary had compromised her own ambitions, accepted her husband's chronic infidelities, and taken on the role of protecting him from the consequences of his often self-destructive behavior. To justify these decisions to herself, she needed to have faith in his special destiny. Thus, keeping Bill's career on track invariably took precedence over resolving the problems of their private life.

And, ironically, even as the federal investigation into Roger Clinton and Dan Lasater threatened to destroy the Clintons' future, they were achieving recognition on the national scene. In 1984, *Esquire* magazine listed Bill and Hillary Clinton among 272 outstanding representatives of the baby boomers, and the National Association of Social Workers honored them as Public Citizens of the Year. Their position on teacher testing, however calculating it might have been, played well in the national media, earning the governor an appearance on *Face the Nation*. The possibility of a Clinton presidency, which Hillary had predicted to Bernie Nussbaum back in 1974, was now being taken seriously by the world at large.

Like a number of other Democratic politicians of his generation,

Bill Clinton had succeeded in reinventing himself. The children of the Great Society and Watergate were now middle-aged Atari Democrats. With a conservative Republican in the White House and the taxpayers no longer willing to fund visionary new programs, smart, energetic Democrats in state and local government had begun to look toward other sources of financing, such as bond issues and public employee pension funds. They were getting their ideas from theorists like Belden Hull Daniels, Lawrence Litvak, and Bill and Hillary's old friend Robert Reich, who advocated something called socially responsible investment. Basically, the idea was that capitalism wasn't going away anytime soon, but since it was a dumb and inherently inefficient system, it could benefit from a little enlightened tinkering.

Among the governors who bought into the trend were Michael Dukakis of Massachusetts, John Carlin of Kansas and Bill Clinton of Arkansas. Of course, fostering economic development wasn't a new idea, even in Arkansas. The state already had an industrial development agency as well as a housing authority, the Arkansas Housing Development Authority. Clinton's plan was to replace the latter with a new entity, the Arkansas Development Finance Agency, which would issue securities directly. The pension funds of state employees— teachers, state police, highway crews, clerks—would also be used to finance ADFA bond issues, advancing money until the bonds sold on the open market. In addition, the pension funds would be required to invest between 5 and 10 percent of their assets inside Arkansas, with ADFA reviewing the investments to decide which of them were creditworthy.

The problem with socially responsible investing is that enormous pools of cash—much of it other people's retirement money—are placed in the hands of elected officials and their appointees in the expectation that they will go forth and do good things with it. The temptations are many, the guidelines few. Clinton's critics charge that ADFA degenerated into a front for money laundering, graft and even narcotics trafficking. The evidence is provocative, though controversial. But even when ADFA worked more or less as expected, the line between good works and corruption was often a matter of perspective. ADFA bond issues generated business for primary and secondary bond underwriters, law firms and trustee banks, all of which could be expected to contribute to Clinton's campaigns. ADFA was also used to reward favored businesses, giving them an edge on their competition. Meanwhile, pensioners' money—by law—was no longer being invested for the maximum benefit of its owners, who were the workers, not the state.

Even while the legislation creating ADFA was being debated by the assembly, testimony in the trial of Sam Anderson, Jr., was revealing that Clinton's favorite broker, Dan Lasater, and a number of his associates were the targets of a narcotics investigation. In April there was another crisis. Over the Easter weekend, Bevill, Bresler and Schulman, a New Jersey dealer in government securities, went bankrupt, losing $52 million in state pension funds lent to it by the trustee, the Worthen Bank. Worthen had taken an unacceptable risk with the pension money, but the ultimate responsibility lay with the state for its failure to supervise the bank—a bad omen for its ability to run ADFA. In fact, Worthen's contract was so poorly written that it may not have been legally obligated to make good on the loss. Clinton's political future hung in the balance until Jack Stephens flew in from his home in Georgia and agreed to cover the losses.

Thanks in part to Webb Hubbell, who figured out that Worthen could recover part of the money from its insurers, the losses were capped at $32 million. Major shareholders, including Joe Giroir and the Riady family, would be stuck with the rest of the losses. Meanwhile, an investigation by the comptroller of the currency turned up a slew of questionable loans, many of them in Indonesia. Worthen Bank stock went into a free fall, and the Riadys would eventually sever their ties to the institution. But the Clintons now had reason to be grateful to both Jack Stephens and James Riady, who was Worthen's president at the time of the debacle and had declined to pass the losses off onto the state. In October 1985, Bill and Hillary traveled to Asia on a two-week trade mission, and in Hong Kong they were escorted by a Riady group executive named John Huang, who would later play a key role in Bill Clinton's political future.

Meanwhile, on the same day that Bevill, Bresler went broke, Jim McDougal was holding a fund-raiser for Bill Clinton at the headquarters of Madison Guaranty Savings and Loan, on the northern edge of the Quapaw Quarter.

After living for several years in a trailer home in the tiny hamlet of Kingston, Arkansas, where they operated a small bank in partnership with former Clinton aide Steve Smith, the McDougals were back in Little Rock and, to all appearances, prospering. Madison's art deco headquarters in a former laundry building was a symbol of the Quapaw's renaissance, and Susan was conspicuous on television, dressed in tight shorts and a work shirt that showed more than a hint of cleavage as she delivered an advertising pitch for a low-end development called Maplewood Farms. She and Jim were living in the Heights now, in a house Susan had redecorated herself, and they drove classic cars—his and hers Jaguars and, for a time, a sky-blue Bentley.

But the McDougals were floating in a sea of debt. They didn't own the cars, which belonged to a troubled leasing company Jim had done business with, and they owed at least half a million dollars, including approximately $200,000 borrowed from the Worthen Bank to buy Madison Guaranty's nominal parent, a beaten-down savings and loan in the small Arkansas town of Augusta. Vince Foster, who had represented Jim McDougal in a case involving his bank in Kingston, considered him a prime example of the hucksterism that was ruining Little Rock.

Despite Foster's low opinion of Jim McDougal and the Clintons' own experience with the Whitewater land deal, Hillary and Bill were opening a second chapter in their relationship with the McDougals. Clinton always needed financial angels, and it may be significant that he turned to McDougal even as he had become wary of dealing with Lasater and his friends. Clinton was carrying a $50,000 personal loan from the Bank of Cherry Valley that he had taken out to pay for a last-minute ad blitz in his 1984 campaign, and he asked McDougal to help him pay it off.

McDougal's fund-raiser took place on the eve of the Easter holiday weekend and attracted about fifty people, who sipped wine and munched canapés in the thrift's lobby. Clinton and Betsey Wright didn't arrive until the party was almost over, and former senator William Fulbright, touted as the guest of honor, failed to show up at all. Still, McDougal handed Wright $35,000 in checks, each for no more than the maximum legal contribution of $3,000. A complete list of the donors no longer exists, but some of the checks were from individuals who later denied making any contribution at the fund-raiser. Fulbright, whose signature was on one check, said he was never even invited to the event.

Where did the money come from?

Interestingly enough, one day before the fund-raiser, McDougal withdrew $30,000 from the Whitewater Development Company account, money identified as a "loan repayment." The transaction left the account with a $28,000 overdraft. McDougal covered the overdraft by having Madison Guaranty pay him a $30,000 "bonus," which he deposited in the Whitewater account.

Shuffling money around had become a way of life for McDougal. In December 1980, he had decided that the Whitewater Estate lots would sell better if there were a model house on the property. McDougal was running the Madison Bank in Kingston at the time, but regulations prohibited him from lending any more money to himself. He got around the prohibition by transferring the title of Lot 13 to Hillary, who took out a $30,000 loan from Madison to pay for the house. McDougal, acting on behalf of the Whitewater Development Company, arranged the construction of a prefabricated house, sold it to a man named Hilman Logan and used his money to cover Hillary's loan. The company also covered taxes and incidental expenses. Later, after bank examiners questioned the loan, Bill Clinton borrowed $20,000 from Security Bank in Paragould, Arkansas, to pay back the Kingston bank, but Whitewater continued making payments on the new loan.

Hillary, McDougal would recall vaguely, "seemed irritated" by this arrangement even though it was "merely paperwork and cost the Clintons nothing." Perhaps this was because McDougal had gotten her involved in a sham transfer of title to evade banking regulations. By 1984, Whitewater had begun to miss payments on the loan with the bank in Paragould, increasing the possibility that auditors would begin to ask questions. Unfortunately, Hillary's claim to be an innocent victim of McDougal's bad advice was undermined by her and Bill's decision to deduct the interest payments on the loan on their personal income tax return. Moreover, it would seem that Bill Clinton may have taken out other loans from McDougal's bank. In February 1982, Clinton wrote a check to Madison for $20,744.65. The Clintons originally claimed this check was related to Whitewater expenses, but by March 1994 had remembered that the loan was to help his mother buy her cottage on Lake Hamilton.[15]

Although Hillary may have been irritated with McDougal, she took him on as a client in April 1985, just days after the Easter weekend fund-raiser. McDougal gave different versions of how this came about, the common theme being that early one morning in the summer of 1984 Bill Clinton interrupted his usual jog through the Quapaw Quarter to pop into Madison Guaranty. Settling his sweaty body into a brand-new leather chair, Clinton complained that Hillary was having trouble bringing new clients into the Rose firm. McDougal noted, "I hired Hillary because Bill came in whimpering they needed help." This story does not explain why McDougal waited until the following spring to put the Rose firm on retainer. In his last attempt to account for the gap, he would say that he had begun paying Clinton $2,000 a month in graft. Giving Hillary's law firm business merely legalized the setup.

Hillary's version of how she came to represent McDougal was that she had merely stepped in to act as name partner, accommodating a young associate, Rick Massey, who solicited Madison Guaranty's business on his own initiative. Massey, however, said his involvement came after Hillary met with McDougal to discuss taking on his business.

However the arrangement came about, McDougal had an obvious motive for wanting to hire Hillary. He had been milking Madison Guaranty's assets to support his real estate ventures. The S&L was at the center of a pyramid scheme, supported by kiting checks from one company to another and doling out government-guaranteed loans to associates in exchange for various favors. As early as January 1984, the Federal Home Loan Bank Board had issued the first of several scathing reports on Madison Guaranty's financial condition. The board was threatening to shut down the S&L unless McDougal could find a way to recapitalize. He needed the help of a politically connected attorney like Hillary Clinton, and he got it. In her first act as McDougal's attorney, Hillary called the state securities commissioner, Beverly Bassett, who gave preliminary approval for a sale of preferred stock.

Acting on behalf of his sometime partner Jim Guy Tucker, McDougal had lobbied to get Bassett appointed to her job, apparently on the theory that since she was young, blond, female and a former partner in Tucker's law firm, she would be easy to manipulate. All this would come as news to Bassett, whose decision appears to have been made on technical grounds. But the stock issue soon ran into other problems—fortunately for Hillary, who had cited a fraudulent audit in her attempt to persuade Bassett that Madison Guaranty was solvent.

Less than a month after Hillary's call to Bassett, in May 1985, McDougal sold off the remaining Whitewater lots to Chris Wade, the local real estate broker who had handled the sale of the land to McDougal. Wade put up no cash, but he agreed to pay $35,000 toward the Citizens Bank mortgage and gave McDougal a Piper Seminole airplane. The Piper was later purchased by Webb Hubbell's father-in-law, Seth Ward, who financed the purchase with a loan from Madison Guaranty. The Clintons may have been unaware of this sale at the time.

Later in the year, however, Hillary had a warning that Whitewater was about to become an embarrassment to the Clintons when their names appeared on a list of property owners who were delinquent in paying real estate taxes in Marion County. Voters do not look kindly on politicians who fail to pay their own taxes, and Hillary was furious. According to James Stewart's book *Blood Sport*, McDougal saw this as an opportunity to get the Clintons out of Whitewater. He'd been paying their share of the bills anyway, and he reasoned that by signing over their interest, they could save themselves future embarrassment. In Stewart's account, McDougal called Clinton and suggested that he and Hillary sign over their interest in the investment; in exchange, he and Susan would assume the company's liabilities. Clinton agreed, and since by this time Jim McDougal and Hillary were at odds, he sent Susan over to the Rose firm with a stock certificate for Hillary's signature.

Hillary's reaction was icy. She stared at the certificate, refusing to touch it. "Jim told me that this was going to pay for college for Chelsea," she said. "I still expect it to do that."[16]

Assuming this story is accurate, Hillary's comment could be read as an indication that she was ignorant of the true state of the Whitewater investment. On the other hand, she may have known the truth all too well and been determined to squeeze the McDougals. Jim McDougal could not unilaterally assume Whitewater's debts, since the bank had Bill's and Hillary's signatures on the mortgage loan, but he obviously wanted the Clintons out of Whitewater—probably because, as his ventures went, it had a good credit rating and he could use it as a shell company to launch other projects. By withholding her signature, Hillary might have hoped that she could motivate McDougal to shuffle some money in the direction of the Flippin bank.

A few months after this acrimonious meeting, Hillary represented Madison Guaranty's interests on a project called Castle Grande, a land flip scheme that stands as a paradigm of the savings-and-loan scandals of the 1980s. Jim McDougal owned a parcel of land just south of Little Rock that he called Castle Grande Estates. He wanted to develop the land into a trailer park, but to do this he needed to build a road to the site. Seth Ward, who by this time was acting as a land scout for McDougal, was sent to negotiate an easement. The owners weren't interested, Ward reported, but they were willing to sell the entire property for $1.5 million.

The larger tract—called Castle Grande, as opposed to Castle Grande Estates, which was the trailer park—was described by federal assessors as "low and swampy" as well as generally undesirable. Nevertheless, McDougal wanted it. Unfortunately, Madison Guaranty was barred from providing more than 6 percent of the financing for property purchased by its real estate subsidiary, Madison Financial. To get around this regulation, Seth Ward became the "straw buyer," purchasing most of the 1,100 acres with an unsecured loan from the S&L. McDougal then subdivided the land and began selling parcels at greatly inflated prices, guaranteeing Ward a 10 percent commission on every parcel sold.

The beauty of playing with federally insured money was that the more inflated the land prices, the more attractive the deal. The buyers got loans from Madison Guaranty, often for much more than they were actually paying for the property. When they defaulted, and most of them did default eventually, the taxpayers were left with the debt. Castle Grande would add some $3.8 million to the negative side of Madison Guaranty's balance sheet.

Unfortunately for Seth Ward, he didn't get his commissions, probably because the bank examiners were already keeping tabs on the S&L's books and the fraud was too transparent. (If Ward owned the land, why would he get a commission for selling it?) At this point, Hillary stepped in to negotiate an agreement giving Madison an option to buy 22.5 acres from Ward for $400,000. A fair price for the parcel would have been about $47,000, and the option was apparently part of a complicated scheme to justify paying Ward for his role as a straw buyer. Hillary represented both Ward and Madison in drawing up the agreement.

In a 1994 interview with investigators from the Resolution Trust Corporation, Hillary would describe Seth Ward as "a persistent, demanding client, someone who pushed very hard for lawyers to respond to him, to get his work done . . . someone who wasn't shy about showing up at your office unannounced and demanding that you give him all the time he wanted right then, no matter what else you were involved in."

This was a fair description of the overbearing Ward, the bane of his son-in-law Webb Hubbell's existence. Hubbell, however, would get revenge by overbilling the Ward family firm, Park-O-Meter, by hundreds of thousands of dollars. It may be significant that Hillary took care of the land option at a time when Hubbell happened to be out of town. Perhaps she was in a hurry to get rid of an obstreperous client and pushed the option through, shrugging off a warning from Madison Guaranty executive Don Denton, who told her that the agreement was drawn in a manner that might be illegal.

The problem with this scenario is that Hillary billed Madison Guaranty for a two-hour meeting with Ward on the day the option was signed, a long time to discuss an option agreement that posed no special problems. She also met with Ward on February 29, 1986, the day Jim Guy Tucker purchased the Castle Grande sewer system from him. This deal was part of a complicated loan swap that included the now famous $300,000 loan made by municipal judge David Hale to Susan McDougal.

When Hillary was questioned under oath by representatives of the Resolution Trust Corporation in May 1995, she denied doing any work on Castle Grande. After her billing records mysteriously surfaced on a workroom table in the private quarters of the White House, she was hard-pressed to explain why she had billed thirty hours on the project, at a rate of $120 an hour. She claimed that about half of the time was spent on low-level research on a plan to build a microbrewery on the site—another project of which she had earlier denied any knowledge. It seems doubtful that a senior partner would have been billing for such work.

With hindsight, there are three possible explanations for Hillary's conduct: She worked on the Castle Grande deals but was too dense to realize that they were a scam on the taxpayers. She did very little work but padded her bill. Or she actively participated in setting up the deals because she had a motive for keeping McDougal's real estate ventures from collapsing. At the very least, she did not want herself and Bill to get stuck with the Whitewater mortgage, of which $56,000 was still outstanding as late as 1986. When finally placed in a position where she could not avoid answering questions, Hillary would opt for the stupidity defense.

The Hale/Tucker/Madison Guaranty loan swap was a last-ditch

effort to keep the pyramid scheme afloat. By the time Susan McDougal's $300,000 loan came through, the Federal Home Loan Bank Board had already begun another audit. On July 11, FHLB regulators and Beverly Bassett met with the Madison Guaranty board and forced them to oust McDougal. The governor's aide, Sam Bratton, had been warned by Bassett that the coup was coming, and Betsey Wright wrote Clinton a note asking him if he still had shares in Whitewater, adding, "If so, I'm worried about it." He wrote back, "No, do not have any more." Whether Clinton actually believed that Hillary had signed the papers getting them out of Whitewater is a mystery.[17]

Just three days after McDougal's ouster, on July 14, Hillary sent a hand-delivered letter to Madison Guaranty, informing the remaining directors that the Rose firm would no longer be representing them. By coincidence, on July 10, Rose had been put on notice that the firm would have to sever its ties with savings and loans if it wanted to represent federal regulators.

The loss of Madison Guaranty was enough to push McDougal over the edge. He and Susan had been living apart for some years, and she had been having an affair with one of McDougal's real estate salesmen, Pat Harris, who is now her fiancé. One afternoon, McDougal was sitting in his living room at the Riviera Apartments when he suddenly "went berserk," climbing out the window and down the fire escape. Doctors diagnosed a condition that cut off blood flow to the brain, as well as long-standing manic-depressive illness.

Stabilized on medication, McDougal still had hopes of turning one last deal. In October he and Susan closed on an 810-acre tract south of Little Rock purchased from International Paper. McDougal called the property the Lowrance Heights, a made-up name he considered quite witty. At least some of the money for this purchase appears to have come from Susan's $300,000 loan. Also, the McDougals bought the land in the name of the Whitewater Development Company. In his posthumously published memoir, *Arkansas Mischief*, McDougal says that he thought Whitewater's losses would create a tax write-off against their profits from lot sales. But another reason suggests itself.

Governor Clinton's connection to Whitewater may have been the key to getting International Paper to provide financing. McDougal, kicked out of his own S&L for cooking the books, was certainly a bad credit risk, but Bill and Hillary Clinton were not. Moreover, International Paper had just that summer received a $22 million tax break from the state.

As bad as this looked for the Clintons, it seems that they were never intended to share in the profits from selling lots in Lowrance Heights. A few days before the deal closed, the McDougals wrote to Hillary once again asking her to sign papers transferring her and Bill's interest in Whitewater to them. Once again, she refused.

The Citizens Bank in Flippin was now known as First Ozark, and the new managers weren't inclined to keep rolling over the Whitewater mortgage loan indefinitely. With a balloon payment looming, Susan McDougal talked the bankers into refinancing under terms that would have directly assigned payments from contract sale customers to the bank. Hillary resisted this arrangement, refusing to provide the necessary personal financial statement and nearly throwing the loan into default. Apparently she was using the statement as leverage to get as much information as she could about Whitewater's finances. But the move was too late, and badly timed. The officers of First Ozark were nervous about foreclosing on a property partly owned by the governor, and when Bill Clinton came to Marion County to deliver a speech, the bank president took him aside and tried to impress upon him the importance of providing the financial statement. Hillary finally came through with the paperwork in March 1987. The third phase of Whitewater—the damage control—had now begun.

EIGHT

The Gary Hart Factor

I N the Clintons' career plans, 1987 was to be the year that Bill advanced from the minor leagues of Arkansas politics onto the national scene. The governor had speaking engagements that spring in eighteen different states, including New Hampshire, and in March he even made a trek to the West Coast, where he was introduced to major Hollywood donors at a dinner given by producer Norman Lear.

And yet even as his day in the national media spotlight approached, the rules that Bill Clinton—and, less willingly, Hillary Clinton—had lived by were rapidly changing. On May 8, front-running Democrat Gary Hart was forced to drop out of the presidential race following the release of photos taken of him with a young woman named Donna Rice while on an overnight excursion to Bimini aboard the aptly named pleasure craft *Monkey Business*. Hart's bad fortune at first appeared to be a lucky break for Bill Clinton. The most formidable competition for first place on the Democratic ticket was gone, a victim, the pundits said, of an ill-advised challenge issued to a *Miami Herald* reporter inviting him to investigate rumors of sexual infidelity, an area normally beyond the bounds of press scrutiny.

Around this time, Dolly Kyle Browning, whose occasional assignations with Clinton had continued through the years, warned him that there was no way he could pass muster under the new standard of media scrutiny. His affairs with women were simply too numerous and too well known in Arkansas. To enter a national race now without setting his private life straight would be political suicide.

In an effort to change the direction of her own life, Browning had entered therapy and become familiar with concepts like sex addiction and the adult-child-of-an-alcoholic syndrome. Asked if he had ever considered his behavior in this light, Clinton conceded that he had, but intellectual insight had failed to motivate him to change.

Whether there really is such a thing as "sexual addiction" is a matter for the psychologists and the moralists to debate among themselves. Certainly there have always been politicians who discover that the aura of power creates certain opportunities that they come to regard as a perk of office. Clinton's childhood hero JFK was undoubtedly the prime example of the statesman as stud, a leader whose affairs with Marilyn Monroe, Judith Exner and others too numerous to contemplate only added to the aura of glamour his name evoked. One could argue that Bill Clinton's behavior had been redefined into deviancy when the rules changed on him in midcareer. From another point of view, the Clintons' marriage had settled into a pattern of denial all too familiar to veterans of AA and their families. Clinton tended to excuse his behavior by blaming his wife's coldness, overlooking the question of cause and effect. Hillary, on her part, had achieved a remarkable ability to ignore the obvious. The woman who decides to stay married to an alcoholic, for whatever reasons, eventually loses interest in hearing the details of his latest binge. For the sake of her peace of mind, it was easier not to know; and when the truth could not be ignored, it was often easier to assume that he had simply been led astray by bad company.

The Clintons marked off their married life in terms of electoral cycles. There was always a campaign going on, or money to be raised, or some crisis with the legislature that needed to be addressed. Ironically, considering the troubles they would later endure over the Whitewater development, Bill and Hillary spent the first seventeen years of their marriage within a few hours' driving time of the scenic Ozark and Ouachita mountain regions without acquiring a weekend cabin where they and Chelsea could enjoy quiet time together. Hillary enjoyed unscheduled vacations, and she and Chelsea tried to take a week every year to get out of town together or in the company of Arkansas-born actress Mary Steenburgen, a close family friend. Chances to enjoy such free time with Bill, away from the distractions of official business, seem to have been all but nonexistent. In Little Rock, their social calendar was filled with official functions, and the trips they took together were usually official trade missions, financed by the Arkansas taxpayers. Clinton was often out of town on business, and when he wasn't away, Hillary was, attending board meetings or events sponsored by the Children's Defense Fund.

And now the Clintons were looking forward to a challenge that would keep them even busier. Hugh and Dorothy Rodham had sold their home in Park Ridge, Illinois, and moved into a Little Rock condo, in part so that they would be able to look after seven-year-old Chelsea while Bill and Hillary were on the campaign trail. With July 15 set as the date for Clinton's formal announcement, old friends were flying in from around the country, and Carolyn Staley, Bill's high school friend, had planned a party in his honor.

Meanwhile, a few close advisers were hearing disturbing rumors. The byline of Meredith Oakley, a frequent Clinton critic, had been missing from the pages of the *Democrat* for some days, inspiring fear that she was closeted somewhere working on a devastating exposé. Another story had it that someone from *The New York Times* had called the governor and asked, "We hear that you are screwing a judge in Arkansas. Is it male or female?" Oakley later denied that she was contemplating any exposé, and the story about the *Times* inquiry was doubtless apocryphal, too. (The rumor concerned municipal judge Beth Coulson, who was appointed by Clinton to the state court of appeals. Coulson has denied an affair.)

A few days before the fifteenth, Betsey Wright confronted Clinton with a list of women's names, challenging him to tell her the truth about his relations with each of them. The experience was a sobering one for Clinton, though he did not make a final decision until the eve of the press conference. By then, representatives of the national media were already in Little Rock, as were a group of old friends, including Mickey Kantor, who was staying at the Governor's Mansion along with Carl Wagner, whom the Clintons had known since he was a McGovern organizer in 1972. The Clintons had no choice but to appear at the rented ballroom of the Excelsior Hotel the next day, where Bill made a statement attributing his decision not to run to his concern that a presidential campaign would force him to spend too much time away from Chelsea. Hillary, standing nearby, appeared teary-eyed and, some thought, stunned.

With so many friends in town, Clinton didn't have an opportunity to fully explain himself until sometime after his nonannouncement. There are varying accounts of the major blowup that took place then, obviously all speculative. It is very unlikely that he made anything approaching a detailed confession of his infidelities. Clinton had at least two good reasons not to go into specifics: guilt, and fear of giving his wife ammunition should she decide to file for divorce. According to one source, Hillary was bitterly angry, primarily because Clinton had panicked and decided to withdraw from the race. She thought his problems could be "handled."

Meredith Oakley quotes a comment Hillary made to a friend suggesting that she was relieved. Observing that she and Bill had been through fifteen contests, primaries as well as general elections, in thirteen years, Hillary had expressed relief that the governor's term had been lengthened to four years, meaning that Bill wouldn't have to run again until 1990. "When we got the four-year term, we were looking forward to something like normal life," she said. "He could never say 'this is what a normal person would do.' "[1]

Any hope of the Clintons enjoying a period of normality would be short-lived. Entering their forties, Bill and Hillary were no longer prodigies. The days when they were the nation's youngest governor and first lady, with ambitious plans for transforming Arkansas into a model state, were long past. The Rose Law Firm was in turmoil and in danger of breaking up. Bill was now a four-term governor who had neglected his duties in order to promote his national ambitions, and the legislature, catching the anti-tax mood, was in a state of near rebellion. Far from heading off to Washington in triumph, they faced a very real danger that they would both find themselves at career dead ends.

Even so, Bill Clinton could not stay away from national politics for long. Throwing his support to Massachusetts governor Michael Dukakis, Clinton found himself at the Democratic convention in Atlanta after all—not as the nominee but in the role of nominator. Delivering the speech that nominates the party's anointed candidate is normally a plum assignment, an opportunity for a relative unknown to enjoy a showcase on prime-time TV. With an audience of pumped-up delegates cheering his every utterance, who doesn't look good?

But Clinton, though often a brilliant extemporaneous speaker, has an unfortunate tendency to ramble on in formal speeches, and Dukakis and his aides had added extra material to an already lengthy draft, which lasted a seemingly interminable thirty-two minutes. Frantic stage managers worked the podium light, signaling Clinton to cut his speech short. Restless delegates booed and shouted, "Get off! Get off!" Still, Clinton stuck to his prepared text until the long-awaited words "And now, in conclusion" caused the hall to erupt in applause.

Characteristically, the Clintons were quick to suspect sabotage. For some reason, the lighting director had failed to dim the house lights, as had been done for Jesse Jackson and Ann Richards, and therefore Clinton had been forced to compete with a lot of distracting activity on the floor.

Hillary recalled:

"As we were going up to the podium I said to the woman who was leading us up, 'Are you sure that the lights are going to go down?' She said, 'Well, I suppose so.' I said, 'Well, would you double-check for me?' And she came back and said, 'Yeah.' So Bill started his speech. Lights did not go out. The Dukakis floor people were not telling the delegates to listen. In fact, in several instances, they were telling their delegates to yell every time Bill mentioned Dukakis' name, which was often, since it was a speech about Dukakis. . . . If the speech had been given the way we were told that it was to be given, it would have done exactly what we thought it was supposed to do and what Bill had prepared for it to do. . . . It was one of the most agonizing moments of my life because I knew that we had been misled and I couldn't figure out why. I couldn't understand what the problem was."[2]

Hillary may have known "exactly" how things were supposed to proceed, though it would seem that both she and Bill knew the purpose of a nominating speech is to whip up enthusiasm for the candidate. The operant question was not whether—or why—the Clintons had been "misled," but why Bill didn't abandon his prepared text and work the crowd's emotions.

Judging by his later statements, Clinton had been paralyzed by fear that cutting the speech would offend Dukakis, who had approved the text in advance, a reaction that betrayed a lack of self-confidence curiously out of sync with his usual demeanor. But whatever points he might have gained for his wrongheaded display of loyalty Clinton quickly squandered by throwing a temper tantrum backstage, denouncing Dukakis as "that little Greek mother-fucker." Hillary, according to a journalist who happened to be standing near her, retreated into a shell of silence.

Mollified by a successful appearance on the *Tonight* show, arranged by Harry Thomason, the former Arkansas basketball coach turned Hollywood producer, and his wife, Linda Bloodworth-Thomason, Clinton returned home to wrangle with a legislature that was in a poisonous mood. Voters across the country were resisting higher taxes, and Arkansas, one of the poorest states in the nation, already had the thirteenth highest tax burden. Nevertheless, Clinton's legislative program for 1989 called for the largest tax increase in his tenure as governor.

Worse yet, the court case that had inspired the 1983 Clinton education initiative remained unsettled. A plan to merge the school district of Little Rock with two largely white suburban districts, North Little Rock and Pulaski, was encountering fierce resistance, and there was a very real chance that a federal judge would seize control of the schools, a step that would have meant political disaster for the Clintons. Thirty years after the first African-American students walked through the doors of Central High, the federal government would once again be intervening to enforce desegregation. The symbolism would obliterate Bill Clinton's claim to have led his state into a new era of racial progress and educational reform.

Back in 1982 the Little Rock school district, increasingly black and underfunded, had sued in federal court to force unification with schools in the fast-growing bedroom communities north and west of downtown. Parents who in many cases had moved precisely to get away from schools that were decrepit and dangerous wanted no part of consolidation and did everything possible to obstruct implementation. After five years of frustration, in 1987 federal judge Henry Woods decided to turn the problem over to a three-person committee, made up of one representative from each district. Walter Smiley, president of the Rose client Systematics, was the member for Little Rock, and Judge Woods appointed Hillary Rodham Clinton as the committee's legal adviser.

By all accounts, Hillary did far more than give legal advice. She was instrumental in getting the citizen committee to agree to a plan, announced in January 1989, that essentially backed away from mandatory integration. Little Rock schools, which had continued to deteriorate in the years since the suit was filed, would receive extra funds, though far less than would be needed to create the kind of magnet schools that have worked elsewhere to attract a diverse student body. The state of Arkansas would be able to settle its financial obligations under the suit for as little as $73 million, far less than its liability would be if the court stepped in to run the schools.

Critics of the agreement—and there were many—noted that Governor Bill Clinton had emerged as a clear winner in an otherwise disappointing settlement. Even Judge Henry Woods, a friend and admirer of Hillary's, agreed that the plan was inadequate. Rejecting the work of his own committee, he asked the Eighth Circuit Court of Appeals to overturn the agreement.

Reportedly, Hillary next went to Judge Woods and tried to persuade him to allow her to argue against the very deal she had brokered. Woods refused to countenance such an obvious conflict of interest. Exactly what Hillary hoped to accomplish is unclear, though of course she could have claimed to be on the winning side, no matter what the court decided. In the end, Hillary did not get her way and the Eighth Circuit allowed the agreement to stand.

NINETEEN eighty-seven was turning out to be a stressful year for Hillary. The Rose firm's decision the previous year to represent government regulators in suits against S&L operators had thrown the firm into turmoil. A number of partners had been opposed to taking the regulators' cases from the beginning. The S&L cleanup wouldn't last forever, but the firm's decision to do the government's work would not be easily forgiven by Little Rock businessmen caught up in the scandal. Vince Foster, on the other hand, had seen an opportunity for the litigation department, so often criticized by other partners for failing to pull its own weight, to bring in lucrative new business. Problems arose however, when the Federal Savings and Loan Insurance Corporation discovered that Joe Giroir had failed to fully disclose work done by Rose attorneys in connection with a loan he had received from the troubled First South Bank. Vince Foster was mortified, and Rose eventually had to pay the FSLIC $3 million to settle the matter.

Never admirers of Giroir, Foster, Hillary and Webb Hubbell now came to see him as a "rogue partner" whose private business dealings were threatening the future of the firm. Their agitation led to Giroir's replacement as the firm's chairman by William Kennedy III and eventually to Giroir's resignation in January 1989. Some of Giroir's friends inside the firm suspected that he had fallen victim to a power play engineered by Hillary and Webb in particular, two partners long criticized for devoting more time to political activities than to profitable cases, and the two of them came under pressure to prove that they could take Giroir's place by playing rainmaker. Hubbell may have worked harder, but he also soon began to improve his billings to clients by charging for work that had never been done. Hillary increased her visibility in the business community by accepting seats on a number of corporate boards, including those of Tyson Foods, Wal-Mart and TCBY. Hillary's board work brought business into her firm, but some found it unseemly that the first lady of Arkansas was making speeches and personal appearances on behalf of corporations.

However, Hillary, who had no trouble recognizing Giroir's conflict-of-interest problems, proved somewhat myopic when it came to her own activities. In 1987, the Rose firm represented the Federal Deposit Insurance Corporation in a $3.3 million suit against Dan Lasater involving his bond firm's work for the First American Savings and Loan Association of Oak Brook, Illinois. While most of the legal work was done by Vince Foster, Hillary billed at least two hours on the case, without revealing to the FDIC her husband's personal and professional ties to Lasater. In a February 3, 1994, article, the *Chicago Tribune* would reveal that the confidential settlement negotiated by Foster had allowed Lasater to discharge the claims against him for a mere $200,000. Hillary was no friend of Dan Lasater's, having concluded years earlier that he was a bad influence on her husband. Still, one could argue that she had a vested interest in disposing of the lawsuit, which threatened to revive publicity about her husband's connections to a convicted narcotics user and his by now scandal-ridden brokerage firm. Despite a March 1994 ruling by the FDIC that Hillary had done only minimal work on the case and therefore violated no federal regulations, the quiet resolution of Lasater's legal problems certainly worked to Bill Clinton's benefit.

While Hillary was preoccupied with the problems of the Rose firm, Bill Clinton seemed to be drifting through a midlife crisis. The decision not to run in 1988 had given him an opportunity to put four years between his presidential ambitions and his history of womanizing, yet he appeared to be intent on continuing as before. By 1989, Clinton was carrying on at least five simultaneous affairs. For starters, there was a married woman in the Heights whom he visited when her husband was away, and another woman who lived in a condo in the Sherwood district, on the outskirts of the city. Clinton assigned one trooper to each woman, charging him with handling phone calls and purchasing and delivering the occasional gift, often lingerie from Victoria's Secret.

Clinton's lovers were primarily married women close to his own age who had no expectations and their own reasons for keeping their affairs private. But he didn't stop there; he also pursued a series of brief assignations while on the road. According to trooper Larry Patterson, Clinton even disappeared into a closed room with an attractive young woman on the night of his disastrous nominating speech at the Atlanta convention, after Hillary had returned alone to their hotel.

At times, Clinton almost seemed to court discovery. Patterson, who joined Clinton's security detail in 1986, would tell of an incident that occurred in 1988 or 1989. On the way to an evening reception for the Harrison County Chamber of Commerce at the Camelot Hotel, Clinton ordered Patterson to make a stop at Chelsea's elementary school. When they arrived, one of Clinton's girlfriends, a young woman who sold cosmetics in a Little Rock department store, was waiting. While Patterson guarded the entrance to the parking lot, Clinton sat with the woman in her car and, according to the trooper, he "saw the lady's head go into his lap." On another occasion, the same woman drove onto the mansion grounds in a yellow-and-black pickup; the governor joined her, and their activities were observed by Patterson via a remote-control security camera.[3]

Clinton also received phone calls late at night, on occasion slipping away from home to join one of his lovers for a rendezvous in the wee hours of the morning. One night, Hillary awoke to find her husband gone and called down to the guardhouse to ask if he had left the grounds. Trooper Roger Perry, who happened to be on duty, alerted the governor by cell phone that he had better get home as fast as he could. When Clinton arrived, he and Hillary could be heard screaming at each other in the mansion kitchen. Afterward, Perry found that someone had kicked the door to one of the kitchen cabinets off its hinges.

Another kitchen argument took place on a quiet Sunday afternoon and was picked up on the intercom located near the outside door. At one point, Trooper Perry overheard Hillary tell her husband, "Look, Bill, I need to be fucked more than twice a year."

It is painful to imagine what this situation must have been like from Hillary's point of view. The troopers who schemed to further her husband's affairs were not just an occasional presence in her life. They patrolled the sidewalk outside the main gate, drove her and Bill to official functions in the Lincoln Town Car and, as just described, occasionally overheard private conversations. This situation dragged on for years, fostering an atmosphere of mutual contempt and hostility. Patterson would recall with particular bitterness one out-of-town trip he made as part of the governor's security detail during which Hillary instructed him not to talk in public because "You sound like a hick."

Patterson found Hillary to be brusque to the point of coldness with family and strangers alike. When Chelsea was younger, he recalls, Hillary arrived home from a long out-of-town trip and, finding Chelsea playing downstairs, gave her a quick pat on the head as she sailed through the room.

On another occasion, preschoolers from the HIPPY program were on the mansion lawn posing for a photograph. Hillary was overheard on the intercom saying, "I want to get this shit over with and get these damn people out of here."

Then there was the morning of Labor Day, 1991, when Hillary noticed as she was driving away from the mansion that the security detail had neglected to raise the American flag. Pulling a U-turn, she came careening back to the guardhouse and screamed, "Where is the goddamn fucking flag? I want the goddamn fucking flag raised every fucking morning at fucking sunrise!"[4]

Larry Patterson obviously had come to dislike Hillary intensely and cannot be considered an objective commentator. Even so, anecdotes from others confirm that Hillary could be astoundingly rude to those who fell outside the scope of her radar screen. An aide of the governor's who had stopped by the mansion to deliver some papers happened to notice Hugh Rodham, Sr., sitting in the kitchen and decided to say hello, only to discover that he had walked in on a Rodham family gathering. Hillary fixed him with a glare so hostile that he fled the room.[5]

Obviously much of Hillary's hostility was deflected from its primary target, the man who recruited uniformed state employees to help him meet women and stand "Hillary watch" to keep his wife from catching on. There could be no doubt that Hillary knew what was going on, even if she was unaware of some specific incidents. Nevertheless, after the troopers' stories appeared in the *Los Angeles Times* and the *American Spectator* at the end of 1993, Hillary told a friend that the accounts were all lies that got started after one trooper's wife found a woman's phone number in his pants pocket and he made up the excuse that he had collected the number for Clinton. To admit otherwise would raise the question of why a woman who had by that time become a feminist icon had put up with so much humiliation.[6]

As 1989 wore on, Clinton took less trouble than ever to hide the evidence of his affairs. He was already the inspiration for "governor jokes," which were understood to refer to Clinton even when told in neighboring states. One of these, told by a college professor who is by no means a Clinton-hater, is perhaps best geared to those with a knowledge of electrical matters: "What do you do with an overheated woman? Put a governor on her."

Moreover, Clinton was allowing his official duties to slide. After all Hillary's work to stave off a federal takeover of the Little Rock schools, he neglected to lobby the legislature, which approved the deal only at the last minute. Even then, the results of the vote were nearly undone by a series of legislative snafus. At one point, Clinton departed on a "trade mission" to Germany, neglecting to call a special session of the legislature that was necessary to approve the annual appropriations bill, without which the school settlement could not be funded.

Betsey Wright had dedicated herself to Bill Clinton since coming to work for him in 1980. Wright controlled access to Clinton, and her bulldoglike ferocity often infuriated legislators and staff. Like Hillary, she cultivated a management style that relied heavily on screaming and cussing—it seemed that this was the only way to get Bill Clinton's attention, and both women sometimes forgot that other people might not respond well to the same tactics. By the summer of 1989, however, Wright was so frustrated that she was on the point of nervous collapse. She later told biographer David Maraniss that she was so angry with Clinton that she wanted to pour boiling water on his head. After a series of bitter arguments, most of them apparently about relatively trivial matters, Wright was banished.

Chelsea was now old enough to be aware of the tension at home. Reportedly, while helping her mother wrap Christmas presents, she blurted out the question, "Mommy, why doesn't Daddy love you anymore?"

For whatever reasons, in December 1989, Hillary persuaded Bill to attend counseling with a Methodist minister. After a few emotional sessions, Bill and Hillary told friends that they had decided they wanted to stay married.

Predictably, the benefits of this short-term counseling would prove fleeting. Clinton appeared to be more focused on his work, but his pursuit of other women, if it was interrupted at all, soon resumed. As she had before, Hillary expressed her determination to make the marriage work by dedicating herself ever more fiercely to the business of getting Bill elected. This dynamic, so obvious to outsiders, inevitably led to speculation that the Clintons had a marriage of convenience, based solely on their shared political ambitions. It wasn't that simple. There was very little that could be called "convenient" about the Clintons' marriage. But Hillary at this point had invested almost as much effort in Bill's career as he had, a consideration that had to weigh heavily against any thought of divorce.

BILL Clinton's immediate problem was deciding whether or not to run for another term as governor. If he stood aside, he would lose his home, his only source of income and his political base. If he ran, he risked rejection by voters irked at having a governor who would spend yet another term pursuing his presidential ambitions. Judge Henry Woods suggested that Hillary run in Bill's place. The Clintons considered the idea and asked Dick Morris to do a poll. He reported back that the voters were not warm to the idea of a Hillary candidacy. In spite of her work on education reform, they still thought that she would be a stand-in, like Lurleen Wallace, who ran for governor of Alabama after her husband became incapacitated. Hillary did not appreciate the comparison, and Bill protested that his wife, with a long list of accomplishments in her own right, was nothing like Lurleen. Nevertheless, a follow-up poll produced similar results, and the idea of Hillary's running was dropped.

Clinton's attention, meanwhile, was distracted by his continuing work for the center-right Democratic Leadership Council (DLC), of which he became chair in 1990, and his ongoing courtship of the Democratic National Committee. Nancy Lieber, a onetime member of the DNC staff, recalls the Arkansas governor's visits vividly: "I remember watching this guy and thinking, 'Wow! Here's a great policy wonk!' The others were so boring you could die. . . . He talked a great game."

Back in Arkansas, the political infighting had never been more nasty. Round one was begun by Attorney General Steve Clark, Clinton's most formidable rival, who announced at an ADFA board meeting that he had been offered a $100,000 bribe to end his opposition to a bond proposal involving nursing homes owned by Beverly Enterprises and a Texas company called Pride House Care. Although it was not generally known at the time, William Kennedy III of the Rose Law Firm had done the legal work on the deal, which generated big profits for Stephens, Inc., and inflated the cost of nursing home care for the chain's elderly residents by as much as 14 percent. Clark's accusation was quickly followed by his arrest for charging restaurant meals and other expenses on his state credit card. According to former Clinton fund-raiser Bert Dickey III, many state officeholders carried such cards and were not always careful about what they charged. Clinton, said Dickey, had looked at Clark's high poll numbers and suggested aloud that someone find some dirt on him, asking, "What can we get that's real good?"

On the Republican side, Tommy Robinson, the former Pulaski County sheriff and congressman known as "Captain Hotdog," had recently switched parties and was running for governor with the support of George Bush and Lee Atwater, a political operative whose penchant for sniffing out the opposition's weaknesses frightened already nervous Clinton supporters. Robinson was a controversial figure, considered an embarrassment by some Arkansas Republicans but popular with the state's socially conservative voters, perhaps the only figure in the state who could match Clinton's charismatic appeal.

Seeing a plot by Bush to deliver a knockout blow to Clinton on his home turf, the Clintonites hired the Investigative Group International (I.G.I.), a detective agency run by former Legal Services Corporation director Terry Lenzner to dig up dirt on the so-called Arkla-Arkoma natural gas deal, best remembered because it made Arkansan Jerry Jones rich enough to buy the Dallas Cowboys. Both Robinson and his Republican primary opponent, Sheffield Nelson, were involved, along with Jack Stephens and his company, which had lost millions on the contract. But Robinson was thought to be the main target of the investigation. Hillary's friend Brooke Shearer was an I.G.I. operative and LSC veteran Mickey Kantor was an old friend of Lenzner's, suggesting this twist on traditional Arkansas electioneering tactics may have been Hillary's idea.[7]

Whether Lee Atwater actually planned to use the issue of Clinton's sex life to help Robinson is uncertain. Spooked by the Gary Hart–Donna Rice scandal, the Clinton inner circle now saw plots everywhere. As the Bible says, "The wicked flee where no man pursueth." Atwater's intentions became a moot point when he was diagnosed with a brain tumor. Robinson then lost the primary campaign after an unusually large number of Democrats crossed party lines to vote for his opponent, Sheffield Nelson. Rex Nelson, a political columnist for the *Gazette* who worked on Robinson's campaign, charged that Sheffield Nelson's primary victory was stage-managed by Hillary and Betsey Wright, who had come out of her enforced retirement to head the state Democratic Party. Wright soon resigned her post, following an argument with Bill Clinton over his attempts to raise funds for the DLC in Arkansas, siphoning off donations that otherwise might have gone to the state party.

Fiscal shenanigans at ADFA once again became an issue in the matter of Larry Nichols, a onetime Clinton supporter whose On Air production company had done some Clinton political ads as well as promotions for the Pepsi Challenge for Better Education, part of Hillary's educational effort. Clinton had placed Nichols in a job with ADFA, but he was forced out in 1989 after being charged with making 642 unauthorized phone calls, many of them to Adolfo Calera and other leaders of the Nicaraguan Contra movement. Nichols claimed that the complaints about his phone calls were an attempt to discredit him because he had discovered that ADFA was nothing more than a money-laundering scheme for key Clinton donors, including Dan Lasater.

In September 1990, Nichols filed a lawsuit that charged, among other things, that Governor Clinton maintained an ADFA slush fund, which he used to finance dalliances with women. A judge who was a Clinton appointee sealed the court papers, and later another judge dismissed the suit with prejudice—failing to notify Nichols, who had no idea the suit had been thrown out until July 1991. Nichols's information was by no means always accurate; he was a guy with a grudge. Nevertheless, by making his charges in a formal lawsuit he had upped the ante. Governor Clinton's sexual behavior was no longer strictly a private matter.

BY 1989, Hillary Clinton was chairman of the board of the Children's Defense Fund and in that capacity, ironically enough, committed to opposing the Bush administration's plan to provide tax credits that parents could use toward the day-care arrangement of their choice. George Bush was more open to the idea of spending federal tax dollars on day care than Michael Dukakis, who had shown little interest in the subject. But as in 1971, Marian Edelman was not prepared to compromise. The CDF had its own proposal, known as the Act for Better Children, or the ABC bill, which called for $2.5 billion in spending for the first year and would have placed day-care centers under federal regulation. There followed a bitter two-year battle, during which Edelman denounced two erstwhile congressional supporters as traitors and called the Bush bill a plan "to rob millions of children."

With the battle over alternative views of day care still pending, Hillary joined a fourteen-member study panel on a two-week tour of French day-care facilities. Sponsored by the French-American Foundation, the study tour received financial support from American Express, Citicorp, Du Pont and *The New York Times,* among others.

Advocates of government-sponsored day care inevitably cite French day care as their model. Instituted after World War II as part of a program to combat falling birthrates, the system has become such a treasured entitlement program that many French parents now believe they are incapable of raising their own young. As reported by *The New York Times* in 1997, French mothers believe that day care does the necessary work of "civilizing" their children and "awakening" them to the demands of society and produces teenagers who are "less rambunctious." Despite an unemployment level of over 12 percent, a rate that would be considered a calamity in the United States, and the country's difficulties meeting the requirements of the European financial union, expenditures on day care keep rising—up 60 percent between 1989 and 1994, and another 9 percent between 1994 and 1996, and this despite efforts of the conservative government to control the budget. Even so, Parisian parents typically wait up to a year to get their child a place in the system, and then "sometimes it takes connections." In the meantime, of course, the excluded parents still pay taxes to support the system, so it is hardly surprising that they are demanding to the point of hysteria. "Young parents now approach us as if it is our duty to take their child," says one teacher interviewed by the *Times.* "Some become quite aggressive."[8]

That said, studying the French system remains an excellent idea. French bureaucrats do seem to do some things better. Visitors from the States, for example, are always amazed to discover that French day-care centers feed children steamed fish and mashed fresh vegetables, as opposed to peanut butter and crackers. Nevertheless, the term "study" suggests some exercise of the critical faculties that does not appear to have played a role in the 1989 study panel's work. The group's official report, prepared by Gail Richardson and Elisabeth Marx, appears to be designed to sell Americans on the superiority of the French system, with no mention of the drawbacks.

Hillary did her part on her return from the tour by contributing an op-ed piece to *The New York Times*—one that "we helped her write," according to an institute spokesperson. Entitled "In France, Day Care Is Every Child's Right," the article complains that the American focus on family values "assumes that parents alone can always determine and then provide—personally or through the marketplace—what's best for their children and, hence, society." Rather, writes Mrs. Clinton, "To do our children justice, we need to develop a nationwide consensus on how to best nurture our children." She suggests that the French standard, which views children as "a precious national resource," should replace the American view that "the cost and disruption of child bearing is . . . a private choice and responsibility."[9]

At the risk of quibbling, one notes from the Richardson and Marx report that the child-to-teacher ratio in the *Ecoles Maternelles,* a program for three- and four-year-olds, is twenty-eight to one. And the child-to-staff ratio for infants is twenty-two to one. (True, the teacher has help from an aide, whose services are shared between two classrooms, and there is extra help at lunch and nap time.) This implies a degree of regimentation that most Americans would consider shockingly inappropriate for children so young. Hillary herself made an impassioned plea to the Arkansas legislature in 1983, in which she stressed that limiting *kindergarten* classes to no more than twenty children must be a top priority of the state's education reform package.

Hillary does say in passing in her *Times* article that Americans would not wish to adopt the French system "wholesale." But teacher-to-child ratios are not a detail. Smaller classes would enormously increase the costs of the system, which is already expensive enough to prove a burden on the French economy. In addition, there are factors in the United States with which the French don't have to contend, including our federal system of government and laws such as the Americans with Disabilities Act, which would make it difficult for supervisors in a government-funded system to fire workers for reasons such as past substance abuse. All this helps to explain why legislators who by no means wished to "rob children" eventually compromised with the Bush administration to pass the bill opposed by Hillary Clinton and the CDF.

Government-run day care, while a goal close to Hillary Clinton's heart, was by no means the most ambitious part of her vision for America's future. Her commitment to a plan that would meld the public schools with an industrial policy for American business dates back to 1986, when Bill Clinton was scheduled to take over as chairman of the National Governors Association. Tom Kean of New Jersey and Lamar Alexander of Tennessee had pushed education to the top of the NGA agenda, and Clinton was to assume the chairmanship at the organization's summer conference on Hilton Head Island, delivering a speech that would stake his claim to succeed Kean and Alexander as a national spokesman on education issues.

Arkansas Gazette columnist John Brummett gives an interesting aside on this conference. A group of Clinton friends, arriving on the island to cheer his speech, found the governor on the eve of his big day prostrate with a stomach virus and looking miserable. Concerned, they asked Hillary what she was doing for her husband. Had she given him medicine? Called a doctor? Hillary shrugged off the questions. "I'm not his nurse," she said.[10]

Clinton recovered to give his speech, and during the course of the conference became acquainted with Marc S. Tucker, whose work on the Carnegie Corporation study *A Nation Prepared: Teachers for the 21st Century* was the talk of education reform circles. As so often happens in this field, the panacea of a few years back was now being defined as the problem. An earlier Carnegie Corporation study, *A Nation at Risk,* had inspired Arkansas as well as a number of other states to pass reform plans that in one way or another scapegoated teachers and limited their control over lesson plans and teaching methods. Testifying before an NGA working group panel the previous April, Margaret J. Lathlaen, a former member of the Teacher in Space program, vented frustrations shared by many of her colleagues: "I no longer have autonomy over what is taught. I don't even have a lot of choice about how I teach."

At the Hilton Head conference, Lathlaen's complaint was being echoed by frustrated educationists, including Al Shanker of the American Federation of Teachers, who fumed, "I've got curriculum reform coming out of my ears." Tucker's approach appeared to be the perfect antidote. He observed that the American high school diploma had become a meaningless piece of paper, and he stressed high pay and recognition of teachers' professionalism, less school bureaucracy and an emphasis on "standards."

The Rochester, New York, school board was so impressed with Tucker's ideas that it invited him to oversee the remaking of the city's schools. The Carnegie Corporation provided funding that allowed Tucker and a nonprofit advisory group, the National Committee on Education and the Economy, to take on the project. The people of Rochester soon learned, however, that what education experts mean by terms like "standards" and "professional autonomy" was not quite the same as what we ordinary folks mean. The NCEE defined a standard as a level that almost every child could reach, so in practice, the class could not move forward until the most recalcitrant members had mastered the material. Instead of rewarding students who complete their work on schedule by allowing them to move on to more challenging material, the "standards" concept in effect punished them by assigning them to spend their time tutoring the laggards. Rochester teachers, moreover, were to be held personally responsible for the progress of their students in all areas of life, an unrealistic goal that required them to function as amateur social workers.

While the Rochester school experiment was foundering, Hillary, who had met Tucker on Hilton Head, advocated his ideas within the Children's Defense Fund. Advocates of "standards" also enjoyed a vogue at such forums as the 1989 "Education Summit" convened by George Bush for the nation's governors—this despite the warning of Chester E. Finn, Jr., the former head of educational research under Reagan, who told the governors, "These people are the blob, the ones who gave you the schools the way they are now."

Around this time, Tucker was collaborating with Ray Marshall, secretary of labor during the Carter administration, on a book entitled *Thinking for a Living*. The U.S. economy was in recession, and Tucker and Marshall quoted with approval David Halberstam's quip that the Cold War was over and Germany and Japan had won. Tucker and Marshall proposed gearing American schools toward vocational training and apprenticeship programs so that they could be integrated into a national "labor market system."

Thinking for a Living was published in 1992, by which time the bloom was off the Japanese and German miracles. As the *New York Times* reviewer noted, the "utopian" nature of Tucker and Marshall's plan called for a system of expert boards and commissions that would enforce educational standards and tell businesses how to reorganize themselves to employ more skilled labor: "The price would be billions of dollars and require a nearly unimaginable level of integration and cooperation among interests." Nevertheless, the book received the Sidney Hillman Prize, and the paperback edition appeared with a generous blurb from Bill Clinton. Hillary Rodham Clinton was thanked in the acknowledgments for reading a draft of the manuscript.

In fact, though her work attracted little notice, Hillary had been hired by Tucker in 1989 as a lobbyist for the NCEE's Workforce Skills program. (Technically, the NCEE was a client of the Rose Law Firm, and the firm received moneys totaling $101,630 in return for Hillary's services.) Hillary was listed as the Workforce Skills task force's cochair for implementation, and her work consisted of selling governors and state education departments on the merits of the NCEE approach. Nowhere, it seems, was she more successful than in lobbying the governor of Arkansas, who was so impressed by her ideas that he would include the NCEE program of school-to-work apprenticeships in his "invest in America" platform for the 1992 presidential race.

According to an NCEE spokesman, Hillary resigned her position three days before Bill Clinton announced his run for the presidency. Nevertheless, Hillary and Ira Magaziner were still listed as program cochairs and coauthored an article that appeared in the March 1992 issue of the journal *Educational Leadership*. In this article, Magaziner and Clinton warn of a dire crisis facing the U.S. economy: "What we are facing is an economic cliff—and the frontline working people of America are about to fall off. If we are to avert catastrophe, we must make drastic improvements in our rate of productivity growth."

In order to stave off the impending disaster, Magaziner and Clinton propose a system that would make the individual's progress in the workplace contingent on obtaining a series of certificates. The student would begin by earning a Certificate of Initial Mastery at the age of sixteen (tenth grade). The national standard for awarding this certificate would be high—"benchmarked to the highest [standards] in the world," the authors assure us. On the other hand, "virtually all" sixteen-year-olds would receive certificates. Figuring out how to make this happen would be the job of the states, who would create "Youth Centers" and other "alternate learning environments" to serve school dropouts.

But this is just the beginning. Magaziner and Clinton go on to advocate that "a comprehensive system of technical and professional certificates should be created for students and adult workers who do not pursue a baccalaureate degree. National committees of business, labor, education, and public representatives should be convened to define certification standards."

To pay for this effort, employers would be expected to "invest" 1 percent of their payrolls in worker training. In return, panels of experts would advise them on how to reorganize their workplaces to utilize skills being taught in the certificate programs. Finally, our present "fragmented" approach to employment would be replaced by a national data bank that would track all workers and match them to available jobs.

All this, the authors promise, would save us from the stagnation that results from our lack of a "coherent" labor market: "The system we propose provides a uniquely American solution. Boldly executed, it has the potential not simply to put us on an equal footing with our competitors, but to allow us to leap ahead, to build the world's premier workforce."[11]

A Great Leap Forward, if you will.

Readers who believe that the changes recommended by Mrs. Clinton and Magaziner in this article would make our economy more productive are welcome to their opinion. As for "standards," they have so far failed to transform American youth into an army of highly skilled technicians. In Cottage Grove, Oregon, which instituted the country's first Certificate of Initial Mastery program, students received credit for demonstrating their ability to assemble a peanut-butter-and-jelly sandwich, and tenth-grade instruction in Shakespeare consisted of having students read two comic books and watch a video of the Steve Martin film *Roxanne*. (The movie is based on a play by Rostand, not Shakespeare, but never mind.)

Standards-based education also implies a large component of social control. Students who have trouble coping with the demands of the classroom, for whatever reason, would no longer be merely disadvantaged; they would be virtually unemployable, since without the proper certificates they could not be placed by the national data bank, where all jobs are listed. And since "standards-based" curricula tend to include elements of politically correct values instruction, students who resist having their attitudes adjusted may also find themselves among the permanently unemployed. For example, in Oregon, one of the first states to embrace standards-based education, each student has an "electronic portfolio," which includes not only grades and psychological evaluations but information on whether he or she has "assimilated the desired behavior or attitude required for a given outcome, such as understanding diversity."[12]

Hillary Clinton's employment as a lobbyist for a program her husband was advocating attracted some interest in the aftermath of the 1992 campaign. Observing that some of the money paid to the Rose Law Firm came from grants by the state of New York, state attorney general Dennis Vacco opened an investigation but abandoned it in June 1997, announcing that he had found "no evidence that Mrs. Clinton's time records are inaccurate or that the services described in the records were not performed."

Mrs. Clinton was not asked, either during the campaign or later, to defend the ideas expressed in the *Educational Leadership* article. Instead, and despite her repeated desire to concentrate on issues, she would find herself besieged by questions about chocolate chip cookies and other weighty matters. Little wonder if she feels contempt for the media, and for all of us.

Everyone Has a List

CHELSEA Clinton was just six years old when her mother and father began to prepare her for the rigors of political campaigning by engaging in role-playing sessions. One of them would pretend to be a political opponent and "say some bad things about Daddy," such as "Governor Clinton has done a terrible job."

When Chelsea became upset and wondered why anyone would say such a thing, Bill would tell her, "I don't know." That was just the way politics worked. Bad people (Republicans? journalists?) said hurtful things for their own selfish reasons.

As time went by, Hillary's warnings would become more shrill. During the 1992 campaign she told her daughter, "By the time this is over, they'll attack me, they'll attack you, they'll attack your cat, they'll attack your goldfish."

In a 1993 interview, Hillary explained her approach to preparing Chelsea for the political life as part of an effort to help her develop an "internal compass." As she told *Redbook*'s Roxanne Roberts, "We can't control anybody else. The only thing we can do is try to develop our own values and our own sense of who we are. That's what Bill and I were raised to believe about ourselves, and that's what we can do for our daughter. And I think that's the best gift you can give a child."[1]

With hindsight, these words are painful to read.

As we now know, the nastiest things Bill Clinton's opponents said about him were essentially accurate. And Hillary, who once criticized Virginia Kelley for being in denial, was now passing this legacy on to her only child, training her to believe that the truth was a lie, all in the name of developing her "moral compass."

Dr. Paul Fick explains in *The Dysfunctional President,* a book about the impact of alcoholism on Bill Clinton's life, that children in families where a parent's addictive behavior is consistently denied often appear to cope very well. They get good grades in school and develop outgoing personalities even as they internalize feelings of shame and guilt that will cause them problems in later life. One can only hope that this will not be the case with Chelsea Clinton, who has many friends and appears to have inherited her father's gregarious personality. It is not surprising, however, to hear that she has a low opinion of politics and does not plan to follow her parents into public life.

AS the 1988 presidential elections approached, the threat that Bill Clinton's private life would be exposed had been at least somewhat theoretical. But this was by no means true as Clinton prepared for the 1992 race.

On September 4, 1990, Clinton received the first of a series of disturbing calls from Gennifer Flowers. He had not seen Flowers in more than a year, except in passing at the Hot Springs racetrack. Flowers had a new boyfriend, Finus Shellnut, a former Lasater & Co. broker, previously married to Webb Hubbell's sister-in-law. ("Fineus," as he was called in Arkansas, got his name because his mother, on giving birth to her fourth son, had announced that this was *finis*.) That day, Finus had been at work when a press release came over the fax machine, announcing the filing of Larry Nichols's lawsuit in which Flowers was one of five women named as lovers of Bill Clinton. She was concerned—concerned enough to call Clinton and to tape the conversation.

Clinton told Flowers the lawsuit was a setup, arranged by Republican gubernatorial candidate Sheffield Nelson, and he promised that it would all be taken care of. He also dismissed Nichols's list of his alleged lovers as a "pack of lies": "I told you a couple of years ago, when I came to see you, that I'd retired—and now I'm glad I have because they really scoured the waterfront." Running down the names, he conceded, "At least Lencola Sullivan is a friend of mine. I've never slept with her, but she does come to see me."

A few weeks later, Flowers was on the phone again. The Excelsior Hotel, where she had been appearing as a lounge singer, had cut back her hours because of the recession. Flowers was broke but reluctant to look for work that would take her out of the area and away from her mother, who had breast cancer. She asked Clinton to find her a day job, something with state government.

Clinton passed Flowers's request on to his administrative assistant, Judy Gaddy, who called Flowers in and gave her an application for a job doing multimedia presentations for the Arkansas Historic Preservation Program. Flowers was eventually called for an interview, only to learn on February 22, 1991, that another candidate had been hired. Furious, she wrote Clinton a letter, reminding him that if her financial situation became desperate, she might file a defamation suit against a Little Rock radio station that had allowed a talk show caller to name her as one of Bill Clinton's paramours.

A few days later, Gaddy called Flowers about another job, this time for a newly created Administrative Assistant II position, working for Don Barnes, the head of the state Board of Review. According to Flowers, Gaddy advised her to stress her background in office work, so Flowers changed her application, claiming experience in public relations and computer skills that she did not have.

Flowers also placed a call to Clinton, reminding him that she couldn't afford another screwup.

"No problem," he said. "I talked to Barnes. It's a done deal."

On June 11, Flowers was hired.

Whether Bill Clinton thought of Flowers's job request as a form of blackmail is unclear. During their conversations, which Flowers was preserving on tape, he often sounded wary. At other times, however, the discussions were friendly enough. Flowers was excited about the prospect of his running for president and repeatedly assured him that she would do nothing to harm his chances. She would later say that she saw nothing unusual about the way she had come by her $17,520-a-year state job: "This was the way things were done in Bill Clinton's Arkansas."

As it turned out, not everyone saw it that way. When Flowers showed up for work, a woman named Charlette Perry, who happened to be at the top of the promotions list, refused to speak to her. Perry filed a formal grievance, and soon Clinton was receiving more calls from Flowers, who was panicky over the prospect of having to testify under oath about how she had been hired.

During this series of conversations, many of them placed to the Governor's Mansion late at night, Flowers and a mildly wary-sounding Clinton reminisced about their affair. At one point, Flowers reminded Clinton of some pillow talk that occurred during an earlier campaign—presumably in 1987, when he was thinking of announcing for the presidency. Clinton had been rehearsing Flowers, asking her to tell him what she would say to a reporter who asked her about their relationship.

"I said, 'You eat good pussy.' "

"What!" exclaims Clinton.

Flowers laughs. "I said, I had to tell them you ate good pussy and you said, 'Well, you can tell them that—if I don't run for president.' "

Given the situation with Flowers, one can well understand why Clinton continued to have some doubts about running. On the Fourth of July, it was customary for him and a group of aides and reporters to fly around the state, showing up at various parades and civic celebrations. During the 1991 tour, Clinton got into a conversation with Joan Duffy of the Memphis *Commercial Appeal* about his presidential prospects. He was worried about the effect his candidacy might have on Chelsea, he said, and he wondered if it wouldn't be better for him to wait four years until she was almost out of high school. Clinton added that his chances of winning in 1992 weren't great in any event. But, "if lightning strikes and I win, she loses me."[2]

Clinton would later credit Hillary for being the first to see that 1992 could be his year. As early as the Gulf War, when the pundits were calling George Bush's reelection a sure thing, Hillary had seen that the economy and the public's desire for a change after twelve years of Republican control of the White House would present an opportunity for a Southern governor who was young, personable and positioned as a moderate on the issues.

But it wasn't quite that simple. Back in April 1989, Al From of the Democratic Leadership Council had come to Arkansas and made Clinton an offer. If he would become chairman of the DLC and adopt its centrist message, the organization would back him for the presidency. Hillary, fortified by their recent sessions of pastoral counseling, had decided that she was prepared to cope with whatever scandals might be dredged up in the course of a national campaign. Neither of the Clintons could anticipate just how ugly things would get, but Hillary, despite an ingrained suspicion of the press, was perhaps the less realistic because she had a tendency to see the rumor mill as an Arkansas phenomenon. Who in the national press was going to pay attention to a crank like Larry Nichols?

Immediately after winning his fifth term as governor, Clinton began acting like a presidential candidate. Despite his promise to the voters that he would not run, Clinton spent much of his time traveling around the country, giving speeches and conferring with Democratic strategists. Meredith Oakley of the *Arkansas Democrat* kept a running tally of the governor's whereabouts and found that during one fifty-two-day period he was in the state on only fourteen days. On a trip to a Democratic Leadership Conference convention in Seattle, he was accompanied by a fourteen-person security detail. The *Democrat* wondered why the state should continue to subsidize such activities.

By the spring of 1991, Hillary had begun to spread the word of Clinton's plans to his home-state supporters. Skip Rutherford, an Arkla executive and close friend of Mack McLarty's, got the message during a softball game of the Molar Rollers, a softball team on which his daughter and Chelsea Clinton both played. While he sat with

Hillary in the bleachers, she took advantage of a lull in the action to pass on her opinion that she thought Bill could prevail against Bush.

"If she's telling me that, you know she's telling the governor that," Rutherford said to his wife that evening.[3]

By the summer, Mickey Kantor, Hillary's old friend and a former Legal Services Corporation board member, was putting together the nucleus of a national campaign organization. The network that Kantor assembled included a few longtime Clinton enthusiasts, such as media consultant Frank Greer, who had been watching Clinton's progress since 1988. Other key recruits were professional operatives like the Ragin' Cajun James Carville and his partner Paul Begala, who came recommended by the governor of Georgia; George Stephanopoulos, another Rhodes scholar; Gene Sperling, Mark Gearan and Dee Dee Myers. It was no secret that Bill Clinton had a reputation as a womanizer, but few in the national organization knew any details.

In September, after the Associated Press ran a brief item about the Larry Nichols lawsuit, Kantor and Greer urged Clinton to deal with the issue and put it behind him. The Sperling Breakfast, a gathering of Washington, D.C., political reporters, was coming up, and it would be an ideal occasion to make a statement. Clinton knew at this point that Charlette Perry's grievance over being passed over in favor of Gennifer Flowers would be the subject of a formal hearing in October, and he had to realize that nothing he said on this occasion would bury the woman question once and for all. At first, he balked, citing the advice of his golfing buddy and confidant, Washington attorney Vernon Jordan, who had said, "Screw 'em! Don't tell them anything." Kantor wasn't impressed, and Clinton did show up at the breakfast, with Hillary at his side.

Candidates usually didn't bring their wives to these events, and given Hillary's presence, the discussion of Clinton's private life was limited to an awkward question or two. Clinton told the journalists that it was unfair that politicians who resolved their marital problems through divorce were able to wipe their slates clean while he, who had made the effort to keep his marriage going, seemed to enjoy no statute of limitations. Affirming his high regard for his wife, he added,

"We intend to be together forty years from now, regardless of whether I run for president or not."[4]

Clinton hadn't lied in so many words, but his talk of a "statute of limitations" implied that the adultery issue was an old story, many years in the past. Back home in Little Rock, in a phone conversation with Gennifer Flowers, he had a slightly different take on the situation. Indignant over the embarrassing questions about his private life, he went on about how unfair it was that as a married man he couldn't date without being criticized for it: "I might lose the nomination to Bob Kerrey because he's got all the Gary Hart/Hollywood money and because he's single, won the Medal of Honor, and since he's single, nobody cares who he's screwing."

By contrast, Hillary, when pressed to explain her marriage, would talk vaguely about Bill's appetite for life and his "enormous" heart: "He is an incredibly loving and compassionate and caring person." Presumably, she had managed to persuade herself that Bill was just too weak to resist the women who continually took advantage of him.[5]

Obviously, Clinton could not be put before the public without a major damage control operation. The groundwork had been put in place as early as the spring of 1990, when Clinton was running for governor and worried that Tommy Robinson would use the woman issue against him. At the center of the effort were Clinton's close friend Bruce Lindsey and Hillary's confidant Vince Foster. Neither man seemed quite cut out for running a political mop-up squad. The trim, preppy Lindsey came from a socially prominent Heights family, and by general agreement his only apparent character flaw was an utter devotion to Bill Clinton. Foster appeared to be carrying on in the role of Hillary's champion and protector.[6]

Lindsey and Foster began their research by having a heart-to-heart talk with Darrell Glascock, a Republican public relations operative who had managed Frank White's campaign in 1986. (Glascock, who was running for lieutenant governor against Clinton's Democratic nemesis, Jim Guy Tucker, also happened to have been the recipient of many of the unauthorized phone calls that had cost Larry Nichols his job at ADFA.) The goal of the conversation was to figure out just how much Clinton's enemies knew about his affairs, and it turned out to be a lot. "I told them Sheffield had a list, and Tommy had a list. Hell, everybody had a list," Glascock would recall.[7]

Quite soon, Lindsey and Foster had a list, too.

Just how Foster and Lindsey went about collecting information on Clinton's affairs is an interesting question. Despite his famous photographic memory, Clinton had a tendency to be forgetful on this subject. Like an alcoholic who had been through many relapses, he found the topic too painful to discuss.

According to British journalist Ambrose Evans-Pritchard, Foster farmed out the job of investigating Clinton's affairs to a Little Rock private detective named Jerry Luther Parks. Jerry Parks is not around to confirm this, having been ambushed and shot dead in September 1993, while he was driving home from his usual Sunday afternoon meal at the El Chico Mexican restaurant in Little Rock.

Evans-Pritchard's account is based largely on the word of Parks's widow and his son, Gary, who have been notably reluctant to talk to other investigators. As told in his book *The Secret Life of Bill Clinton,* the story is full of melodramatic and improbable details: Foster making pickups of laundered money in a Kmart parking lot, Foster calling Parks in July 1993 to say that Hillary wanted the files that Parks had been keeping in his home, and so on. Evans-Pritchard cites Gary Parks, the son, as saying that his father had been looking into Clinton's affairs since 1988—"working on his infidelities," he called it—and that the raw files, which included photographs, were stolen from the Parks home in 1993, after which Jerry Parks began to fear for his life. While some of this reads like fiction, Hillary had hired a private detective once before, and it is logical that when she needed one again she would have worked through Foster, who knew Parks because he had handled a case for the West Memphis security firm Parks worked for in the mid-1980s. Parks had other connections to the Clintons. He was appointed by Governor Clinton to the Arkansas Board of Private Investigators, and in 1989, when he decided to start his own security firm, he received a $47,959 loan from the Arkansas Teachers

Retirement Fund. In 1992, moreover, his firm received the contract to handle security for the Clinton campaign headquarters in Little Rock.

Wherever Lindsey and Foster got their information, they did nothing with it for the time being. However, there were other problems to be addressed. After Larry Nichols was fired for making personal phone calls at state expense, it occurred to someone that Bill Clinton's phone records could also be obtained under state freedom of information laws. Trooper Larry Patterson was told by Raymond "Buddy" Young, the head of Clinton's security detail, that he would have to be prepared to take responsibility for calls made by Clinton from a hotel room in Charlottesville, Virginia, where he had stayed while on a state-financed trip. One call, placed at 1:23 A.M., lasted ninety-four minutes; another, made at 7:45 the next morning, lasted eighteen minutes. Despite the warning to Patterson, Clinton eventually wrote a personal check for $40.65 to cover the calls. After the spring of 1990, however, pages from cell phone and hotel bills listing the numbers the governor had called were no longer included in state records.

Meanwhile, James McDougal was on trial in federal court in Little Rock for thirteen felony counts involving the Castle Grande development. McDougal had been living in a trailer on the Rileys' property in Arkadelphia, surviving on $591 a month in Social Security disability. On the eve of the trial his memory was so foggy that he checked into a Little Rock hospital, where doctors performed another surgery to clear blocked arteries that were restricting the flow of blood to his brain. His attorney asked the court to order a psychiatric evaluation, but the doctors ruled that McDougal, though manic-depressive, was competent. As it happened, when the time came for McDougal to take the witness stand, he was in a manic phase. He put on a brilliant performance, presenting himself as a small-town banker straight out of *It's a Wonderful Life*. The jury acquitted McDougal, his wife and her brother, James Henley, on all counts.

After the verdict, Bill Clinton called McDougal to congratulate him, then brought up a problem. Hillary, who now had control of the Whitewater books, had spent about $3,000 on various expenses and she wanted McDougal to reimburse her. This infuriated McDougal. The Clintons were still 50 percent partners in the venture and had lost much less than he and Susan had. He also had the impression that Hillary was listening in on an extension and wasn't willing to talk to him directly.

A few days later, Clinton met with Sam Heuer, McDougal's court-appointed lawyer, and told him Hillary wanted to close the books on Whitewater. Heuer wrote a letter to Hillary laying out the steps that would have to be taken to dissolve the company, but Hillary replied that she needed to gather more information about Whitewater's finances. Hillary never called Heuer back. Undoubtedly, McDougal's acquittal made the matter seem less urgent.[8]

ON October 3, 1991, a crisp, sun-filled autumn day, Bill Clinton stood on the steps of the historic Old State House in downtown Little Rock and formally announced he was a candidate for the Democratic nomination for President. Clinton's rambling thirty-two-minute speech was a kind of Rorschach test for Democrats. If you were a centrist New Democrat, you could take heart from his references to the "forgotten middle class" and his calls for an ethic of personal responsibility. If you were a liberal, like many of the political operatives recruited by Mickey Kantor, you were likely to come away with the impression that Clinton intended to expand entitlements to the middle class. One of the striking features of the Clinton candidacy was that the press rarely challenged him to clarify where he stood. Of the seven candidates in the Democratic primaries, a group sometimes derided as the seven dwarves, Clinton was the most personable and articulate. He was a proven fund-raiser with broad support in his own party and a family man, in contrast to the bachelor Bob Kerrey, who had allowed his girlfriend, actress Debra Winger, to stay over at his residence with him when he was governor of Nebraska.

Nevertheless, there was evidence that Clinton was not universally popular in his home state. A few demonstrators had turned out, protesting Clinton's now broken promise to serve out his four-year term as governor, a theme echoed by an editorial in the *Democrat* mocking the candidate as "Billy the Eternal Wunderkind Clinton." The day's papers also featured ads paid for by a group called ARIAS, challenging the media to expose the myth of the "Arkansas Miracle." One of the group's organizers, attorney Cliff Jackson, who had known Clinton at Oxford, was already trying to interest reporters in his claim that Clinton was a draft dodger.

But the most bizarre feature of the celebration was the presence of a picketer, Robert "Say" McIntosh, distributing handbills featuring a photograph of a boy about seven years old who he claimed was Bill Clinton's illegitimate black "love child," discovered by one Righteous Reverend Tommy Knots "after a six month investigation." Little Rock natives knew Say McIntosh as the owner of a local diner that specialized in barbecue and sweet potato pie and a political provocateur with a flair for street theater. McIntosh was known to dress up as the devil or Santa Claus; he had also draped himself over a wooden cross in a mock crucifixion, in the pursuit of a cause no one could remember. Reporters who asked for more information about the so-called love child found McIntosh vague and evasive.

Say McIntosh's charges became the inspiration for the plot of Joe Klein's roman à clef, *Primary Colors*. The actual story is more sordid. Bobbie Ann Williams was a Little Rock prostitute who worked out of an apartment near Seventeenth and Main streets, a seedy neighborhood a few blocks from the Governor's Mansion and not far from Clinton's regular morning jogging route. Williams, then twenty-four, later claimed that she had sex with Bill Clinton about thirteen times in late 1983 and early 1984, on one occasion recruiting two other prostitutes to help him fulfill a fantasy of being with three women at once. When Williams became pregnant and told her mother and sister that the baby was the governor's, they laughed at her. But the little boy turned out to bear a certain resemblance to Bill Clinton, and her relatives began to think she might be telling the truth.

In 1988, McIntosh picked up on talk about the child, whose name was Danny Williams, and handed out the first of his pamphlets. Lucille Bolton, Bobbie Ann Williams's sister and sometime guardian of her son, was upset about the publicity and called the Governor's Mansion, getting Hillary on the phone. During their brief conversation, Hillary asked, "Is it true he has this illegitimate child?" Bolton said it was, and she wanted to know what could be done to stop Say McIntosh from spreading the story all over town. Hillary gave Bolton the phone number of a private security agency that, she said, would help to get the rumor stopped. Sometime later, Bolton brought Danny, then four, to the Governor's Mansion, demanding to see Clinton and discuss the paternity issue, but she was turned away at the gate. Unable to get a meeting, she resorted to picketing a campaign rally, holding up a sign pleading for JUST ONE DROP OF BLOOD, to establish paternity.[9]

Nothing more was heard from Say McIntosh on the subject until 1991, when he filed a lawsuit against Clinton, claiming that the governor had reneged on a promise to pay him $25,000, money he had been counting on to begin retail marketing of his sweet potato pies, in exchange for his help in stemming "negative publicity" about Clinton's affairs. Furthermore, McIntosh charged, Clinton had promised to obtain an early release for his son, Tommy McIntosh, who was serving a fifty-year term for cocaine distribution. McIntosh hinted that the money was supposed to have been a bribe from Hillary. A Clinton campaign spokesman said that McIntosh had been hired to heckle an opposition rally but was never promised anything like $25,000.

McIntosh had made so many wild and scurrilous charges over the years that no one took him seriously. Clinton, however, did address the lawsuit in a July 1991 press conference. Vowing that the allegations about his personal life were "not so," he added, "My lawyer is over there having a heart attack that I would say anything else." Clinton acknowledged, however, that he had approached the Yarnell Ice Cream Company about manufacturing McIntosh's pies, though nothing came of the talks.

In January 1999, the editor of the supermarket tabloid the *Star* announced that tests comparing Danny Williams's blood with a DNA profile of President Clinton obtained from papers released by Independent Counsel Kenneth Starr had established that the boy was not in fact Clinton's son. In hindsight, Bobbie Ann Williams's allegations showed just how vulnerable Clinton was to rumors from the most unlikely sources. Right up to the day the *Star* revealed the test results, the White House failed to issue a flat denial of the allegations. Either Clinton did have relations with Bobbie Ann Williams at some point or, perhaps more likely, even his closest aides were never entirely sure that they could trust his protestations of innocence. Indeed, the "Clinton love child" story would take another strange turn, suggesting that the Clinton camp struck a deal with Say McIntosh in exchange for his silence.

On January 20, 1993, the day Bill Clinton was inaugurated as President, Tommy McIntosh was pardoned. The papers were signed by Dr. Jerry Jewell, president pro tempore of the state senate and acting governor for the four days during which Clinton's successor, Jim Guy Tucker, was in Washington attending the inaugural festivities. Dr. Jewell, a dentist and one of Little Rock's most prominent black political figures, later said he was just executing paperwork prepared by the governor's office. Three other prisoners were pardoned or granted clemency at the same time, and the Little Rock chapter of the NAACP defended the pardons, dismissing any criticism of them as being racially motivated. A few weeks later, Jerry Seper of the *Washington Times* encountered Say McIntosh on a downtown street. McIntosh told Seper that he had struck a deal with Clinton immediately after the election. "Those who question my credibility should ask themselves, 'If there was no deal, how did this happen?' " he said. "How did my son get out of prison eighteen years before he was eligible for parole?"[10]

The Clintons, of course, knew even before Bill entered the primaries that Say McIntosh's scandalmongering pamphlets would pose a problem. The same goes for Gennifer Flowers, whose hearing before the State Grievance Committee took place on October 9, just six days after Clinton's formal announcement. As instructed by Clinton, Flowers avoided implicating either him or Judy Gaddy, testifying that she learned of her job through a notice in the newspapers. Afterward, she reported her lie in a phone call to Bill, whose reaction was, "Good for you!"

The Grievance Committee found in favor of Charlette Perry based on misrepresentations that Flowers had made on her résumé; however, Don Barnes overruled the decision and kept Flowers in her job. Even so, Flowers's problems were far from over. Steve Edwards, a reporter for the *Star*, had arrived in Little Rock to track down the women named in Larry Nichols's lawsuit. Deborah Mathis, who was temporarily working in Mississippi, firmly denied an affair with Clinton, and Nichols himself said he had been wrong about a second woman, Clinton aide Susie Whitacre. Edwards tracked down Lencola Sullivan, a third woman named by Nichols, at an NAACP luncheon, but she didn't want to discuss Bill Clinton.

Edwards and other reporters found Flowers an easier target, since her grievance hearing was a matter of public record. On October 14, Flowers called Clinton to report that she had been fired from her night job at the Flaming Arrow, a private club in the Quapaw Tower, after a crew from *A Current Affair* staked out the parking lot. Moreover, her mother, who was remarried and living in southern Missouri, had received a phone call from a menacing stranger who told her that her daughter would be "better off dead."

In early December, Flowers twice came home to her condo—she had moved from the Quapaw Tower—and found the deadbolt locked, even though she hadn't left it that way. Several days later, she discovered the door ajar and her apartment ransacked. The intruder had rifled through several boxes filled with old photographs. When Flowers reported this to Clinton in a December 11 phone call, he blamed "Republican harassment" and asked, "Do you think they were looking for something on us?"

Something about Clinton's tone of voice made Flowers suspicious. "It was you," she thought.

According to Flowers, she had been contacted by a locally prominent Republican named Ron Fuller, who offered to put her in touch with people who would pay $50,000 if she would sign an affidavit attesting to her affair with Clinton. (Fuller denied this.) Informed of the offer, says Flowers, Clinton asked her to sign an affidavit accusing Fuller, then suggested ways she might embellish her account. Flowers knew that a former Quapaw Tower neighbor, attorney Gary Johnson, had been beaten up and gravely injured after word got out that he had a surveillance tape showing Clinton arriving at her apartment. Although Flowers had originally told Edwards that she wanted to tell her story to *The Washington Post,* if at all, she began to reconsider when she was informed that the *Star* had pictures of her and was going to name her as Clinton's mistress with or without her cooperation. Worried that she was no longer safe in Little Rock, she hired an attorney and bolted to New York, where she signed a six-figure deal to sell her story.

Even before the *Star* could publish Flowers's story, the latest issue of *Penthouse* had arrived on the newsstands featuring an article called "Confessions of a Rock 'n' Roll Groupie," based on the diaries of Connie Hamzy, a Little Rock woman who had achieved underground notoriety for having sex with Keith Moon, Don Henley, David Lee Roth and numerous other rock-and-roll stars. Hamzy's ghostwriter had discovered an item in her diary about a meeting with Bill Clinton, which took place one day as Hamzy was sunning herself by the pool at the North Little Rock Hilton. Clinton, who had just delivered a speech in the auditorium, sent one of his aides to fetch Hamzy and told her, "I want to get with you." There followed some consensual groping in a laundry room off the pool area.

Hamzy, who had nothing against Clinton, passed on a friendly warning through Rose firm partner William Kennedy III. When the *Penthouse* story appeared, the Clinton campaign countered by releasing an affidavit from a state legislator who had been with Clinton at the hotel that day. The legislator's version of the meeting was that Hamzy had approached the governor and dropped her swimsuit top to give him a good view of her breasts before he fled in embarrassment. "I'm a slut, not a liar," fumed Hamzy. But a projected book based on her diaries never appeared, the rights having been bought up by a group of Little Rock brokers.

While Connie Hamzy's Clinton anecdote proved to be a one-day story, the *Star*'s accusation that Flowers had been Clinton's mistress for twelve years had to be answered. Hillary had a theory that the

Gary Hart campaign collapsed because Hart's wife, Lee, had allowed herself to be portrayed as a victim. (Indeed, the media's treatment of Lee Hart had been cruel; she was seen as pitiful, and yet in some unspecified way culpable for remaining married to a philanderer.) Hillary was determined to be at Bill's side when he responded to the *Star*'s allegations, and campaign pollster Stan Greenberg agreed, pointing to opinion surveys that said that Americans would overlook a candidate's infidelity if he had confessed them to his wife and obtained her forgiveness.

Prepared to tough it out, Hillary personally chose the setting for this all-important interview, vetoing an arrangement by campaign headquarters for an appearance on *Nightline* in favor of *60 Minutes* with interviewer Steve Kroft. The *60 Minutes* producers were uneasy about the subject matter but offered the Clintons the chance to appear on an abbreviated edition of the show, immediately following the Super Bowl. The Clintons' ground rules were that they would speak generally about their marriage but would not go into details on specific allegations. This did not stop Kroft from doing some gentle probing in an effort to get some details about Flowers:

KROFT: Was she a friend, an acquaintance? Does your wife know her?
HILLARY: Oh, sure.
CLINTON: Yes. She was an acquaintance, I would say a friendly acquaintance. . . .
KROFT: She is alleging and has described in some detail in the supermarket tabloid what she calls a twelve-year affair with you.
CLINTON: That allegation is false.
HILLARY: When this woman first got caught up in these charges, I felt as I've felt about all of these women: that they . . . had just been minding their own business and they got hit by a meteor. . . . I felt terrible about what was happening to them. Bill talked to this woman every time she called, distraught, saying her life was going to be ruined, and . . . he'd get off the phone and tell me that she

said sort of wacky things, which we thought were attributable to the fact that she was terrified.

KROFT: I'm assuming from your answer that you're categorically denying that you ever had an affair with Gennifer Flowers.

BILL CLINTON: I've said that before. And so has she.

KROFT: You're trying to put the issue behind you, and the problem with the answer is it's not a denial. And people are sitting out there—the voters—and they're saying, "Look, it's really pretty simple. If he's never had an extramarital affair why doesn't he say so?"

Clinton, looking shaken, launched into a rambling answer, but Hillary interrupted him, saying, "There isn't a person watching this who would feel comfortable sitting here on this couch detailing everything that ever went on in life or their marriage. And I think it's real dangerous in this country if we don't have some zone of privacy for everyone. . . ."

At this, Kroft backed off. The interview did not heat up again until he commented, "I think most Americans would agree that it's very admirable that you've stayed together—that you've worked your problems out and come to some sort of understanding and arrangement." The word "arrangement" drew protests from both Clintons, especially Hillary. Speaking with a slight Southern drawl that would never be heard from her mouth again, she said:

You know, I'm not sitting here—some little woman standing by my man like Tammy Wynette. I'm sitting here because I love him, and I honor what we've been through together. And you know, if that's not enough for people, then heck, don't vote for him.

Shortly thereafter, the interview drew to its close with a final thought from the candidate:

CLINTON: I have absolutely leveled with the American people.

KROFT: You came here tonight to try to put it behind you. . . . Do you think you've succeeded?

CLINTON: That's up to the American people and to some extent to the press. This will test the character of the press. It is not only my character that has been tested.

This was a classic example of what Larry Nichols calls the Bill Clinton "rope-a-dope" strategy. Confronted with embarrassing questions, he may appear sheepish, even panicky. It takes a while to realize that he has conceded almost nothing, and then only as a prelude to delivering a counterpunch: Shame on all of you for asking these sleazy questions.

As *The Washington Post*'s Tom Shales observed, Hillary had remained "impressively impervious" even while Bill Clinton "looked like a scared kid." Moreover, she had made a point of saying that she was well aware of Gennifer Flowers's late-night calls to her husband. Indeed, according to *Salon* magazine's Murray Waas, Hillary actually listened in on some of them. (Presumably these were the conversations in which Clinton appears to be guarding his language while counseling Flowers to deny their affair.)

Leaving the Ritz-Carlton Hotel in Boston, where the *60 Minutes* segment had been taped, the Clintons were pumped up with nervous energy, almost ecstatic over how well the interview had gone. A few well-wishers had gathered on the sidewalk, and they cheered as the couple got into their car, Hillary in the front seat and Bill in back. As the car pulled away, Bill reached over the top of the seat and Hillary held tight to his arm.

The euphoric mood did not last long. While Clinton returned to New Hampshire, Hillary flew back to Arkansas to be with Chelsea when the segment aired later that day. What she saw did not please her. She felt that some of the best moments had been cut, including an incident when a piece of lighting equipment tipped over and Bill instinctively reached out to protect her.

Of course, Hillary was wrong about this. Clinton's semiconfession of adultery on national television made his candidacy, thanks in large part to an editing job that winnowed down ninety or so minutes of tape to a brief interview presenting the Clintons in a distinctly favorable light. The Clintons "came to us because they were in big trouble in New Hampshire," producer Don Hewitt later recalled. "They were about to lose right there and they needed some first aid. They needed some bandaging. What they needed was a paramedic. So they came to us and we did it. . . . As I said to Mandy [Grunwald, the Clinton media consultant], 'You know, if I'd edited it your way you know where you'd be today? You'd still be up in New Hampshire looking for the nomination.' " Hillary never did come around to seeing it that way, however, and after Clinton became President, Hewitt was persona non grata at the White House.

What some might interpret as evidence of a pro-Clinton bias on the part of Hewitt and CBS perhaps had more to do with the media's thirst for drama. Democratic candidate Tom Harkin of Iowa complained that he and his wife would have loved to have fifteen minutes after the Super Bowl to meet the public and talk about their marriage. But there was no Gennifer Flowers in their lives and hence no "story" worthy of network airtime. "Tom is trying to get anyone to admit that they slept with him," quipped Rob Schneider of *Saturday Night Live*.

While the press focused on the drama of Bill and Hillary's marriage, they largely missed another, equally compelling story. The Clinton campaign, from the very beginning, was a saga of lies, false affidavits, perjury and intimidation. Skeletons kept slipping out of the closet of Bill Clinton's past, and every time his advisers cracked open the door to shove one back in, two more escaped.

On the Monday afternoon following the Clintons' *60 Minutes* appearance, Gennifer Flowers appeared at a press conference arranged by the *Star*. Looking beautiful but blowsy, Flowers portrayed herself as a longtime mistress, in pain because her married boyfriend was denying their love—not most people's idea of a sympathetic figure. But Flowers also announced that she had tapes of her phone conversations with Bill Clinton. (The Clintons may have suspected this al-

ready. Why else would Hillary have seized the opportunity to tell
Steve Kroft that she knew there had been phone calls during which
certain "wacky" things were said?)

After the Flowers tapes were released, Clinton flew back to Little
Rock, where, it was said, he was needed to escort eleven-year-old
Chelsea to a father-daughter dance. Hillary, left alone to handle the
fallout, went on *Prime Time Live* and suggested with a straight face
that the Flowers tape might be faked. On Thursday evening, she
attended a roast for Democratic National Committee chairman Ron
Brown, and managed a stoical smile when emcee Larry King teased
her, "It's ten o'clock. Do you know where Bill is?"

Back in New Hampshire, the Clinton spinmeisters had gone into
overdrive, attacking Flowers's credibility by pointing out errors in the
Star's story, a few real but most imaginary. (Flowers had not been
crowned Miss Teenage America, as the *Star* claimed. She *had* appeared
on the TV show *Hee-Haw,* though whether this made her more or
less believable was anybody's guess.) The *Star* had distributed a sixty-
minute tape of excerpts from the original recordings, which James
Carville immediately denounced as "doctored." Carville would continue to use this term to describe the unedited tapes, later sold by
Flowers through an 800 number. But the tapes had been authenticated
by an independent laboratory, and one can be sure that if the Clinton
campaign had evidence that the tapes were doctored, it wouldn't have
hesitated to get specific.

As feeble as some of this spin-doctoring seems in retrospect, it
worked. Political reporters and their editors were grappling with the
question of just when a candidate's private life becomes the public's
business. While it may not be true that "everyone does it," many
people have some corner of their lives that they would not care to
have featured on the front page of the *Star,* and no one wants to see
the electoral process reduced to a game of competitive character assassination. It didn't hurt, also, that mainstream reporters liked the
Clintons and were generally appalled by the *Star*'s checkbook journalism. Moreover, the notion that there was a lot of money to be
made out of women "coming forward" to tell the stories of their

extramarital flings with the candidate Clinton turned out to be a fantasy. Flowers, who eventually earned several hundred thousand dollars from going public, would be the exception to the rule. The *Star* had paid her its highest fee ever but was disappointed by its newsstand sales and would not be so free with its money again. For the next several weeks, pundits and editorial writers indulged in a ritual of self-criticism over their decision to report Flowers's claims. In the process, they overlooked the message of the tapes: Clinton had been caught in the act of urging a state employee to lie under oath.

Although it had been granted a reprieve by the press, the Clinton campaign organization was still in a panic mode. Like Flowers, Dolly Kyle Browning had been offered a six-figure deal to talk about her long relationship with the candidate. Browning, now a Dallas real estate lawyer and happily married, had no plans to take the money, but she wanted to talk to Clinton personally and get his reassurance and advice on how to handle the overtures from reporters. Browning left messages for Clinton, who was back in New Hampshire by this time, but he failed to return her calls. Eventually, she heard from her brother, Walter Kyle, who was one of the Arkansas Travelers, a group of Clinton supporters who paid their own way to New Hampshire to work for the campaign. Kyle told his sister that Clinton couldn't take her call because he now feared his conversations might be taped, and he added a strong warning. If Browning went public she would regret it. "We will destroy you," he threatened. Browning had the impression that Clinton was in the room, listening in on this conversation.

Miami attorney John B. Thompson, a conservative and longtime antagonist of Janet Reno, happened to be an old friend of Bruce Lindsey's law partner, M. Samuel Jones III, having graduated from Vanderbilt Law School with him in 1976. Around this time, Jones called Thompson from New Hampshire to congratulate him and his wife, who had recently learned that they were about to become parents after seventeen years of marriage. Sounding exhausted, Jones complained that he was worn out from his assignment of tracking down Clinton's ex-lovers. Clinton, he groused, was a pathological womanizer on the scale of JFK—"The man has simply got to have it." Jones, who had been Clinton's attorney of record in handling the Larry Nichols suit, said that the campaign's attitude toward Clinton's women was to do whatever was necessary "to make them go away." In a later call, Jones said that he had even located the woman who claimed that Clinton was the father of her son, adding, "I tracked her down but I don't think there is anything to it." After Thompson discussed these conversations with a reporter, Jones denied saying such things and cut off their two-decade-old friendship.[11]

In California, meanwhile, Elizabeth Ward Gracen, the former Miss America, was finding that her acting career was suddenly picking up. *Playboy* magazine had featured Gracen on its cover, and she was getting calls from reporters, including the *Star's* Steve Edwards. Gracen also heard from Roger Clinton, who invited her to visit the set of *Designing Women,* the sitcom produced by the Clintons' good friends Linda Bloodworth-Thomason and Harry Thomason. Gracen, who had no desire to become the next Gennifer Flowers, agreed to make a statement that the allegations about her and Clinton were false. Around this time, Clinton campaign director Mickey Kantor and Harry Thomason got together with Gracen's manager, Miles Levy, in a deli on Ventura Boulevard. Gracen, who knew nothing of the meeting at the time, soon received an offer to fly to Croatia, where there was a role waiting for her in Michael Viner's miniseries *The Sands of Time.* No sooner was the shooting finished than she was offered another acting job, this time in Brazil, followed by a starring role in the *Highlander* series, also filmed abroad.

All went well for Gracen until late 1997 when she, her parents and her friends began receiving phone calls from people who warned that she had better duck subpoenas from the lawyers of Paula Jones and Independent Counsel Ken Starr, or her reputation would suffer. Gracen went into hiding, but mysterious callers had no trouble reaching her and she concluded that she was under surveillance. While she was staying at a resort on the Caribbean island of Saint Martin, her room was ransacked and an employee told her that he had admitted two men in suits, assuming they were friends. "Physically scared," Gracen hired Bruce Cutler, best known as the attorney of John Gotti, and gave an interview to NBC's *Dateline* in which she acknowledged a single sexual encounter with Bill Clinton in 1983, an assignation that took place in a Quapaw Tower apartment belonging to one of Clinton's friends.[12]

BUT inconvenient former girlfriends were hardly the Clinton campaign's only worry. By late February, Clifford Jackson, a member of the group of Arkansas Republicans calling themselves ARIAS, had been trying for months to get the press interested in Clinton's draft history. *The Wall Street Journal* finally brought up the issue twelve days before the New Hampshire primary, and soon the press had a copy of Clinton's December 1969 letter to Colonel Holmes. Clinton's first reaction was that Bush campaign operatives had unearthed the letter from official files. "I want you to ask the President to call the Pentagon and find out who leaked this!" he angrily told reporters. (Later, it was established that the letter came from the personal records of the National Guard commander Lieutenant Colonel Clinton D. Jones.) While issuing a series of revisions of his draft history, Clinton swore to his staff that the *Journal* had it wrong when it claimed he had actually received an induction notice. Then, on April 5, AP reporter John King handed George Stephanopoulos a copy of the actual induction notice. Confronted with the evidence, Clinton told Stephanopoulos, "I forgot about it." Stephanopoulos was so depressed that he retreated to his bed and spent the morning hiding under the covers. But he made what he called "a choice to believe" Clinton and soldiered on.[13]

Clinton's denials about the draft were clumsy in comparison with the web of obfuscation Hillary was weaving around Whitewater. When investigative reporter Jeff Gerth of *The New York Times* began asking questions about the Clintons' relations with Jim McDougal, Hillary appointed her close friend Susan Thomases as spokesperson on the issue, at the same time retaining Thomases as her lawyer so that their discussions would be covered by attorney-client privilege.

A veteran Democratic operative and partner in the New York firm of Wilkie, Farr & Gallagher, Thomases had known Bill Clinton since 1970, and she met Hillary four years later when Thomases spent some time in Fayetteville helping out with Clinton's congressional campaign. Hillary became friendly with Thomases, brought her onto the board of the Children's Defense Fund, used her as her personal attorney in drawing up her partnership agreement with the Rose firm and confided in her about subjects that were difficult to discuss freely with anyone in Arkansas. On at least one occasion, Thomases also played the role of a "mean out-of-town lawyer," scaring off a woman who had become a threat to Bill's career.

Thomases is memorably characterized in Joe Klein's novel *Primary Colors* as Lucille Kauffman, a charmless woman in power suits who radiates "a disconcerting sense of ownership about the campaign." Long single, Thomases had recently married the artist-cum-carpenter who was renovating her apartment and, at forty-six, had given birth to her first child. Career and family obligations kept her from joining the campaign full-time until the final months, but she sat in on strategy sessions anyway, often acting as Hillary's surrogate. Thomases shared many of Hillary's more abrasive qualities, except that unlike Hillary she made little attempt to keep them under control. She monitored other people's smoking and eating habits, had an unjustified faith in her own snap judgments and used up a lot of the energy in whatever room she happened to be in at the moment. "I'll kill you for that" or "You'll never work again in this business" was her idea of a mild rebuke, and she was soon embroiled in feuds with Democratic National Committee chairman David Wilhelm, campaign counsel David Ifshin, Dee Dee Myers and even Al Gore.[14]

Nevertheless, one can only sympathize with the dilemma Thomases faced as the point person on Whitewater. Despite the way the press had dissected the finances of Geraldine Ferraro's husband in 1984, Hillary seemed to be surprised that she was being asked detailed questions about her personal finances and legal career. She even complained that the *Times*'s reports were instigated by the paper's Washington bureau chief, Howell Raines, a Southerner who she claimed was jealous that Bill Clinton was headed for the White House while he remained a mere journalist. To the horror of the professional campaign operatives, she suggested launching a counterattack against the *Times*.[15]

Moreover, when it came to Whitewater and work for Madison Guaranty, Hillary's famous ability to master complicated subject matter and focus in on the essentials seemed to fail her. Thomases had to fly to Little Rock herself to interview Yoly Redden, the Clintons' personal accountant, and she assigned lawyers attached to the campaign to talk to others involved in Madison Guaranty dealings and discover their versions of what transpired. Two of these attorneys, James Lyons and his associate Kevin O'Keefe, grilled a wary Vince Foster and Webb Hubbell about Hillary's legal work for Madison. Hubbell writes of feeling that he and Vince were "Butch and Sundance, trying to find footing in a dangerous world."

And no wonder. Hubbell, in particular, had been the lead attorney for the FDIC in a case against Frost and Company, the accounting firm that had issued a report declaring Madison Guaranty solvent. The government's charges covered some of the very transactions involving Hubbell's father-in-law that the Rose firm had handled for Madison, with Hillary as the billing partner. Needless to say, the FDIC was not aware of this conflict, and the firm's files on Castle Grande had been purged. It turned out, however, that Hillary's billing records still reflected her work for Madison Guaranty, even if it was impossible to say exactly what she had done. Foster drew the assignment of getting Rick Massey to confirm Hillary's story that he was the one responsible for bringing in Madison Guaranty's business. Massey verified the story, at least according to Hubbell, but he clearly was not happy. Thomases was frustrated by her inability to give Jeff Gerth more specific answers and hinted that the Rose firm wasn't cooperating fully with her efforts.

On March 8, James Lyons issued a report showing that the Clintons had lost $68,000 on their interest in Whitewater. The figure included a $20,000 check signed by Bill Clinton and made out to the Madison bank of Kingston. Only after his mother's death would Bill Clinton recall that the check was actually a repayment of money he borrowed to help his mother buy her cottage on Lake Hamilton. The Clintons hardly seemed wealthy enough to have forgotten the purpose of a $20,000 loan.

Hillary's billing records, meanwhile, had disappeared from the files of the Rose firm, which would be unable to locate them when they were subpoenaed in 1994. One copy of the 116-page printout would turn up much later in Vince Foster's attic. Another, with notes in Foster's hand addressed to HRC—Hillary Rodham Clinton—would not be seen again until January 1996.

With even top campaign aides becoming frustrated by the Clintons' inability to provide straight answers about Whitewater, it was time to recall Betsey Wright from Harvard, where she had been doing a turn as a guest lecturer. With the aid of a few volunteers, including Hillary's friend Diane Blair, Wright set up her own operation separate from the inner circle of the Clinton campaign. Without irony, she called it the "truth squad."

Part of Wright's mission was to collect the files from Hillary's major legal cases. Webb Hubbell's recollection is that Wright needed these files because Clinton's opponents were claiming that Hillary was not a "real lawyer." But the reality of Hillary's lawyering was never the issue; rather, Democratic challenger Jerry Brown had suggested that her work for the Rose firm involved serious conflicts. For example, at the request of her husband, Hillary had represented the state in setting a dispute with Arkansas Power & Light that allowed the utility company to pass on to consumers 80 percent of its cost overruns from the Grand Gulf nuclear plant, popularly known as "Grand Goof." Hillary's position was that there could be no conflict because she had not been awarded any part of the Rose firm's income from the case; technically, she had worked for nothing. On the other hand, the settlement did help Bill Clinton politically. While appearing to be a victory for the public, it did not unduly hurt Arkansas Power and Light and kept Clinton in the good graces of some major political donors.

No one can be sure exactly what other subjects were covered by the "Betsey files," as they came to be called. The files were intended to provide answers on issues that might be raised by the press, so presumably there was material on Bill Clinton's medical history. Recent presidents and presidential candidates have voluntarily sacrificed their personal privacy in the belief that the public has a right to know the truth about their health. Jimmy Carter released information about his hemorrhoid problems; Ronald Reagan's doctors showed the world pictures of his colon polyps; George Bush revealed that he took the powerful sleeping pill Halcion. Bill Clinton turned down requests from *The New York Times* for an interview with its health correspondent Dr. Lawrence Altman, and in December 1992 his appointment with a heart specialist, Dr. Andrew Kumpuris, was omitted from his published daily schedule. In January 1993, just weeks after his inauguration, a package from Clinton's allergy specialist, Dr. Kelsy Caplinger, arrived at the White House containing a vial with a note identifying the contents as "President Clinton's allergy medicine." Dr. Burton Lee, who had been George Bush's official physician, refused to administer the medicine by injection without more information.

"I was not happy about how the serum came to us, delivered to the White House gate, no covering letter, no idea what was in the bottle or why it was mailed," said Dr. Lee, who called the President's Little Rock physician, Dr. Susan Santa Cruz, to request his full medical records. Dr. Santa Cruz said she'd have to get Hillary's permission to release the records. Two hours later, Dr. Lee was fired. For whatever reasons, Clinton has continued to be sparing in disclosing details about his health.

Another of Betsey Wright's duties during the campaign was riding herd on L. D. Brown, the state trooper who'd had a falling-out with Clinton after serving as his bodyguard and confidant during the mid-1980s. Brown has been the source of several controversial allegations about the Clintons, notably his claim that in 1984 Clinton guided him through the process of applying for work with the CIA. Brown went along on two covert operations flights to Central America, only to become disillusioned when he discovered that the pilot, Barry Seal, was also smuggling drugs. During the late 1980s, as head of the 500-

member association of Arkansas state troopers, Brown came into conflict with Clinton again, after the governor reneged on a promised pay raise. In 1990, he gave information about Clinton's womanizing to his primary opponent Jim Guy Tucker, in the hope that Tucker would use the charges in his campaign. (Tucker did not.) Brown now began to notice that when he stopped by his favorite haunt, Diego's bar, for a drink, a woman named Kathy Ford, one of Betsey Wright's assistants, just happened to be there. Ford was alone, not drinking, and eager to draw him into conversation. Suspicious, he followed her out of the bar one day and found her in a huddle with Wright at a convenience store half a block away.

Wright also had assistants keeping track of the daily movements and contacts of out-of-town reporters, a move she justified on the grounds that they were liable to fall into bad company. "Most of these people have no concept of Arkansas or what a small state worked like," she said at the time. "One of the things I'm determined to do is protect the record from unfair distortion because of things like that."[16]

But the effort to keep tabs on L. D. Brown was amateurish compared to the ongoing search for potentially troublesome females. By now, the campaign had several private detectives on retainer, including Anthony Pellicano and Jack Palladino, the latter a former protégé of San Francisco private eye Hal Lipset, who did the investigative work for Black Panther lawyer Charles Garry. Palladino specialized in digging up dirt that could be used to discredit Clinton's ex-lovers. "Every woman I investigated was either manipulated or had come forward as a result of the media throwing money out there, or out of ideological craziness," he boasted to *The New Yorker*'s Jane Mayer, though when threats of character assassination failed, Palladino could resort to outright intimidation.

Interviewing Lauren Kirk, a woman who had been Gennifer Flowers's roommate in Dallas, Palladino asked, "Is Gennifer the type to commit suicide?" When such questions are asked by a surrogate for a man soon to be President of the United States, people get nervous.

A prime target of the "truth squad" was Sally Miller Perdue, who claimed to have had a brief fling with Bill Clinton in late 1983. A fun-loving blonde seven years Clinton's senior, Perdue was crowned Miss Arkansas in 1958, worked as a reporter for the local public television station and also briefly hosted a radio call-in show. Perdue, in short, fit the profile of so many other women connected with him— the ex–beauty queen working for the local media. As Perdue told it, when Clinton visited her condo he would have his driver park in a wooded area and then signal him to come pick him up by flashing the patio light when he was ready to leave. She recalled that Clinton liked to drink a Bud or two, then don her black negligee, just for fun, and play fifties rock-and-roll hits on his sax—"badly"—while she accompanied him on the piano.

Sally Perdue was the woman Betsey Wright was talking about when she coined the immortal phrase "bimbo eruptions." Perdue had taped an episode of the *Sally Jessy Raphael* show that was scheduled to run in July, during the Democratic National Convention. She had also agreed to appear at a press conference sponsored by Lenora Fulani's New Alliance Party. Forewarned, Jack Palladino and his partner/wife Sandra Sutherland called up Perdue's relatives until they found one who was willing to say negative things about her character. Wright herself used Perdue's connection to the New Alliance Party as evidence that Clinton was the victim of a politically inspired vendetta.

The Washington Post's Howard Kurtz, in *Spin Cycle,* a book about the Clinton administration's press operations, dismissed Perdue as a part of the lunatic fringe of the Clinton scandals, a woman who "couldn't even convince the *National Enquirer* to publish her story." It is true that the *Enquirer* backed out of negotiations with Perdue because she, unlike Gennifer Flowers, had no hard evidence to back up her story, though she had told a couple of friends about it at the time. (In any event, the *Enquirer* later got its Perdue story for nothing, headlining it "Weird Cult Wants to Destroy Clinton.") In retrospect, however, Perdue's story of a couple of "no strings" dates with Clinton at her condo is a lot more believable than Bill Clinton's word, delivered via Betsey Wright, that he had never so much as met Sally Perdue.

It wasn't enough to discredit Perdue. On August 19, after returning home to Clayton, Missouri, Perdue received a call from a man who claimed to represent the Democratic Party. He asked Perdue to meet him at a local restaurant called the Cheshire Inn. Nervous, Perdue had a coworker station himself in the bar area, where he could listen in on their conversation. The man told Perdue that if she were "a good little girl and didn't kill the messenger," he could arrange a federal job at "level 11 or 12" (about a $60,000 salary level); but if she refused, she had better be careful when she went jogging because, as Perdue would recall, "he couldn't guarantee what would happen to my pretty little legs. Life would get hard."

Perdue refused the offer and was soon fired from her job in the admissions office of Lindenwood College. She also began to receive threatening letters, including one that said, "Marilyn Monroe got snuffed." She found unspent shotgun shells on the front seat of her Jeep and, on another morning, discovered the rear window shattered, apparently by a shotgun blast. Perdue, of course, had no way to prove who was doing these things. She gave at least two more interviews, one to a correspondent from the London *Daily Telegraph,* then dropped out of sight. She is said by acquaintances to be living in China.

The "bimbo patrol" detectives cost the Clinton campaign about $100,000 in all, paid via James Lyons's Denver law firm. Some of this may have been drawn from taxpayer-supported matching funds. Reacting to an earlier figure of $28,000, which Betsey Wright mentioned in her "bimbo eruptions" conversation with *The Washington Post's* Michael Isikoff, Bush political director Mary Matalin noted, "I controlled the money in the [Bush] campaign, and Betsey Wright announced that she was spending $28,000 on the bimbo patrol. . . . And $28,000 was to me—as the political director—that was equal to the campaign budget for four states in the Rocky Mountains. I thought— how could they spend this much money, because there is a limited budget? How could they basically give up four states to track down bimbos? And that's what I found shocking. . . ."

Needless to say, Republican operatives were doing their best to get some of Clinton's ex-girlfriends to come forward. This was an unsavory business, perhaps, although there is a difference between coaching people to lie and urging them to tell the truth. In any event, they weren't getting cooperation from the Bush campaign. Copies of threatening letters Clinton's ex-girlfriends had received from private investigators working for the Clinton campaign were forwarded to Mary Matalin. Moreover, she recalled, "I had handwritten letters from the women. I got one letter from one woman's father saying, 'This is so horrible. Here's what they're going to do to us.' " Much to the disgust of the conservative partisans, Bush and his advisers refused to take up the issue.

This history is worth retracing because it gives the lie to the argument that Bill Clinton is a victim of "sexual McCarthyism," hounded by a puritanical prosecutor and his right-wing allies. If anything, in 1992 the Clintons and their surrogates were the "sexual McCarthyites," engaged in a desperate campaign to get his detractors to stop talking about sex. The motives of the individuals involved may have been unattractive—tabloid cash, fifteen minutes in the spotlight, political enmity—though, in several cases, the Clintonites' harassment and threats were also a catalyst, causing them to come forward. Dolly Browning didn't care to hear her brother talk about destroying her. Sally Perdue was amazed by Clinton's "I didn't inhale" denial of pot smoking, since she said she had seen him smoke joints. Similarly, state trooper Larry Patterson, later one of the first of Clinton's bodyguards to go public against him, began to question his loyalty the day he realized he might end up taking the blame for the personal phone calls Clinton had charged to the state.

IRONICALLY, considering all the energy that was going into spin and counterspin on the "bimbo" issue, Hillary, not Bill, emerged from the *60 Minutes* interview as the Clinton with an image problem. Hillary's comment that she wasn't "some little woman standing by my man like Tammy Wynette" riled the country music queen, who fired off an open letter to the press demanding an apology on behalf of "all the little women in the country who 'stand by their man' and are potential supporters of Bill Clinton."

In Hillary's defense, she was just trying to defuse an awkward situation with a flip remark, but her words were unintentionally revealing. Released in 1968, "Stand by Your Man" was ridiculed by the women's liberation movement and trend-conscious commentators as *the* anthem of female submissiveness. But the women who thought this weren't listening closely. There was nothing whiny or submissive about the lyrics or the delivery of the Tammy Wynette. Moreover, Wynette, whose hits also included "D-I-V-O-R-C-E," broke into the country music business at a time when DJs were reluctant to play cuts by women vocalists and was regarded by many fans as a feminist heroine. Hillary Clinton could hardly be expected to know all this, but having lived most of her adult life in Arkansas, where country music has more than a few fans, she might have realized that a great many women did not see either the song or Tammy Wynette as a joke.

In a blunder that Bill Clinton would have been unlikely to make, Hillary had alienated the very people who were most likely to cheer her spunky defense of her husband—women who know that "Stand by Your Man" is a song about the need to support a flawed and occasionally embarrassing husband whom you happen to love. No one could justly fault Hillary for being a loyal wife. It was the suggestion that she had other, superior, motives that made some people nervous. With hindsight, knowing that Clinton was lying about Gennifer Flowers, and Hillary was lying, too, her *60 Minutes* performance becomes even more problematical. The very appearance that introduced her to a national TV audience was an exercise in deceit, not a promising beginning for a woman who hoped to be taken seriously as a political presence in her own right.

The Clintons had not anticipated this reaction. In New Hampshire,

Bill Clinton used the tag line "Vote for one, get two," and he continued to return to this theme in the early months of the campaign, predicting that he and Hillary in the White House "would be an unprecedented partnership, far more than Franklin Roosevelt and Eleanor." When asked if he would appoint his wife to a cabinet post, Clinton said, "I wouldn't rule it out." (Could it be that the Clintons were unaware of the anti-nepotism law, passed after Jack Kennedy named his brother Bobby attorney general? The law made it illegal for the President to name his wife to his cabinet.)

In many respects, Hillary was an enormous asset to the campaign, and to Bill Clinton's image. She had her own all-female staff of eighteen and kept a full schedule, speaking at fund-raisers and schools, before women's groups and professional associations. Long a tireless networker, she had a solid base of support within the liberal wing of the Democratic Party, the very people who otherwise would have been squeamish about supporting her husband.

Hillary originally planned to get back to Little Rock every four days to spend time with Chelsea, a resolution that became impossible to keep. As the campaign became more frenetic, she resorted to helping Chelsea with her homework by fax. Professional women empathized. They knew what it was like to juggle work and family obligations, and they tended to bristle at any criticism of Hillary. Women journalists, in particular, were familiar with the strain of long hours on the road, and they rushed to file articles describing Hillary Clinton as emblematic of her generation, a woman striving to balance work, marriage and home life. Many working women came to see any criticism of Hillary Clinton as a veiled attack on them. And they weren't always wrong. Too often, the Republican mantra of family values translated into "Let the women do the work."

But when it came to the double standard, Hillary usually benefited. Because she was a woman, dedicated to good works, and not a candidate herself, she escaped tough questioning. When *The American Spectator* ran a piece asserting that Hillary believed in children suing their parents, publications including *The Washington Post* dutifully repeated Susan Thomases's spin that Hillary's ideas had been taken out of context, and that Hillary had advocated such suits only in extreme cases of abuse and neglect. This sounded plausible; it just didn't happen to be true. Similarly, when Evan Gahr of *Insight* reported that Hillary had served on the board of the New World Foundation, which gave a grant to a Boston-based group that funded PLO-affiliated organizations on the West Bank, Hillary's press secretary, Lisa Caputo, said the money had been earmarked for antiapartheid work. Hillary herself commented, "If the money was diverted, I was not aware of it." But the money was part of a general grant, it didn't have to be diverted, and it was common knowledge that the organization, Grassroots International, supported PLO affiliates. Of course, Hillary was free to disavow her past views or defend them. Instead, too often, she and her staff implied that the very act of bringing up her record was a smear tactic.

Hillary did well in scripted appearances but got into trouble when she had to improvise answers to tough questions. Campaigning in Chicago for the Illinois primary, she was confronted by a reporter wanting a response to Jerry Brown's charges that her legal work may have involved her and her husband in conflicts of interest. Hillary began by denying that she had done work dealing with Madison Guaranty's relationship with the state of Arkansas—a statement that happened to be untrue. Exasperated, she also added, "I suppose I could have stayed home, baked cookies and had teas, but what I decided was to fulfill my profession, which I entered before my husband was in public life."

Most papers didn't mention Hillary's response about Madison Guaranty. Nor did they report her further comment that "this campaign is about issues that I've worked on for twenty years." Instead, they focused on the "I could have stayed home" phrase. Doubtless, Hillary never meant this as a slam on stay-at-home wives. More likely, she was thinking of her onetime nemesis, Gay White, a notorious cookie baker. (Indeed, Hillary held many politicians' wives in low esteem, including Kitty Dukakis, Lee Hart and Barbara Bush.) But the remark created an impression that could not be erased, even when the campaign countered by sponsoring a bake-off with Barbara Bush.

For what it was worth, Hillary, who liked to bake cookies at Christmastime, probably put in as much time in the kitchen as Mrs. Bush did. But cookies were beside the point. As William Safire observed, Hillary seemed unable to learn a key lesson of politics: "Self-definition by exclusion is needlessly defensive."

By spring, Hillary's image had become a divisive issue within the Clinton campaign. Polls showed that 67 percent of voters were cool to the idea of a president sharing power with his wife and about a quarter of all voters viscerally disliked Hillary. A campaign memo known as the "interim report" showed that focus groups saw Hillary as unaffectionate, unfeminine and power hungry. Among the Clinton campaign's suggested fixes were having "Bill and Chelsea surprise Hillary on Mother's Day" and arranging events in which Hillary could be seen relaxing with friends and "where Hillary can laugh, do her mimicry." This last phrase inadvertently hit on a problem: Hillary's humor consisted of making fun of people. Some of her bits were harmless. Others, while genuinely funny, like her take on a witness from a case many years before who said he had found a mouse's behind in a can of beans, had an edge that some might see as denigrating the common folk.

Hillary's sense of humor was destined to remain under wraps, but she responded to the polls by dropping the co-presidential "we" from her vocabulary and giving reassuring interviews to the women's magazines—*Redbook, Ladies' Home Journal* and *Working Woman.* The pundits, meanwhile, analyzed Hillary's image problem and concluded that the trouble was . . . her headbands. Hillary's black velvet and tortoise-shell fashion accessories were thought to make her look too schoolteacherish, or perhaps too matronly. When the Clintons arrived in Los Angeles to campaign for the California primary, Linda Bloodworth-Thomason invited her to attend a taping of the Thomason production *Hearts Afire.* Instead, the day turned into a makeover session with Chris Chally, who did costumes for *Designing Women,* and the hairdresser Cristophe. Both men were eventually dispatched to Arkansas for further consultations. Additional makeovers would follow after the election, with Hillary going through so many coiffures that they inspired a Hillary's Hair-do website.

Whatever one might say about Hillary Clinton, there was nothing wrong with her looks. She had given herself a makeover before the campaign started—lightening her hair, dieting, investing in a wardrobe of St. Johns knits in the bright, cheerful colors she enjoyed. Later efforts only succeeded in erasing the characteristics that reflected her individuality—her Southern matron's clothes sense, her preference for casual looks, her color choices. The remade Hillary looked like a TV anchorwoman, a generic blonde in a pastel power suit. Not surprisingly, the media pronounced her perfect and left her alone.

The appearance police are so relentless that one can hardly blame Hillary, or any public figure, for doing everything in her power to appease them. Still, one wonders, is this feminism? How can it be that Barbara Bush, the traditional homemaker, was so comfortable with herself, while Hillary, a forty-five-year-old Yale lawyer and policy wonk, could not brave criticism of her headbands?

IN a bizarre twist of fate, 1992, the year of suppressing "bimbo eruptions," also happened to be the Year of the Woman. In July, the Democratic National Convention devoted its entire Tuesday evening to a prime-time showcase for women candidates. The DNC also put out a special women-in-politics newsletter entitled *Getting It*—presumably no double entendre intended. A month later, in August, Hillary was the principal speaker at an American Bar Association luncheon in San Francisco honoring Anita Hill. In her remarks, Hillary saluted Hill's "courageous testimony" before the House Judiciary Committee, which had "transformed consciousness and changed history" on the issue of sexual harassment. "All women who care about equality of opportunity, about integrity and morality in the workplace, are in Professor Anita Hill's debt," she said. Hill, in her response, urged the attorneys present to educate colleagues on the issue and "look out for potential victims."

Surely, it must have crossed Hillary's mind as she listened to these words that her husband was a sexual harassment suit waiting to happen. But for several years—probably since 1989, when she decided to back Bill's run for President—Hillary made a deliberate decision to adopt the posture of a defense lawyer. Though she might know, in a general sense, that Bill Clinton fooled around, she had decided to take his word for it that he was innocent of specific allegations. To think otherwise would put her in an untenable position.

Hillary had no reason to feel anything but bitterness toward the likes of Gennifer Flowers and Connie Hamzy. Still, campaign aides who happened to overhear her and Bill discussing the "bitches" who were making his life miserable were shocked by the level of vitriol. Hillary had almost worked herself around to believing that Bill was the sexual harassment victim, beset by predatory females who were being used by his enemies to bring him down.

It was a lot easier to go on thinking this way, because she was not spending a great deal of time in her husband's company. On the campaign trail, she maintained her own schedule, her own staff and her own airplane. This distaff side of the campaign was known to aides as "HERC and the Girls," a play on Hillary's monogram, HRC.

Bill Clinton's behavior on airplanes had been notorious during his years as governor of Arkansas—one story had him being masturbated by a female companion while he ogled the shocked daughter of a state official, who happened to be sitting across the aisle from him. Such goings-on continued on his campaign plane *Longhorn One,* staffed at his request by attractive blond flight attendants. Clinton aide Bruce Lindsey did his best to play chaperon. Lindsey would quietly but firmly tell the attendants to avoid photographers and ignore Clinton's invitations to work out with him at the YMCA in Little Rock or switch to the same hotel he was staying in during campaign stopovers. But there was only so much he could do.

Christy Zercher, a regular crew member, later told tales of Clinton's lewd behavior: He dipped his finger in his tea and sucked on it, mimicking oral sex, and invited her to join him in the lavatory. Zercher also claimed that Clinton eagerly read the *Star*'s exclusive interview with Gennifer Flowers, calling Flowers's effusions over his sexual techniques "pretty accurate," and adding obliquely, "A lot of people make mistakes in marriage."

On the rare occasions when Hillary came aboard *Longhorn One,* she ignored the attendants, failing even to say hello. After one such flight, the women were called in to have the length of their skirts measured. During another, according to Zercher, Clinton waited until Hillary was asleep and snoring, then moved to the seat beside Zercher and began stroking her breast. Afraid of making a scene that would wake Hillary and cost Zercher her job, she let him.

According to Zercher, Clinton offered all three attendants jobs in the White House, though only one, Debra Schiff, accepted. Moreover, in July 1994, after the White House learned that she was getting calls from a *Washington Post* reporter, Zercher's house was burglarized at two A.M. while she lay asleep in an upstairs bedroom. The thief left behind jewelry, money and other valuables. Zercher later told the *Star* that the only items taken were her diary and most of the photographs she had from days on the crew of *Longhorn One*. She kept quiet until 1998, after the Lewinsky story broke, then got her revenge by selling her story to the tabloid.

The atmosphere of the "HERC and the Girls" side of the campaign could not have been more different, at least superficially. Brooke Shearer, an old friend of Hillary's who was traveling with her, recalled: "At seven in the morning, we would all gather in Hillary's room, and while she was on the phone talking to a local radio station, somebody was doing her hair, somebody else was ironing her dress, and a press person was dialing the next radio station. Sometimes at the end of the day, when Bill telephoned, we'd be laughing and doing imitations and carrying on."[17]

But it wasn't all lighthearted fun. Shearer had joined Hillary's staff after resigning her job as an operative for Investigative Group International, the private detective agency run by Terry Lenzner, who had presided over the Office of Legal Services during its radical heyday until he was fired by Richard Nixon. (Lenzner went on to become an assistant to Sam Dash on the Senate Watergate Committee and got his revenge when he was tapped to serve the President with a subpoena.) Hillary kept a certain distance from the work of Jack Palladino and the "bimbo patrol," though if she didn't know the details of their work, it was only because she chose not to ask. But through Brooke Shearer, who was married to Clinton's Rhodes scholar roommate, *Time* writer Strobe Talbott, she had a direct link to I.G.I.

During the campaign, Shearer's twin brother, Cody, a freelance journalist who had been a "subcontractor" for I.G.I. on at least one other case, was seen around the firm's offices, dipping into data banks and bragging indiscreetly about his conversations with the Clintons. According to Judy Bachrach, who wrote about the Lenzner-Clinton connection in *Vanity Fair,* Cody Shearer was looking for dirt on Republicans, trying to revive the old charge that Dan Quayle had purchased marijuana and the decade-old gossip that Bush had been involved in a romantic relationship with a State Department employee named Jennifer Fitzgerald.

Hillary brought up the "Jennifer with a J" question while being interviewed by Gail Sheehy for a magazine profile. Hillary told Sheehy that she'd been having tea with Anne Cox Chambers of the Cox newspaper chain when Chambers wondered aloud why the press got so excited about Gennifer Flowers when "everybody knows about George Bush." Her husband, she suggested, was being singled out because he was an outsider—"the establishment—regardless of party— sticks together. They're gonna circle the wagons." When Sheehy's article appeared, Hillary said she had thought the interview had been off the record, an excuse few believed.[18]

"He did it first" is a lame excuse that rarely works for people over the age of ten. The problem was that the Jennifer Fitzgerald story was tired gossip that had been investigated several times before but never substantiated; Fitzgerald denied the affair and George Bush didn't have any old girlfriends tattling on him in the pages of the *Star, Playboy* or other publications. Nevertheless, Hillary appeared to buy the argument that since others in this world may not be perfect, her husband should be exempt from any moral judgment. Murray Waas, the author of many an attack on "Clinton haters" in the on-line magazine *Salon,* has confirmed that Hillary continued to encourage journalists to pursue the Fitzgerald angle. Sidney Blumenthal, then with *The New Yorker,* eventually convinced a writer for the satirical magazine *Spy* to do a piece on the allegations. Blumenthal, writes Waas, "then publicly questioned the ethics of *Spy* for publishing the story."[19]

I.G.I.'s connection with the Clintons did not end with the 1992 campaign. Terry Lenzner later met in the White House with Clinton aide Harold Ickes, and I.G.I. did work for the Clinton legal defense fund and for the Clintons' personal lawyers at Williams and Connelly, investigating Paula Jones, prosecutors on Ken Starr's staff and the Republican lawyers Victoria Toensing and Joe diGenova, among others. Other accounts have Lenzner's operatives snooping into the backgrounds of Kathleen Willey, Monica Lewinsky and Linda Tripp. In a deposition connected with a suit by Larry Klayman's Judicial Watch, Lenzner claimed privilege, refusing to answer the question of whether Hillary Clinton had been a client.

Dick Morris has called I.G.I. the "White House secret police," and judging by the revolving door between I.G.I. and the administration this is not an exaggeration. Former I.G.I. operatives who found their way onto the government payroll include Brooke Shearer, who ran a White House intern program and then moved to a job in the Interior Department. Ricki Seidman, who reportedly worked on the Arkla-Arkoma matter, joined the Justice Department. Raymond Kelly, who left I.G.I. to head the New York City Police Department, became chief of the Secret Service. Even Terry Lenzner's daughter found a place in the West Wing, as a White House intern assigned to George Stephanopoulos. Meanwhile, Howard Shapiro, the FBI general counsel who presided during the Filegate scandal and gave the White House an advance peek at Gary Aldrich's manuscript for *Unlimited Access,* became Lenzner's attorney, and Larry Potts, censured by the FBI for mishandling the shootings at Ruby Ridge, left the Bureau to join I.G.I. as an executive.[20]

One can argue that sex should be a private matter, but keeping Bill Clinton's sex life secret had become a cottage industry that led the Clintons to bring private eyes into their administration and turn "opposition research" into a permanent part of their lives.

TEN

The Transition That Wasn't

I STILL cannot believe you won," Marc Tucker wrote to Hillary a few days after the election. "But utter delight that you did pervades all the circles in which I move. I met last Wednesday in David Rockefeller's office with him, John Sculley, Dave Barram, and David Heselkorn. It was a great celebration. Both John and David R. were more expansive than I have ever seen them—literally radiating happiness. My own view and theirs is that the country has seized its last chance."

Tucker's letter reflected the high hopes that he and the board members and supporters of the National Committee on Education and the Economy held for the Clinton administration. And with good reason. Not only had Bill Clinton endorsed their ideas, but "Hillary Rodham Clinton, First Lady of the United States" would continue to be listed in NCEE materials as co-chair for implementation of the organization's school-to-work program. Not surprisingly, therefore, Tucker's letter went on to report that he and a group of NCEE staffers had come up with a plan on how implementation should proceed. The remainder of Tucker's single-spaced eighteen-page letter laid out what he called "a very large leap forward in terms of how to implement the agenda on which you and we have all been working—a practical plan for putting all the major components of the system in place within four years—before Bill has to run again."

These ideas included:

- Universal early childhood education, fully financed for the poor and subsidized for the middle class.
- A reorganization of the education system to emphasize training for the work force of the future. The government would guarantee three years of schooling beyond the basic competence certificate awarded at about age sixteen. However, educational institutions receiving grants and loans under the new policy would be required to adhere to a "uniform format."
- A 2 percent levy on business to pay for the program. To overcome the resistance of the business community to mandatory participation in the training program, Tucker suggests, "Bill could gather a group of leading executives and business organization leaders, and tell them flat out" that if they don't spend 2 percent on training voluntarily he will submit legislation to Congress imposing a tax of that amount on their revenues.

A number of labels could be applied to this government-directed melding of the education and labor systems. One that would not be inappropriate, though doubtless it had never occurred to the people who thought up the NCEE plan, was corporate fascism—a partnership between government and big business to create a planned economy—with, of course, multiculturalism replacing the more traditional nationalism, racism and anti-Semitism as the unifying ideology.

Tucker expressed no doubts that Bill and Hillary were in essential agreement with his vision, though he recognized that they might not be prepared to move as quickly as he hoped they would. "Radical changes in attitudes, values and beliefs are required to move any combination of these agendas," he conceded, proposing that Clinton undertake a public relations campaign to sell the program. "After six months or so, when the public has warmed to the ideas," Clinton could then hold a summit of "experts" to formalize his agenda.

Along these lines, Tucker passed along a tip for making the apprenticeship component of the program more palatable. Focus group research, he noted, had established that vocational training is the last thing parents want for their children—"parents everywhere want their kids to go to college, not to be shunted aside into a non-college apprenticeship 'vocational' program." Tucker's solution was a name change that "reframes the Clinton apprenticeship proposal as a college program." In other words, the program would remain the same, but it would be called "college" in order to assuage public opinion. And, indeed, as President, Bill Clinton would drop his commitment to apprenticeships in favor of guaranteeing at least one year of college for all students.[1]

TUCKER'S letter reflected one side of the conflicting expectations that tugged at the Clintons as they prepared for their move into the White House. Certainly, the majority of Americans, who had just elected Bill Clinton in the belief that he was a middle-of-the-road, fiscally responsible Democrat, would be surprised to learn that any such dramatic increase in federal control over the nation's schools and workplaces was being contemplated. (They were, of course, wrong. The controversial Workforce Investment Partnership Act, incorporating the essence of the NCEE plan to link a federally regulated education system with data banks maintained by the Department of Labor, became law in August 1998.) In the short run, however, the Clintons were caught in a squeeze between their activist allies, whose expectations for bold action would be difficult to satisfy, and other segments of the Democratic Party, who had supported them in the belief that Clinton would build on the party's appeal to moderates.

Bill Clinton's problem was especially acute because the first constituency he had to satisfy was his wife. The Clintons had never resolved between themselves what Hillary's role would be in the new administration. Since learning she was barred by the antinepotism law from a cabinet post, Hillary had begun to think in terms of serving as her husband's chief of staff. Dick Morris was among those who did their best to explain to Hillary why this wouldn't work: a President had to be able to fire his chief of staff if the occasion arose, and it was difficult to imagine the First Lady being fired. Hillary's next suggestion was that she might serve as chief domestic policy adviser. Although this was a relatively low profile position, George Stephanopoulos and other key Clinton aides dreaded the prospect of the positive press coverage traditionally accorded to a new presidency being sacrificed to a national debate over Hillary's job title.

There were also likely to be practical consequences. Hillary's ties were to the activist community, and she tended to share their enthusiasm for bold, comprehensive programs. Bill Clinton may have been intrigued by the ideas Hillary advocated, but he was more realistic about what could be accomplished in one year, or even in four.

Much to the surprise of Democratic Party leaders, the Clintons decided not to move to Washington immediately. Instead, they would lay plans for the new administration from their home base in Little Rock, while Warren Christopher and Vernon Jordan, the official heads of the transition team, ran a small office in the nation's capital. The official explanation for this decision was that staying at home would be more economical, an excuse served up with a dash of anti-Washington paranoia. Although George Bush had vowed to do his part to make the change in administrations "the best transition ever," some of Clinton's people continued to suspect sabotage. In mid-November, when the Clintons traveled to Washington to attend a dinner hosted by Democratic fund-raiser Pamela Harriman, the transition organization asked to stay at Blair House and use a presidential jet for transportation. Informed of the costs—$1,500 a night per person, including food for Blair House, and $27,000 an hour for the jet—they issued a statement rejecting the terms and announcing that they had negotiated a special rate of $120 a night at the Hay Adams Hotel, as well as a bargain air charter. The press release implied that the Bush administration's cost figures, if not intentionally padded to soak the Democrats, were evidence of Republican profligacy.

In reality, the decision to remain in Little Rock reflected a sense that, as one friend of Hillary's put it, "some of these people were not

quite ready for prime time." At the former Clinton campaign headquarters in the fortresslike former *Arkansas Gazette* building, now the
official headquarters for the transition, the disarray was all too obvious.
No one had thought to supply the staff with wearable ID documents
or name tags. Offices were designated by makeshift signs scrawled on
sheets of 8-by-11-inch paper stuck to the doors with Scotch tape.
After a day or two, the sheets began to sag, rendering most of the
signs unreadable.

One visitor from out of state recalls arriving at the *Gazette* building
for the first time and stopping at the metal detector just beyond the
main entrance to empty his pockets of keys and other metal items. A
security guard seated nearby watched his efforts with benign amusement. "Don't worry," the guard advised, waving the visitor on in,
"that thang doesn't work anyway."

These were the visible symptoms of an organization that lacked
direction from the top. Mickey Kantor, originally slated to head the
transition effort, had been forced out immediately after the election,
dogged by press reports questioning whether he had used his relationship to the Clintons to attract clients to his law firm, Manatt,
Phelps, Phillips and Kantor. Kantor had not been popular inside the
campaign, and he had also run afoul of Susan Thomases, whose influence with Hillary made her a figure to be reckoned with.

With Kantor gone, the power vacuum was filled by George Stephanopoulos, James Carville and Paul Begala, aides who had done an
often brilliant job of developing a campaign strategy but who lacked
experience managing large organizations. Clinton's failure to appoint
a strong manager undoubtedly reflected a sense that Hillary, despite
her lack of an official title, remained his most important adviser. Had
the transition been run from Washington, Hillary's influence would
surely have been more visible and controversial. In Little Rock the
issue was muted, because the Clintons were conducting interviews
for cabinet-level positions at their kitchen table, an informal setting
where Hillary's active participation in the interviews attracted less attention.

Nevertheless, Hillary's influence was substantial, especially when it came to appointments affecting education, social programs and the Justice Department, areas of special interest to her. A strong transition chief with Washington experience would have recognized that the résumés of some Clinton appointees were going to be politically controversial. The Clintons, especially Hillary, appeared to believe that because the Cold War was history, the tensions that divided moderates, liberals and radical leftists no longer mattered. This would prove to be not quite the case.

One of the first and most controversial Clinton appointments was the choice of Johnnetta Cole, the president of Spelman College, in Atlanta, and a friend of Hillary and Marian Wright Edelman, to head the transition team's planning in the areas of education, the arts and the humanities—a post that simultaneously made Cole the likely choice for secretary of education. A former leader of the Venceremos Brigades, which recruited American students to cut sugarcane in Cuba, Cole had long been a Communist fellow traveler, serving as an executive board member of the U.S. Peace Council, a Soviet front, and former president of the U.S.–Grenada Friendship Society, which supported the Marxist regime of Grenadian president Maurice Bishop. Cole, for example, had joined the chorus condemning Joan Baez after the folksinger had the temerity to say that former antiwar activists should protest human rights violations by Communist Vietnam. Cole signed a petition, later published in *The New York Times,* condemning Baez and asserting that Vietnam "now enjoys human rights as it never has in history."

Cole's affiliations had not hurt her in academia; nor did they prevent her from serving on the boards of a number of our largest corporations, including Coca-Cola—after all, multinational corporations do business with Communist regimes when they can. It was another matter to imagine the U.S. Senate confirming a longtime defender of totalitarian regimes as secretary of education. More alarming than the appointment itself was the obvious surprise in Little Rock when an article on Cole's pro-Soviet views appeared in the *Forward,* a nationally circulated Jewish newspaper. Press secretary Dee Dee Myers initially dismissed the *Forward*'s criticisms as "silly," and the Reverend

Jesse Jackson opined that Cole was the target of "Jewish complaints." When commentators in other publications joined in questioning Cole's fitness, the Clintons simply allowed her to drop off their radar screen. As a result, Cole was subjected to embarrassment and her supporters in the civil rights establishment felt they had been cheated out of a cabinet appointment. Cole, meanwhile, proclaimed herself a victim of "right-wing extremists," but it was an unusual definition of right wing that could embrace publications like the *Forward,* which had endorsed Bill Clinton for the presidency.[2]

The naming of Lani Guinier, a friend of the Clintons from Yale, as chief of the civil rights division of the Justice Department would raise a similar flap, even though the controversy would not gather steam until a few months into the new administration. Guinier was a law professor with an impressive résumé, but her nomination would run into trouble when it was revealed that in her academic writings she had advocated a form of proportional representation. Guinier, whose ideas are taken seriously by her colleagues, would be shocked when *The Wall Street Journal* described her as a "quota queen." Not only was the label arguably offensive, but the *Journal* and other publications critical of Guinier at times appeared naive. Civil rights enforcement, as practiced under both Republican and Democratic administrations, has often taken the form of tinkering with the electoral system in order to produce results that are considered socially desirable. What Guinier refused to acknowledge, willfully or not, was that in the political arena proportional representation has a bad odor because it promotes, or at least institutionalizes, factionalism.

The best-remembered experiment with proportional representation in the United States took place in New York City between 1937 and 1947, when a plan designed to weaken the influence of clubhouse bosses enabled the American Labor Party, at the time a Communist front, to gain a bloc of seats on the city council, wielding power far out of proportion to its public support. Although it was rarely mentioned during the controversy over Lani Guinier's nomination, her father and personal hero, Ewart Guinier, had been an ALP functionary and candidate for Manhattan borough president during this period.

Once again, when criticism of Guinier's appointment mounted, Clinton backed away, declaring that he had only just got around to reading her articles and didn't agree with them. This raised the question of why Guinier had been appointed in the first place, and why her name was released to the public before she had been thoroughly vetted. No doubt the answer was that Hillary had wanted Guinier, though she, too, washed her hands of the nomination once it began to go sour. Guinier would later recall running into Hillary in a corridor at the White House. Hillary tried to brush past her with a chirpy "Hi, kiddo." When informed that Guinier had come to discuss strategy for getting her nomination through Congress, she muttered vaguely, "I'm thirty minutes late to a lunch," and bolted.

Like Cole, Guinier would attribute the unraveling of her nomination to a "right-wing" cabal, in this instance involving not only *The Wall Street Journal* but *The New Republic* and the American Jewish Congress, as well as Democratic senators Joe Biden and Patrick Leahy.

The Clintons' treatment of Lani Guinier has often been cited as an example of their lack of loyalty to old friends. But this criticism misses the point. Politics isn't about doling out rewards to one's friends. No doubt the nomination was handled clumsily, causing Guinier needless embarrassment, but she also violated political etiquette by refusing to withdraw quietly when it became obvious that she had become a liability.

If anything, the Clintons—and Hillary in particular—put too much stress on personal loyalty. Presidents-elect normally call on the FBI to run background checks on their appointees. In the past, the process of clearing cabinet-level nominees had often been completed by the time the new President took office. The Clintons relied on old friends like Webb Hubbell and Mickey Kantor to do their vetting for them. Hubbell's friendly questioning of Zoë Baird, a friend of Susan Thomases and Hillary's first choice for attorney general, did not uncover Baird's failure to pay Social Security taxes for her children's nanny, a problem reporters ferreted out even before the inauguration.

It was doubtless inevitable that Hillary would offer jobs in the administration to her close friends Vince Foster and Webb Hubbell. She decided to ask Foster to serve as assistant White House counsel under Bernie Nussbaum, her former colleague from the Watergate Judiciary staff. Neither Foster nor Nussbaum—nor, for that matter, Rose partner William Kennedy III, who also joined the counsel's office—had experience with the federal regulations and security issues that are a large part of the work of the counsel's office. They didn't understand how agencies of the executive branch, including the FBI and the Secret Service, normally function, and Foster and Kennedy weren't likely to learn a great deal about these matters as long as they remained in Little Rock. More to the point, perhaps, Foster's close relationship with Hillary had already been a topic of speculation among some reporters during the campaign. At this point, whether there was any substance to rumors of an affair—past or present—was almost beside the point. With Foster serving as the First Lady's personal legal adviser inside the White House, the chances that the gossip would become public at some point were increased, with devastating consequences for Hillary, Foster and their families.

Meanwhile, Webb Hubbell was placed at Justice, where he eventually became associate attorney general, officially the third most important position in the department. Hubbell's only obvious qualification for this post was his personal devotion to Hillary. Indeed, when the volatile Betsey Wright was nudged into retirement, Hubbell became the custodian of the "Betsey files," ten storage boxes containing the paper trail on various sensitive questions about the Clintons' past, as well as the remaining files on the Rose–Madison Guaranty relationship, which Vince Foster had obtained from Rose associate Rick Massey. (Hubbell would insist that he was never curious enough about the contents of these files to glance at the index prepared by Betsey Wright, a story somewhat contradicted by his statement that on one occasion he was asked to go through the boxes and locate a file on the tangled marital history of Clinton's father.)

When Zoë Baird withdrew her name from nomination, Stuart Gerson, an assistant attorney general in the civil division, became acting attorney general. But Gerson, a Republican, would be a nonperson as far as the White House was concerned, while Hubbell, the highest-ranking political appointee, became the de facto head of the Justice Department, a situation in which Hubbell appeared to be completely out of his depth. During this unsettled period, the position of FBI chief William Sessions was in limbo, creating confusion within the Bureau; the World Trade Center was bombed by terrorists; and the ATF attempted to raid the headquarters of the Branch Davidians in Waco, Texas, setting up an extended confrontation. Hubbell also signed off on the administration's decision to demand the resignations of U.S. attorneys across the county at once, instead of replacing them gradually with political appointees. While this was seen by some as cover for Clinton's attempt to place a friendlier face in the U.S. Attorney's Office in Little Rock, Hillary's eagerness to get more women in influential positions may have been the most important consideration.

Cronyism is, of course, in the eye of the beholder. New presidents naturally want to work with appointees they know and trust. Still, the Democrats had been out of office for twelve years, and the new administration was receiving up to 3,000 résumés a day from party supporters who were eager to go to work for the Clintons. Considering the size and breadth of the applicant pool, the number of appointees who were chosen because they were FOHs—Friends of Hillary—was striking. To mention just a few, Donna Shalala, who became secretary of health and human services, was a former chairman of the board of the Children's Defense Fund, sometimes known as the "queen of political correctness" for instituting a controversial hate-speech code at the University of Wisconsin. Maggie Williams, Hillary's campaign aide and soon-to-be chief of staff, had done publicity for the CDF. Sheldon Hackney, whose wife had served on the CDF board, was tapped to run the National Endowment for the Humanities. The post of secretary of labor went to the Clintons' old friend Robert Reich, who favored large-scale social investments, including a vast job-training program that would subsidize the income of unemployed Americans for up to two years at a cost of perhaps $50 billion annually. Key slots at Justice were eventually filled by Eleanor Acheson, Hillary's Wellesley classmate, who would have control over recommending candidates for the federal bench; former CDF fund-raiser Ann Bingaman; Yale professor Drew Days, an advocate of laws against gender-based discrimination; and Sheila Anthony, the sister of Vince Foster. Margaret Richardson, another FOH, became head of the Internal Revenue Service.

Meanwhile, moderate Democrats were getting short shrift. Even the stalwarts of the DLC, who had done so much to position Clinton for the presidency, were all but ignored. Interestingly, the one area where they had some clout was in economics. Clinton named Senator Lloyd Bentsen as his first secretary of the treasury; Robert Rubin, a whip-smart Wall Street investment banker and multimillionaire, became head of the National Economic Council; and California congressman Leon Panetta signed on as budget director.

Bill Clinton could often be indecisive. During the course of the campaign he had made so many conflicting promises that after the election, pollster Stanley Greenberg was dispatched to New Jersey to run a series of focus groups in the hope of figuring out which ones the voters remembered. Greenberg, one of the more liberal campaign aides, reported back that voters wanted economic stimulus to create more jobs as well as health care and welfare reform. Deficit reduction and the middle-class tax cut were far down on their list of concerns. Clinton was hearing a different story from his economic advisers, and his instincts told him they were right. Bob Woodward's *The Agenda* describes the moment when another adviser, Princeton economist Alan Blinder, explained to Clinton that winning the confidence of the bond market would be essential to keeping the country out of recession. Clinton was appalled but also impressed. The bond market was like a daily poll, delivering instant feedback on the state of his administration.

Clinton understood feedback, and he came to realize that his hopes of being a two-term president might hinge on his ability to pare down the national debt and keep interest rates low. At the same time, paring the deficit wasn't the sort of accomplishment that fulfilled his concept of political leadership. He wanted to do things for the people. Other, contradictory, goals had claims on his attention, including the idea of a stimulus package to create jobs and an ambitious social agenda in-cluding health care reform, which had emerged during the campaign as the number-one concern of the voters, even above education.

Like George Bush and all of his Democratic rivals in the primaries, Clinton had put forth his own answer to the health care dilemma, developed by the so-called Blueberry Donut group, named for the pastries they consumed during their meetings. The group, which in-cluded Bruce Reed, a former DLC policy analyst, and Judy Feder, who had been involved in health policy issues for many years, had come to favor a "pay or play" system. Employers would have a choice between providing private health insurance for their workers or pay-ing into a national insurance fund. The trouble with this plan—and, in fact, every plan designed to extend health insurance to low-income Americans—was that it would cost money, and lots of it.

Enter the magician, Ira Magaziner. When not working with Hillary to advance the NCEE's school-to-work agenda, Magaziner had been preparing a health care reform plan for the state of Rhode Island, and he had become an enthusiast for managed competition—a system that would allow purchasing cooperatives representing large pools of in-dividuals to buy health coverage at the best available prices. Advocated during the primaries by Paul Tsongas, managed competition basically relied on competitive market forces to keep the costs of health care manageable. Magaziner, however, came to favor a variant of the sys-tem that would include government-mandated benefits and spending caps.

Once the government began to dictate the size of the health care budget, the whole character of managed competition changed, and the system began to resemble the single-payer plans in effect in Europe and Canada. But Magaziner made a compelling case. He figured that under his plan it would be possible to cover the 37 million Americans who were without health insurance without new taxes. In fact, by 1997 the government could actually be saving several billion dollars a year.

Magaziner and the Clintons went back a long way. He was a big idea man, and as Bill and Hillary surely knew, he had been confidently but spectacularly wrong on a number of occasions. As an undergraduate activist, Magaziner had been responsible for deconstructing the curriculum of Brown University. Later, after attending Oxford with Bill Clinton, he became part of a group that set out to turn Brockton, Massachusetts, into a model social democracy. By the time the plan was abandoned two years later, Brockton's last two shoe factories had been driven out of business and the locals were bitterly hostile. Magaziner then went on to become a management consultant and a convert to the merits of industrial policy—a "third way" between socialism and capitalism. His company, Telesis, commanded high fees as an adviser to corporations and foreign governments, though some of his biggest contracts went awry. General Electric lost half a million dollars when a particular type of compressor Magaziner recommended turned out to be defective. And when scientists in Colorado reported a successful demonstration of "cold fusion"—nuclear fusion at room temperature—Magaziner testified on their behalf before a congressional committee, urging the government to invest $25 million in research "for the sake of my children and America's next generation," lest the Japanese corner the market first. Cold fusion, however, turned out to be a figment of the scientists' imagination.[3]

Magaziner's cost projections were based on the assumption that capitalism was inherently wasteful and experts like himself could squeeze considerable cost savings out of the health care system without lowering standards of care. Clinton's economic advisers had no faith in his figures, nor did health care analysts like Judy Feder and Bruce Reed, who had devoted years to studying the problem. Magaziner infuriated members of the Blueberry Donut group by telling them, "It's you people in Washington who created this problem." And he talked blithely of moving the entire population into HMOs within a year.[4] Nevertheless, he appeared to offer the Clintons a way to combine fiscal discipline with an ambitious social agenda.

Bill Clinton, meanwhile, could solve the problem of Hillary's role in the new administration by naming her to head the health care effort, with Magaziner as her chief deputy. Hillary didn't necessarily want to deal with health care; she was more interested in day care and education. But Clinton had always trusted her to handle delicate assignments, like working out Little Rock's school desegregation plan. After years of handling political chores in Arkansas, Hillary wanted official recognition and a chance to play a substantive policy role. Clinton was willing to let her have it, but at the price of taking on a difficult and risky assignment.

That Bill Clinton would even consider making his wife responsible for health care reform raised questions about their relationship, which was already a mystery to aides who had observed them during the campaign. One rumor had it that the Clintons had worked out a formal deal through the mediation of Washington attorney Lloyd Cutler, agreeing that Hillary would be given power over domestic policy in exchange for standing by her husband throughout the campaign. The truth was that their marriage was both more traditional and more volatile.

Hillary saw her husband as a brilliant but psychologically vulnerable man, capable of accomplishing great things but in need of her protection. In essence, she was doing what politicians' wives—and, perhaps, wives in general—have always done. It was just that Bill Clinton needed a lot more rescuing than most men. If Hillary sometimes made snap decisions, one had to understand that she was married to a man who saw so many sides to every issue that he often found it difficult to settle on a course of action. If Hillary seemed paranoid at times, and single-minded about needing to be surrounded by old friends who were personally loyal, it was because she knew that her husband had a lot of secrets to protect, far more than most of his top aides suspected. She had helped Bill get elected, and the sense that his presidency might be undone at any moment by some unforeseen scandal had to prey on her mind.

With the move to Washington fast approaching, it was apparent that Clinton hadn't yet committed himself to changing his personal habits. Clearly savoring his last days of relative freedom, he enjoyed driving around town, creating traffic jams with his Secret Service motorcade and playing golf at the Country Club of Little Rock. He was also still having clandestine meetings with female friends. One woman, an Arkansas utilities company executive named Marilyn Jo Jenkins, showed up at the Governor's Mansion on at least four occasions, including once at 5:15 A.M., dressed in an oversize raincoat and a scarf that concealed her face. She and Clinton disappeared into the recreation room area of the mansion basement, while a state trooper stood guard, prepared to give the alarm if Hillary or Chelsea stirred from their upstairs bedrooms. Jenkins later denied in an affidavit that anything sexual had occurred on these occasions. Clinton, who was not required to characterize his relationship with Jenkins, would say in a sworn deposition that he had merely met with Jenkins to give her going-away presents he had purchased for her and her children.

The President-elect's assignations were just one of several reasons why he and Hillary were having serious problems with their Secret Service detail. Unlike the Arkansas troopers who had guarded Bill and Hillary for more than a decade, Secret Service agents were trained to avoid any activities that might distract them from their duty as bodyguards. They didn't make small talk, rate women on a one-to-ten scale, fetch errant balls on the golf course or carry luggage. The Clintons suspected—correctly—that the agents assigned to their detail disliked them. There was some dispute, however, as to which Clinton was the more difficult. Hillary moved in circles where an infantry sergeant's vocabulary was still considered a sign of liberated womanhood. One agent, who politely explained to Mrs. Clinton that his duties did not include toting suitcases from their airplane to their limo, was shocked when she replied, "If you want to remain on this detail, get your fucking ass over here and grab those bags."

Relations between the Clintons and their Secret Service escort deteriorated so badly that some members of the detail groused to journalists about their treatment, both on and off the record, an unheard-of lapse of confidentiality. One former agent interviewed by *Capitol Hill Blue*'s Doug Thompson complained that Clinton would greet female voters with a display of boyish charm, then turn away and speculate under his breath about whether they gave good blow jobs or not. "He was used to swapping dirty jokes and using Arkansas

State Troopers as his personal servants. He learned real quick that Secret Service agents don't play that game."[5]

The Clintons' raucous arguments also alarmed their Secret Service detail. Retired agent William Bell told of an argument that took place in the Clintons' limo. At one point, Hillary, who was in the backseat, became so infuriated that she threw a briefing book at Bill, who was in the front seat, hitting a Secret Service agent by mistake.[6]

Whatever this argument may have been about, the Clintons continued to squabble about Hillary's place in the administration through December and into January, until eventually she was persuaded to assume the leading role in health care reform, with Magaziner as her chief deputy.

So as the price of having her own area of responsibility, Hillary was taking over another thankless job, charged with producing a health care reform bill that would also save the government money over four years. Hillary was the ideal candidate for the job, because she appeared to have faith in Magaziner's projections, which others found fantastic.

In the meantime, right up to the moment the Clintons departed for the inauguration, Hillary was faced with reminders of her husband's infidelities. On the day of the flight to Washington, state trooper Larry Patterson was assigned by Clinton to pick up one of his former lovers and bring her to the airport, where a group of friends had gathered to see the Clintons off. Hillary spotted the woman as soon as she arrived, and threw a fit. "What the fuck do you think you're doing?" she screamed at Patterson. "I know who that whore is. Get her out of here." Clinton stood by, saying nothing, and Patterson did as he was told.

ELEVEN

Reclaiming America

No one could say for sure what the theme was for Bill Clinton's first inaugural. Of several mentioned by the organizers, the one heard most often was "Reclaiming America"—a phrase that rankled some of our capital's civil servants, suggesting as it did that the government of the previous twelve years had been not quite legitimate. The phrase did have a populist ring to it, however, and considering how often Clinton had denounced lobbyists and special interests, many were expecting that he would celebrate a "people's inaugural" like Jimmy Carter's, at which tickets to the inaugural balls went for a flat twenty-five-dollar fee. Instead, the Clintons had chosen to go Hollywood, with a week of celebrations "produced" by their friend Harry Thomason. Inaugural ball tickets were selling for up to $5,000, but the high point of the festivities was the star-studded inauguration eve gala for 18,000 held at the Capital Center, featuring Warren Beatty, Aretha Franklin, En Vogue and Barbra Streisand. Resplendent in a low-cut white Donna Karan dress, Streisand sat on a stool and serenaded the Clintons with Stephen Sondheim's "Children Will Listen," a warning that whatever adults do or say, they are setting an example for the young.

Already there had been a mini–fund-raising scandal involving the first brothers-in-law, known to some Democratic Party activists as "the boys." A lawyer representing Tony Rodham and Hugh Rodham, Jr., had called up Ford, Chevron, Mobil and other major corporations, asking for $10,000 donations "on behalf of the family" of the incoming First Lady. Hitting up corporations for money was not exactly unheard of in Washington, but this was a new twist. Hugh Rodham, however, was amazed by all the fuss. "We obviously need to pay for it somehow. I don't see it as a conflict at all," he told *The Wall Street Journal*. "We're just small fries, that's all we are." Rodham then conceded that he hadn't informed the Clinton-Gore campaign of his project, because the events were to be "personal thank-yous. . . . The less people know about it, the better."[1]

Tony Rodham, heretofore a private detective, soon turned up on the payroll of the Democratic National Committee as an "outreach coordinator," earning a salary of up to $70,000 a year for attending fund-raisers and playing golf with Democratic donors. Hugh junior, a public defender with no political experience, would become a candidate for the Senate in 1994, brushing off suggestions that he was trading on his name by accusing the incumbent, Connie Mack, Jr., of doing the same.

BILL and Hillary Clinton had been aiming toward the White House all their adult lives, yet now that their day of victory had arrived they did not appear to be ready for the spotlight. On the morning of Inauguration Day, the Clintons were half an hour late departing for the customary pre-inaugural courtesy call on the outgoing President and First Lady. TV cameras outside Blair House caught a sharp exchange as Hillary joined her obviously impatient husband. A shocked park police guard later reported that Clinton had referred to his wife as a "fucking bitch," while she came charging out the front door calling him a "stupid motherfucker." Much to the surprise of George and Barbara Bush, when the Clintons showed up, they had two uninvited guests in tow, Harry Thomason and his wife, Linda Bloodworth-Thomason. The Bloodworth-Thomasons were scheduled to spend the night in the Lincoln Bedroom, and after dinner they invited their friends and fellow White House guests, actress Markie Post and her husband, in for a look around. While a jubilant Linda and Markie jumped up and down on the bed, shouting, "We won! We won!" Harry snapped their picture. The photo, eventually leaked to the *National Enquirer,* goes far to explain why many permanent employees of the White House were in a state of shock.

From day one of the Clinton administration, it was apparent that the buttoned-down conservatism of the Reagan-Bush years was over. Mindful that anything they did might end up as grist for an investigation by Congress or a special prosecutor, White House employees had disciplined themselves to communicate through official channels and keep their memos short and businesslike. Meetings followed preset agendas. The dress code was strict, its tone set by Reagan, who thought it disrespectful to remove his jacket in the Oval Office.

The Clintons consciously set out to liberate the West Wing and the Old Executive Office Building, introducing informality, diversity and fun. Political appointees, mostly former campaign volunteers in their twenties and thirties, showed up for work in casual attire. Men wore unconstructed jackets and earrings, or sometimes even T-shirts and jeans; women came to work in miniskirts, pants, bare midriff blouses and—in at least one case, as we now know—thong underwear. They brought with them their personal coffeemakers, radios and take-out food. They referred to the White House complex as "the campus" and treated it like one. George Stephanopoulos chewed bubble gum during press conferences. Annie Leibovitz showed up to photograph the President and was allowed to set up a boom box in the Oval Office, where she worked to the accompaniment of Eric Clapton.

As the Clintonites saw it, their arrival marked the victory of youth, enthusiasm and high IQs over the old fogies. Techno-conscious young staffers were shocked to learn that their predecessors had used outdated personal computers and even on occasion an obsolete machine known as a typewriter. Scornful of Microsoft, they were soon filling out requisitions for Mac Powerbooks, which had a way of disappearing from "the campus" almost as soon as they arrived. "The

Republicans said they wanted to run things like a business," said Paul Begala. "Instead, they were running it like Bartleby the Scrivener."[2]

With the Clintons' approval, White House operations were being reorganized along nonhierarchical lines. As *Time* magazine put it, the new administration believed that "the best way to solve any problem is to assemble the smartest people in a big room and pull a lot of all-nighters." One Clinton aide confided that it was hard to apply an up-to-date managerial style in the West Wing, which had been designed for "fat old men and their secretaries." *Time*'s Michael Duffy was able to find at least one holdover from the Bush years who professed to be enjoying the new regime: "Before, I always felt afraid that I would do something wrong. Now, anything goes."[3]

In some ways, the change was refreshing. Minorities, women and the handicapped were well represented in the new administration. On the other hand, since no one had gotten around to vetting résumés, there were also petty thieves, drug users and individuals who were simply too inept and/or indiscreet to be working on matters of national importance. One Clinton volunteer was arrested on the streets of Washington for possession of crack during inauguration week.

The Clinton transition team never had gotten around to filling out its organization chart, so many volunteers from the campaign were told to just show up and lend a hand until they could be given formal work assignments. Other staffers were working at the White House on political issues while on the payroll of the Democratic National Committee. Even among top-level aides it was often difficult to tell who was officially on staff and who wasn't. Political strategist and speechwriter Paul Begala did not want to give up his private clients and file disclosure forms, so he was sitting in on meetings as an unpaid adviser. Harry Thomason, the President's unpaid "producer," had a small office in the East Wing. Susan Thomases had no official role in the administration, but she held an unrestricted pass to the official residence and was not shy about telling people how to do their jobs; Hillary's staff nicknamed her the "Midnight Caller."

The chaotic first year of the Clinton presidency has been docu-mented from different points of view in bestsellers from Bob Woodward's *The Agenda* to Gary Aldrich's *Unlimited Access*. What is obvious only in hindsight is the extent to which this situation stemmed directly from the tensions in the Clinton marriage. On some fundamental level, Bill and Hillary undoubtedly were in love and always had been. They shared a long history together, valued their roles as parents of Chelsea and each understood the other better than anyone else ever would. But the fact was, they often didn't get along. Their disagreements, at this point, probably had less to do with infidelity—an old story—than with Hillary's desire not to be humiliated in public and her impatience that Clinton often sided with fiscal conservatives like Robert Rubin at the expense of the social agenda she cared about.

At any rate, it was no secret to the press corps that a number of women working in the White House appeared to be there because they were past, present or perhaps future girlfriends of the President. The most formidable was Marsha Scott, the daughter of Miss Arkansas 1945 and Clyde "Smackover" Scott, who played football for the Philadelphia Eagles. A longtime resident of California, where she worked as a Head Start teacher and interior decorator, Scott was known to be a close confidante of Bill Clinton and once described herself in an interview with a hometown paper as his "hippie girlfriend" from the sixties. (A man calling himself Scott's ex-boyfriend later told the *National Enquirer* that Scott joked that she had sex with Bill Clinton once a year when she was in Arkansas for the holidays, in lieu of a Christmas present.) Catherine Cornelius, an attractive blonde in her early twenties who had handled the Clinton campaign account for her travel agency, seemed to enjoy unusual access to the Oval Office; and Debra Schiff, formerly a flight attendant on Clinton's campaign plane, later turned up as a West Wing receptionist, a post normally reserved for candidates with an extensive background in foreign languages and protocol. Robin Dickey, a former administrator of the Governor's Mansion in Little Rock and the mother of Helen Dickey, who was staying at the White House as Chelsea Clinton's live-in nanny and companion, drew attention because of her habit of giving the President back rubs. The press also gossiped about Clinton's friendly relationship with Regina Hopper Blakely, Miss Arkansas of 1983 and later the hostess of the Miss Arkansas pageant. Formerly a lawyer with the politically connected Little Rock firm of Arnold, Grobmyer & Haley and then a reporter for KHTV, the local CBS affiliate, Blakely joined the network in 1991 and was assigned to the White House shortly before Clinton took office.

The President's womanizing was just one of the new administration's ill-kept secrets. Hillary Clinton's active role was another. Although she had lost her bid to have her importance acknowledged with a suitable title, Hillary had established a firm power base for herself. Carol Rasco, who got the title of chief domestic policy adviser, had been Hillary's aide back in Little Rock. Meanwhile, Hillary's chief of staff, Maggie Williams, was also named special assistant to the President, a post that ensured that she would sit in on high-level meetings and see key memos. Hillary, Williams and Rasco all had offices in the crowded West Wing, occupying space on the second floor, next to Bernie Nussbaum and Vince Foster. The rest of Hillary's staff, all women except for press aide Neel Lattimore, was installed not in the East Wing, the traditional realm of First Ladies, but in a suite in the Old Executive Office Building, which quickly came to be known as "Hillaryland."

Operating without portfolio, Hillary interviewed potential hires and oversaw the President's schedule. She attended staff meetings, where she often played the role of the closer, directing the discussion and summarizing opposing points of view. When E. D. Hirsch and Diane Ravitch were invited to the White House to present their ideas on educational policy to the President, they found Hillary sitting in on the session. Even though his visitors represented views at odds with his own, Clinton turned on the charm. "He has got a kind of sex appeal that can get anyone to do what he wants," conceded Ravitch. But it was Hillary who played the tough cop role, cutting off a discussion of whether classroom computers should be a top priority by saying, "Chelsea is on the Internet, and I want every student to have the same opportunity."[4]

A presidential aide speaking off the record told Washington reporter Elizabeth Drew that Hillary "has in many cases served functionally the way a Chief of Staff would in terms of accountability and discipline and sometimes unruly organization. She has made the point openly in his presence. What she does privately I can only imagine."[5]

No one doubted that Hillary had her husband's best interests at heart, or that she often tried to counter his tendency to transform meetings into open-ended policy seminars. But Hillary's position was in itself a disruptive force. The nominal chief of staff, the Clintons' old friend Thomas "Mack" McLarty, was not about to get into a turf battle with the First Lady and thus came to be seen as weak and disorganized. Donna Shalala, a former CDF board president and Hillary's handpicked choice as secretary of health and human services, would soon find herself upstaged; her staff did much of the support work while Hillary made policy and took over the spotlight, at least as long as the publicity was good. Some of the old Arkansas hands, a faction that included Marsha Scott and Bruce Lindsey, made it known that they were not fans of Hillary, but aides who lacked a long history with the Clintons were reluctant to criticize anything she did, for fear of putting themselves in the middle of a marriage whose dynamics they did not understand.

Nevertheless, it was understood that Hillary had been involved in several troublesome decisions. Clinton's order lifting the ban on gays in the military, announced on January 29, ran afoul of the Joint Chiefs of Staff and put him publicly at odds with the very popular General Colin Powell. Zoë Baird's nomination for attorney general had quickly run into trouble when it turned out that she had not paid Social Security taxes for her children's nanny, despite a household income in the neighborhood of half a million a year. Kimba Wood, the Clintons' second choice for the job, turned out to have a similar problem. The selection of a third woman, Janet Reno, would make it obvious that the post of chief law enforcement officer in the land had been reduced to an affirmative action slot. As the chief prosecutor of Dade County, Florida, Reno had built her reputation by pursuing child abuse cases; when the "believe the children" movement peaked, the cases began to unravel, amid charges that Reno aides had used leading questions to elicit lurid charges from impressionable preschoolers.

Despite unmistakable clues that the Clintons took the idea of a co-presidency seriously—such as the prominent display of Hillary's portrait instead of Al Gore's—the position of the Clinton White House was that Hillary was not wielding power. Her active role, which was controversial but not illegal, became another ill-kept secret.

Our modern Presidents and First Ladies live inside a twenty-four-hour security bubble, as confining and almost as artificial as the environment in the movie *The Truman Show*. One might think that the Clintons, after years in the Little Rock Governor's Mansion, would be used to the semi-public life of officeholders. But Bill Clinton hadn't been President for three hours when he was overheard calling the White House "the crown jewel of the federal prison system." As for Hillary, in Arkansas she had been able to shrug off her trooper escort more or less at will, and she hoped to enjoy at least some of the same freedom in Washington. On several occasions, she managed to slip out and walk the streets unrecognized in the company of Diane Blair, a frequent overnight guest at the White House. And once, during a visit to Little Rock, Hillary shook off her Secret Service escort and drove around town for over an hour, later boasting of her success in her weekly newspaper column. Understandable as the impulse to escape into anonymity may have been, these incidents did not endear her to her Secret Service detail, which of course would have taken the blame if anything had happened to her.

Even inside the West Wing offices, Hillary's attempts to be accepted as just another colleague went awry. She used the regular staff bathrooms, to the consternation of female employees who didn't feel comfortable meeting the First Lady of the United States emerging from the stall next to theirs. Unlike her husband, Hillary was not naturally gregarious, and her patience for chatting with strangers who just wanted a few seconds with the First Lady was limited. What she really wanted was to go about her business without attracting attention, which of course was impossible, giving rise to a legend that Hillary expected staffers to hide in their offices when she passed by. White House employees spoke nervously of Hillary's Alzheimer's—a condition that caused the First Lady to erase certain people who had offended her from her memory bank, rendering them permanently nonexistent.

Hillary was by no means the witch portrayed in some hostile accounts of the Clinton presidency. A loyal friend, she made time in her overcrowded schedule to keep in touch with women she had known in grade school, in college and at Yale. She played mother hen to the Hillaryland staff, complimenting their taste in clothes, asking solicitously about their love life and helping to plan their weddings, and her aides repaid her personal interest in them with zealous loyalty. But unlike her husband, who had to go out on the campaign trail and make his case to all the voters, Hillary spent her time with people who shared her activist views. She and her friends took pride in caring passionately about children, the poor and the environment. Republicans were dismissed as people who didn't care.

L. D. Brown tells the story of escorting the Clintons on a trip to the capital during the mid-1980s. As they were driving through the streets of Washington, Chelsea piped up, wanting to know if they could take a tour of the White House. Of course not, Hillary snapped. They would have to wait to visit until "someone decent lived there." Now the Clintons were living in the White House, but they were still surrounded by career employees who had served under the Reagans and the Bushes and for the most part remembered them quite fondly.

Within days of the inauguration, two anecdotes would find their way to the press about the Clintons' dissatisfaction with the White House stewards who attend to their needs in the residence. Hillary told one interviewer about a steward who appeared out of nowhere, hovering over her as she tried to whip up some scrambled eggs for Chelsea, who was ill with the flu. Another story that got around had it that on the morning after the inauguration, a steward had blundered into the first couple's bedroom, catching them still in bed. Asked if it was true, Clinton said yes, adding that the incident was "not as lusty as it sounds."

With hindsight, when the Clintons volunteered glimpses of their private life, there was usually a reason. The information that a steward had surprised the first couple in bed together answered a question that many had wondered about but no journalist would have dared to pose. In the past, some first couples have maintained separate bedrooms, while others have not. Despite the temptation to read a lot into these arrangements, they may say more about a long-married couple's sleeping habits than their sex lives. The Clintons, however, had managed to come up with a new twist. Officially, they shared a bedroom, though according to sources in a position to know, the President was often exiled from it.

Though the Clinton's sex life is their own business, this hasn't prevented their friends, including Hillary's former spokesperson, Lisa Caputo, from calling attention to their frequent public displays of affection as evidence of a loving marriage. Similarly, Don Senich, the manager of the Washington restaurant where Bill and Hillary dined on their first Valentine's Day after the election, commented that the Clintons must have touched each other more in two hours than the Bushes did in four years. In fact, we don't know what these huggy-kissy moments really signify. Married couples who touch each other a lot in public may just be naturally demonstrative, but they may also be taking advantage of a protected situation to express feelings that get tangled up with negative emotions and issues when they are alone, behind closed doors.[6]

The tale of the blundering steward helped lay the groundwork for Hillary's decision to ban domestic staff from the residential quarters while members of the first family were at home—a rule that made it difficult for workers to complete their tasks without overtime. The Secret Service, which had formerly maintained a small post inside the private quarters, was also banished. Hillary explained the new rules as an effort to protect Chelsea's privacy. This made sense, though Chelsea had always lived in a semi-public house with servants, and concern for her privacy didn't keep the Clintons from entertaining 948 overnight guests during their first four years in office, including 128 domestic and foreign dignitaries, 370 friends from Arkansas and scores of major donors, many of them total strangers to the Clintons or, as the White House press office would phrase it, people who were "not friends yet." Chelsea, meanwhile, entertained 72 sleepover guests of her own.

Hillary's concern that domestic staff and the Secret Service might be spying on her and the family escalated soon after the election, when a Chicago newspaper columnist wrote that during a quarrel in the residence she had thrown a lamp at the President. This sounds too much like a recycled version of the briefing book incident to be believed, though some who claim to know insist that it actually happened. At any rate, Hillary gave the story credibility by denying an interview to *Newsweek,* which had printed a brief item about it. She also spoke to Vince Foster about her concerns that Secret Service agents were the source of the leak. (If the story wasn't so, then there was no telling how it got started, and no reason to think that anyone in the White House was the source.)

It was true that some in the Secret Service, not necessarily in the private security details, did harbor a visceral dislike for the Clintons. They saw the President as a draft dodger, prone to childish displays of temper. A few had made up their minds that Hillary was a lesbian, or speculated that Bill Clinton was a victim of domestic abuse. (The same thought occurred to posters on some Internet sites, who swap photos of the President's occasional bruises including, in 1998, a nasty lump on the forehead, which he attributed to "walking into a door.")

Hillary had reason to be concerned that gossip about her and Bill's quarrels was being leaked to the press, but she was too quick to suspect a political motive. This mentality had rubbed off on Vince Foster, who confided to his wife, Lisa, that departing Bush aides had stripped their offices bare, taking with them important files and even their desk chairs. Foster's complaint revealed a profound misunderstanding of the nature of the transition process. Outgoing Presidents always take their files with them. As for the missing chairs, presumably some aides had provided their own, more comfortable seating or requisitioned chairs from other departments. The notion of a Bushite conspiracy to deprive the Democrats of desk chairs was ludicrous.

In the meantime, the Clintonites were letting go of holdover employees, the people who constituted the institutional memory of the

White House and knew how to keep their departments running smoothly from one administration to the next. Ostensibly, the firings were necessary because Clinton had promised during the campaign to cut White House personnel by 25 percent. Clinton had already shelved other promises, like the middle-class tax cut, however, so it wasn't clear why this one had to be honored. Marsha Scott, placed in charge of the White House correspondence office, immediately fired two dozen employees, some of them women close to retirement age. In a sworn deposition before the House Committee on Government Reform and Oversight, Scott would later explain that the firings were necessary to make the office more efficient: "Unlike our predecessor we had a policy of answering every letter." In fact, the opposite was true. It was the Clintons who were bad about answering their mail. (In a completely unscientific test, Socks the cat proved to be the member of the first family most likely to respond to a letter.) The dumping of up to one million letters has been widely reported.

The Secret Service posed a special problem for the Clintons, because its uniformed employees administered the White House security system. Clinton appointees were resisting filling out their SF-86 background questionnaires for processing by the FBI's Special Investigations Unit—ironically, known as SPIN—and the White House counsel's office wasn't insisting that they cooperate. As a result, by September 20, 1993, only two permanent White House passes would be issued and many employees had not passed even a preliminary screening process.

It was no secret that many Clinton appointees, including top-level aides, had problem backgrounds that would have disqualified them from working in previous administrations. One could make the argument that some of the old standards were obsolete. Gary Aldrich, one of the FBI agents detailed to the SPIN unit, considered anyone who had smoked marijuana fifty-two times in his life to be a drug addict. Aldrich's partner, Dennis Sculimbrene, had never heard of designer drugs until he began interviewing Clinton appointees. In the long run, however, it didn't matter what Aldrich and Sculimbrene thought. The President, acting through the counsel's office, had the authority to decide who was fit to work for him. But Clinton either didn't know what his standards were or preferred not to say. For that matter, the counsel's office wasn't sure that the administration had a right to impose standards at all. In a June 10, 1993, memo, assistant counsel William H. Kennedy III would argue that under the equal employment laws drug abuse was a disability; therefore, even current drug abusers might be said to have a right to work in the White House. The memo revealed how little serious thought the Clintonites had given to security issues. Five months into the new administration, a White House lawyer still wasn't sure whether a crack addict could be excluded from the West Wing.[7]

Meanwhile, Foster and David Watkins had been dispatched to confer with Secret Service head John McGaw about Hillary's fear that his agents were gossiping about the first family. Returning from the meeting, the two men encountered an unfamiliar uniformed Secret Service employee at the White House gate, who insisted on patting them down. They wondered if McGaw suspected that one of them had worn a wire to the meeting; however, whether the incident was a setup or not, it did convey a message, since both Foster and Watkins were still using temporary passes.

McGaw, suspected of being too friendly with George Bush, would soon be removed, transferred to head the Bureau of Alcohol, Tobacco and Firearms. And, amazingly, William Kennedy's reasoning on the right to White House employment would prevail. Twenty-one White House employees, including several top aides, eventually took and failed drug tests but were allowed to continue in their jobs as part of an in-house program that required them to pass "non-random" urine screenings twice a year. The program was supervised by Patsy Thomasson, a former Arkansas Democratic functionary and aide to Dan Lasater. For one reason or another, over 100 Clinton appointees were initially denied permanent passes by the Secret Service but approved by the counsel's office. Among this group were several key Clinton advisers, including Robert O. Boorstin and Ira Magaziner.

★   ★   ★

ON January 25, Clinton announced his appointment of Hillary to head his National Task Force on Health Care Reform. Clinton called his wife "a first lady of many talents," adding, "She's better at organizing people from a complex beginning to a certain end than anybody I've ever worked with in my life." The task force was to present its report within 100 days.

Ensuring access to health care insurance is a problem that has bedeviled American Presidents since Harry Truman. For Hillary, however, it was to be just one more item on her ambitious agenda. As the First Lady's deputy chief of staff, Melanne Verveer, would tell *Washington Post* reporter Martha Sherrill, "Health care is just a short term objective."

Even before the task force began its work, Ira Magaziner's managed competition solution had begun morphing into a vast new entitlement program involving the mandatory participation of all Americans, government price controls and the creation of a National Health Board, which would determine exactly what medical procedures patients were entitled to and how much those procedures would cost. The 100-day deadline, much ridiculed at the time, was actually the smartest part of the Clinton strategy. The only real hope for the plan, and it was not a great hope, was to build a groundswell of public support and muscle a bill through Congress before the opposition got organized. Even within the administration, the Clinton-Magaziner plan had almost no support. Prominent critics included Health and Human Services Secretary Donna Shalala, Alice Rivlin of the Office of Management and Budget, Treasury Secretary Lloyd Bentsen, and in fact almost everyone with expertise in economics. This was not a good sign. Daniel Patrick Moynihan, the influential senator from New York, and other Democrats with clout on health issues were also skeptical. Most important, the President was increasingly focused on getting a deficit reduction budget through Congress and listening to advisers who warned against coupling the budget bill with health reform.

Communications director George Stephanopoulos didn't even want public discussion of the health task force until after the budget issue was resolved. Against the advice of Magaziner, Hillary readily agreed to keep the task force deliberations secret. Much to the frustration of the press, the first interview she granted after being named health care czar was with the *New York Times* food writer, Marian Burros. A photograph that appeared with Burros's article showed Hillary examining table settings for a forthcoming dinner honoring the nation's governors.

Ira Magaziner suffered from a severe case of the "let's pull an all-nighter" syndrome that afflicted the Clinton administration. Under his direction, the twelve-member task force expanded to include more than 630 individuals assigned to twelve "cluster teams" and thirty-eight "sub groups." Magaziner had broken down the reform plan into 1,100 separate decisions, which had to run through a series of meetings he called "tollgates." As one of them put it, the task force members were essentially "research rats," who were kept busy eighteen hours a day, seven days a week, churning out memos and debating details of a blueprint for restructuring one-sixth of the U.S. economy. Hillary's ally Senator Jay Rockefeller pronounced it the most complicated effort since the invasion of Normandy.[8]

Gaunt and frazzled-looking to begin with, Magaziner soon developed a hacking cough, persistent enough to alarm his coworkers. Despite his ambitious plans to remake the health care delivery system, he couldn't deliver himself to a doctor, get meetings to start on time or even provide enough chairs for people to sit on. And, of course, much of this activity was completely beside the point, as Congress had yet to approve even the basic concept of the reform plan.

Dr. Joycelyn Elders, soon to become the Clinton administration's nominee for Surgeon General, recalled, "I would fly up from Little Rock for committee meetings and find people milling around wondering where to go and what the agenda was. One Friday night I flew up for a special Saturday meeting that was supposed to be important, but when the time came, we didn't have a room to meet in. Some of us were sitting on the steps in the middle of the hall—this was in the Old Executive Office Building—and others were roaming around. . . . There weren't a lot of people who were versed in health policy or experienced at getting things done in the health world. I couldn't imagine who might have come up with this mix, or what they had in mind." As for Hillary, Elders noted, "It looked to me like everybody thought she was more involved than she really was. . . . I don't think I ever saw her."[9]

Elders was correct to question the extent of Hillary's involvement. Throughout 1993, Hillary would keep up an exhausting schedule of appearances, including foreign travel with the President, not to mention finding time for activities like posing for *Vogue* in a Donna Karan dress. In December her press secretary, Lisa Caputo, would boast that the First Lady had found time to read forty books on Victorian decorating in order to supervise the work of Little Rock designer Kaki Hockersmith, whose redecoration of the White House included a period restoration of the Lincoln Sitting Room. No matter how adept Hillary might be at mastering complex issues, she was at best a well-briefed spokesperson for the Magaziner program.[10]

In retrospect, the most distressing feature of the task force effort was that it appeared to be taking place in a time warp. We are living through the first stage of a revolution in medical technology whose breathtaking implications are just becoming apparent. Diseases that were invariably fatal a few years ago can now be treated, at a price. In the foreseeable future, even cancer may become a manageable condition. A host of common health problems that we all once had to live with can now be corrected. The mildly depressed chase away the blues with Prozac; the sexually challenged resort to Viagra. Organ transplants, breast reconstruction for mastectomy patients, in vitro fertilization, surgically corrected nearsightedness, dental implants—all have become not just accepted but routine. The bad news is that no matter how we finance health care, it may be impossible to make all the benefits of the revolution available to the entire population as soon as they come onto the market. The good news is that what is rare and expensive today may be routine and affordable tomorrow.

In her speeches on health care, Hillary continually put forward arguments that were, at best, irrelevant to the dawning era of high-tech medicine. She stressed, for example, the theme that universal access to preventive care, along with public health measures like reducing smoking, would lower the total amount we spend on health as a society. On the contrary, the workaholic who drinks, smokes, neglects his health and drops dead of a heart attack at sixty may well pay more into the health care system over his lifetime than he ever takes out in benefits. It's the fitness-conscious woman who sees her doctor regularly and lives to be ninety, requiring treatment for the myriad infirmities of old age and perhaps spending her last years in a nursing home, who will rack up the highest bills. All this had been explained to Hillary, but she refused to accept it. "It's been interesting dealing with the actuaries," she said in an address to the Institute of Medicine. "They don't believe in prevention. They think if you let people go to doctors early, they'll just keep going. . . ."[11]

Americans living under Clintoncare would have been well advised to stay healthy. Hillary's warm endorsement of the benefits of preventive care distracted attention from the task force's plans to strictly ration access to high-tech treatments, especially for the very ill and the elderly. The Clinton program called for moving most Americans into HMOs and cutting $124 billion from Medicare in the first year alone. The government would dictate the curricula of medical schools and restrict the number of graduates allowed to become specialists. An "Advisory Panel on Breakthrough Drugs," under the authority of the secretary of health and human services, would have the power to blacklist drugs from the Medicare schedule if the manufacturers refused to give Medicare recipients a rebate on prices.

Hillary believed that every cause needed a villain, and the pharmaceutical companies had been chosen as the designated enemy of the day. On February 12, Hillary traveled to a health care conference in Harrisburg, Pennsylvania, with Tipper Gore to deliver a speech lambasting drug manufacturers for profiteering at the expense of America's children. Charging that the cost of immunizing a child had risen from $6.69 in 1981 to $90.43 in 1991, an increase of 1,250 percent, Hillary told the audience, "Unless you are willing to take on those who profited from that kind of increase and are continuing to do so, you cannot provide the kind of universal immunization system that this country needs to have."

Whether drug companies actually priced their products too high was a matter of opinion. With many new drugs in development, the pharmaceutical business is becoming more competitive than it was in the past. Investors in start-up biotech companies, in particular, regularly risk—and lose—millions in the hopes of backing a breakthrough drug. High-tech drugs developed through free-market investments hold the best hope for conquering scourges like cancer and Alzheimer's disease. Moreover, they are often highly cost effective compared to alternative treatments such as surgery. On the other hand, millions of Americans who obtain health care through their employers or the government have no idea what it actually costs. Many of them do pay out of pocket for drugs, and therefore they are ripe for the argument that their prescriptions cost too much.

At any rate, the charge that drug companies were responsible for low vaccination rates was as bogus as Hillary's statistics, which did not take into account the addition of new shots to the recommended vaccination program. Manufacturers already donated free vaccines for needy children, and public health specialists, including Dr. Joycelyn Elders, who had supervised childhood vaccination programs during the Clinton era in Arkansas, agreed that the problem was educating parents to bring their children in for shots, not the cost of the vaccines.

Nevertheless, free childhood vaccinations had long been part of the Children's Defense Fund agenda, and the Clintons used the momentum created by Hillary's speeches to push a bill through Congress that gave Health and Human Services the power to bypass the established distribution system by buying up stocks of vaccines at cut-rate prices and storing them in a central warehouse. Government involvement threatened to create a bottleneck in the supply of vaccines and to discourage pharmaceutical companies from developing new ones. Senator Dale Bumpers, whose wife had long been a volunteer in pediatric vaccination programs in Arkansas, denounced the administration's actions as creating a "bureaucratic nightmare." Public health experts agreed, and the administration, faced with overwhelming criticism, abandoned the distribution plan.[12]

Judging from a 1995 report by the General Accounting Office, the pediatric vaccination program that existed in 1993 had been generally on target in meeting its goals. Based on Hillary Clinton's proclamation of a nonexistent crisis, Congress had been stampeded into passing unnecessary legislation. And even though the worst features of the administration plan had been dropped, the country was still stuck with a program that was more costly, cumbersome and wasteful than the one it replaced.[13]

What's more, the alarming statistics Hillary had cited on the rise in prices of prescription drugs were another myth. It turned out that the Labor Department statisticians had gotten the numbers wrong.[14]

This news came too late for investors. The threat of price controls had caused the blue chip pharmaceutical stocks to decline as much as 40 percent, wiping out over $1 billion worth of market capitalization. Some smaller biotech companies were put out of business permanently. Only short sellers profited, among them a private hedge fund called ValuePartners I, run by Smith Capital Management of Little Rock, Arkansas. Hillary Clinton held an $87,000 stake in Value-Partners I, which also owned a block of stock in United Healthcare, an HMO that stood to benefit under the Clinton reform plan. Lois Quam, a United Healthcare vice president, was a member of the task force.

Unlike the Carters, Bushes and Reagans, the Clintons failed to put their assets into a blind trust when they moved into the White House. Hillary resisted the notion that her financial affairs were anybody's business but her own, and she reasoned that since she was not a government employee and the money was in her name, she didn't have to resort to a trust. Vince Foster wasn't sure this was so. After seeing the financial disclosure statement filed by the Clintons the previous December, Foster worried that Hillary's interest in Value-Partners I might pose a conflict of interest.

Moreover, while Hillary was out on the stump denouncing the pharmaceutical companies and the insurance industry, other special-interest groups had been incorporated into the process. For example, the American Dietetic Association had hired Betsey Wright, at the time a lobbyist with the Wexler Group, to make sure that its members' services would be covered under the basic health plan. The working groups were also packed with experts who advocated big-government solutions, such as fellows of the Robert Wood Johnson Foundation, which had funded Arkansas's rural health care clinics as well as other projects in the state during Clinton's tenure as governor.[15]

Reforming health care without some special-interest input would be unimaginable. The problem was that the task force was a closed process. The identities of working group members and consultants were kept secret, so no one could begin to say what guidelines were followed in selecting them. A Watergate-era law known as the Federal Advisory Committee Act existed to prevent exactly this sort of behind-the-scenes influence brokering, and shortly after the health care initiative was announced, the Association of American Physicians and Surgeons had filed a lawsuit challenging the legality of the process on the grounds that the act was being ignored. Because the AAPS was a small organization made up of a few thousand doctors who favored preserving traditional fee-for-service medicine, the press generally dismissed the suit as right-wing obstructionism. Inside the White House, there was a flurry of concern, but no more. In 100 days, when the task force completed its work, the lawsuit wouldn't matter.

HILLARY was soon plunged into her own family health care crisis. Hugh Rodham, Sr., eighty-one, had been in declining health for several years, surviving several mild strokes. Rodham used a wheelchair when he and Dorothy came to Washington for the inauguration, but later in the winter he had been well enough to fly to Los Angeles, where he made a cameo appearance in an episode of *Hearts Afire,* Harry Thomason and Linda Bloodworth-Thomason's situation comedy about a Capitol Hill aide. On March 19, however, Rodham suffered a devastating stroke and was rushed by ambulance to St. Vincent's Hospital in Little Rock.

The Clintons both flew to Arkansas, but the President soon returned to Washington, where Webb Hubbell, showing up at the White House to watch a movie, found that Barbra Streisand was among Clinton's guests for the evening. Hillary remained at her father's bedside for the better part of two weeks, joined by her brothers, Hughie and Tony, and Chelsea.

The hospital set aside a conference room so that the Rodhams could relax and take their meals in privacy. Apart from this concession, Hillary was facing the same hard choices that inevitably challenge one's preconceived notions about hospitals, not to mention the very definitions of life and death. Originally listed as being in serious condition, Hugh senior slipped into a coma, and as his vital signs deteriorated the family was asked to make a decision about terminating life support. They struggled with their emotions and decided to accept the doctors' recommendation to turn off the machines, but Hugh senior didn't die. At this point, the hospital informed Dorothy Rodham that since there was nothing more to be done medically, her husband would have to be transferred to a subacute-care facility.

As usually happens in these situations, the family was told that it would have two days to make arrangements. Few tasks are more distressing than having to locate and apply for a nursing home for a dying family member, especially one who probably won't survive long enough to make the transfer. This was the case with Hugh Rodham, who died at St. Vincent's on April 7, having passed his eighty-second birthday while still in a coma.

While Hugh senior was dying, reporters had kept watch outside the hospital, waiting for news of his condition. Their interest was inevitable, since Hillary was not only the First Lady of the United States but the leader of the health care reform effort, but she had trouble coming to terms with the loss of privacy at this difficult moment for her family. Her father's situation also made her realize that even the comprehensive plan she envisioned had gaps. Like most families, the Rodhams had not carried insurance that covered nursing homes, and Hillary began to think about adding a long-term care option to the package of benefits envisioned under Clintoncare.

Hillary had returned to Washington in early April and on the sixth, the day before her father's death, she flew to Austin, Texas, where she was scheduled to deliver her first major address on health care. She sketched out the themes of the speech herself, surprising the audience with an impassioned and partly extemporaneous description of America as a nation suffering from a "sleeping sickness of the soul"—a country overcome by a "sense that somehow economic growth and prosperity, political democracy and freedom are not enough—that we lack at some core level meaning in our individual lives and meaning collectively, that sense that our lives are part of some greater effort, that we are connected to one another, that community means that we have a place where we belong no matter who we are." The speech culminated with a call for a "new politics of meaning."

The speech came at an emotional moment in Hillary's life, and the content was obviously influenced by her father's situation. She stressed the need for the health care reform effort to address such questions as when life begins and ends, saying "These are not issues we have guidebooks about." Of course, troubling as these questions inevitably are for individuals who have to confront decisions about ending life support, adding them to the agenda of the health care task force was hardly likely to make passage of a reform bill any easier.

As for the philosophical themes Hillary raised, these were by no means new concerns. Both Clintons, it turns out, had been intermittently attracted to New Age guru types, and as early as 1988 they were reading Michael Lerner's magazine *Tikkun.* Founded as a leftist answer to *Commentary,* it puts forth an amalgam of neo-Marxism, Judaism and New Ageism that Lerner calls "the politics of meaning."

A former leftist antiwar activist, Lerner married his first wife in a nonlegal ceremony at which the wedding cake bore the slogan "SMASH MONOGAMY" and the couple exchanged rings fashioned from downed U.S. military planes. According to Lerner's own account, his conversion to the politics of meaning grew out of a 1976 stress clinic in Oakland, California, organized by a group of "psychiatrists, psychologists, social workers, family therapists, union activists and social theorists" concerned because so many of their clients, driven by unsatisfied spiritual needs, seemed to be turning to "the Right." Ironically, the politics of meaning is itself an inadvertent parody of the loonier side of leftism. Lerner gives short shrift to virtues like courage, self-reliance, honesty, patriotism and fidelity—the values he dismisses as exclusionary and judgmental. In his analysis, the problems of present-day America come down to a failure of "empathy." One can only recall Hillary Rodham's 1969 Wellesley commencement speech, in which she said, "Part of the problem with empathy for professed goals is that empathy doesn't do anything." Certainly, many Americans today feel they are up to their eyeballs in empathy; what they want is solutions that work.

Lerner's writings echo the confusion between economics and morality implicit in Hillary's thinking about politics. He dislikes market capitalism, because it rewards "the selfish and the self-centered while putting at a disadvantage those who have taken time off in order to act morally or in a caring way toward others." He wants politics to create incentives "so that people can feel that choosing a moral life does not mean losing all chance for personal advancement, economic reward or social sanction." This is a yuppie's idea of ethics: moral choices without sacrifices.

One of Lerner's proposals is that all bills in Congress be accompanied by ethical impact statements detailing "how the proposed legislation would impact on shaping the ethics and caring and sharing of the community covered by that agency." Hillary liked this idea, according to Lerner, though she worried that words like "caring" and "love" would be misunderstood. Surely the real difficulty is that some of us don't trust politicians to decide what is ethical, much less to redistribute wealth to the virtuous.

After reading about Hillary's speech in the newspapers, Lerner called the White House to ask if the First Lady had drawn on his ideas and was delighted to learn that she had. He was invited to a reception and again for a longer chat with Hillary, who told him, "It's amazing how much we seem to be on the same wavelength."[16]

This meeting of minds would end in mutual disillusionment. Journalists had a wonderful time picking apart what Michael Kelly, writing for *The New York Times Magazine,* aptly called the "unintentionally hilarious Big Brotherism" of the politics of meaning.

Michael Lerner would soon turn against the Clintons, blaming them for failing to articulate a radical vision. But of course it was when Bill Clinton moved toward the center, bringing in a balanced budget and welfare reform, that he was the most successful. Lerner railed against what he called "*New Republic* cynicism," but cynicism seemed to be in order. Who wouldn't be for Clintoncare if we really believed that it could take care of an individual's health needs for $1,932 a year, as promised? But it was hard to believe this, coming from the same politicians who couldn't seem to correct the problems of Medicare, Medicaid and Social Security.

As veteran political junkies, the Clintons surely were aware of the significance of the plaque that President Harry Truman had kept on his desk, proclaiming "THE BUCK STOPS HERE." The slogan meant that the President was held responsible for everything, even though he might be limited in terms of what he could actually accomplish. After just a few months in the White House, Hillary Clinton was already feeling the pressure of being in power. Many of her activist friends, so jubilant over Clinton's victory in November, were already disaffected. Bill Clinton had been forced to compromise on the subject of gays in the military. There would be no 100-day push to reinvent the education system. Even the developing health care proposal, while considered too radical by many Americans, was disdained by some of those working on it as a jerry-built substitute for a Canadian-style single-payer plan.

The press, meanwhile, was already showing signs of the disease of "*New Republic* cynicism." The Hillaryland staff had been delighted when their boss posed in a white dress for the cover of *The New York Times Magazine,* only to see the photo appear with the mocking headline "SAINT HILLARY." No doubt a traditional First Lady would have been more gently treated. During much of her interview with Michael Kelly, Hillary talked earnestly about her Methodist faith, her rather conventional views on child rearing, and her frustration with crime ("I want to live in a place where I can walk down any street without being afraid"). She even expressed agreement with Senator Moynihan's observations on "defining deviancy down." But when she tried to discuss the relationship between individual morality and society, she retreated into confused generalities. She recalled sitting in a class where everyone was intellectualizing about the motives of terrorists, and thinking, "The other alternative is that they are just evil." But this was not quite the great insight she seemed to think it was. Kelly zeroed in on the problem, observing that Mrs. Clinton could talk about her own values but was still too much of a social liberal to feel comfortable about judging the values of others.

Hillary seemed to be groping, very painfully, toward certainties that other people—not necessarily all conservatives—took for granted, and Kelly rightly questioned whether anyone whose ideas on morality were so unformed was in a position to lead the country in a search for "meaning."

TWELVE

A Box with No Windows

HILLARY'S willingness to rush to the bedside of her dying father showed her at her best. On the other hand, no one who plans to reorganize a large segment of the American economy in a hundred days can afford to take twenty days off. Hillary may not have been able to save health care reform if she had remained in Washington, but in her absence it became apparent just how little clout the task force had. Preoccupied with the budget, the President's aides were, at best, indifferent to the health care effort that was proceeding on a separate track in the through-the-looking-glass world of Hillaryland.

The task force discussions were moving so slowly that the bill might not be finished before the end of 1993. In the meantime, Democrat Jim Cooper of Tennessee was working on an alternative bill, Republican opposition was solidifying, and the insurance industry was mobilizing its own lobbying effort. Hillary, moved by her own family tragedy, kept adding items to the health care package. Psychiatric care, drug and alcohol rehabilitation, dental care for children, even a long-term-care option were shoehorned into the plan. All of these benefits were desirable, but they posed cost control problems that couldn't be resolved in a few weeks, or even a few months.

Since the task force would not be finished in May after all, the

AAPS lawsuit also had to be dealt with. Frustrated, Hillary turned to Vince Foster, demanding, "Fix it, Vince." Foster would try to get the suit dismissed, arguing that the 1972 law mandating that federal advisory committees hold open meetings was unconstitutional and, anyway, that it wasn't appropriate to the health care task force because all the task force and working group members were either regular or temporary government employees. The problem with this argument was that only about thirty-five working group members had filed the financial disclosure forms required of government employees. Hillary wasn't among them, and if the court agreed that she was a de facto government employee she might find herself in trouble for failing to place her assets in a blind trust.

Just as troubling, the AAPS had also filed a Freedom of Information Act request for the task force's working papers. Ironically, the people who were supposedly going to slash 28 percent from the costs of medical services could not manage their own budget. The White House originally claimed that the task force effort cost about $300,000, but the true figure exceeded $11 million, a total disguised by charging expenses to various other agencies. Moreover, flights to and from task force meetings were being arranged by WorldWide Travel, the agency used by the Clinton campaign, not the White House travel office. AAPS attorneys would later discover that some travel vouchers were apparently backdated, while others were never located at all.

It isn't clear whether the need to clean up the task force travel accounts had anything to do with the decision to get rid of the White House travel office staff. Most new administrations thought about firing the travel office personnel and replacing them with patronage appointments, only to reconsider when they realized that Billy Dale and his staff did much more than just book charter flights. Dale, who had worked in the White House since the Kennedy administration, and his six assistants were responsible for baby-sitting the members of the press who traveled with the President, getting controversial correspondents past unfriendly passport control agents and moving expensive camera gear through customs.

But Catherine Cornelius, who had handled the WorldWide Travel account during the 1992 campaign, had been led to believe she was in line for Dale's job. Two months after the inauguration, Cornelius talked David Watkins into transferring her to the travel office, where she began snooping around, photocopying bills and vouchers, peering into drawers and listening in on phone conversations. Cornelius soon reported back to Watkins that the staff was casual about filling out expense vouchers. She also suspected them of taking kickbacks—Billy Dale had a summer home, and another staffer had recently vacationed in Europe.

First Friend Harry Thomason and his business partners, Darnall Martens and Dan Richland, were also eager to see the last of Dale, who had rejected the suggestion that the travel office could save money by outsourcing its booking operations to their company, TRM, a broker of charter airline flights. For some reason, the travel office suddenly became an urgent matter during the week of May 10 after Thomason, who happened to be in Washington, had a conversation with Clinton about the situation. Thomason then spoke to David Watkins and Vince Foster, urging them to look into possible graft on the part of Dale and his staff. Weeding out a nest of corruption left over from the Reagan-Bush era would be a publicity coup for the Clintons, he suggested.

On Thursday, May 13, the day after his meeting with Thomason, Foster brought up the travel office situation with Hillary and found her at the end of her patience.

"What's going on? Are you on top of it?" she asked him.

Overburdened, Hillary realized that she wasn't getting the staff support she needed, and she was often bitterly critical of White House aides, especially the Arkansans, whom she saw as slow, overcautious, and indecisive. She had fallen into the habit of venting her frustrations on Foster, whose notes of the meeting show him trying to justify her mood:

I understand—understaffed
(thought more of a combination of
mgment concern, HC [health care] resolution problems
genl frustration, prob aggravated by
grief

Foster had arranged to have an accountant from Peat Marwick conduct an emergency audit of the travel office over the weekend. Associate White House counsel Bill Kennedy also spoke with the FBI about launching an investigation. The call was a mistake. The White House is not supposed to sic the FBI on people. If there are suspicions of wrongdoing, the proper procedure is to inform the attorney general, who will move the request through the proper channels.

Judging by a memo that David Watkins wrote to Mack McLarty but never actually sent, Foster and Kennedy must have felt pressured to act quickly because Hillary was really fired up, telling Watkins, "Harry says his people can run things better. We need to get those people out. We need our people in." Referring to himself and Foster, Watkins continued, "We both know that there would be hell to pay if we failed to take swift and decisive action in conformity with the first lady's wishes." That weekend, Watkins departed on a trip to Little Rock and Memphis. Wherever he went, he was bombarded by phone calls from Hillary, demanding updates on the travel office situation.

Why was Hillary in such a rush to get Dale and his staff out?

Rebecca Borders, who reported David Watkins's side of the travel office story in *The American Spectator*, would take note of an interesting coincidence. On May 11, the day before Thomason began his campaign to get Dale fired, a *Time* magazine reporter had called the White House asking questions about what appeared to be an unusually close relationship between Catherine Cornelius and President Clinton. On a recent out-of-town trip, Cornelius had traveled on Air Force One, and she was allowed to ride in the presidential motorcade instead of the press bus, where travel office employees usually sat.[1]

Although Dale and his staff certainly had a motive to make trouble for Cornelius, there is no evidence that they had been talking to reporters about her. Nor did they need to. Reporters who covered Clinton's trip could see for themselves that Cornelius appeared to be quite friendly with the President, and she herself boasted to coworkers that he called her at least twice a day.

On Monday, May 17, Peat Marwick presented the preliminary results of its investigation. Auditors had found evidence of "mismanagement," and some $18,000 in funds were unaccounted for. Dale's duties involved collecting advances from the networks and publishers to cover the costs of reporters' travel, then issuing reimbursements based on a complicated formula. Dale carried large amounts of cash on foreign trips, as much as $15,000, and as a matter of convenience, he had begun to deposit unspent cash in a separate bank account, held in his own name, using the money to resolve billing disputes. This was an unacceptable practice, but he hadn't stolen any money. No one, however, gave Dale a chance to explain himself. On Wednesday, Dale and his staff were informed that they had ninety minutes to clear out their desks. After thirty-two years on the job and just a few months away from retirement, Dale left the White House in disgrace, riding in a delivery van.

That same afternoon, press secretary Dee Dee Myers released a statement saying that the FBI had opened a criminal investigation of the travel office. This was another violation of proper procedure. The FBI never discloses the names of people who are under investigation unless they are about to be indicted.

The press was immediately suspicious about the motives behind the firings, especially when employees of WorldWide Travel were moved into the office as replacements. Jeff Eller, the pro-Hillary White House director of media affairs, soon came forward to claim that he and Cornelius had been having an affair. Eller also revealed a fact that had the effect of squelching gossip about Cornelius and the President: They were cousins. This was news to everyone, including Vince Foster, who made a note of it: "CC—not know she is a cousin." (Cornelius was certainly not Bill Clinton's first cousin, though he had many distant cousins in Arkansas.)

Their motors racing as usual, Harry Thomason and his wife heaped scorn on the notion that they might have had a financial motive for wanting TRM to get a measly seven million a year in bookings. "Given our salaries," said Linda Bloodworth-Thomason, "setting our sights on the White House business would have been the financial equivalent of us taking over someone's lemonade stand." Calling Washington an "incestuous insane asylum," she added that if she and Harry could be painted as villains, then "Mother Teresa had better stay out of town."[2]

Faced with a storm of criticism, Mack McLarty decided to rescind the firings of the five travel office employees who had no responsibility for financial accounting, placing them on administrative leave. McLarty also removed the WorldWide employees who had taken over travel office jobs and ordered a management review, undertaken by John Podesta, soon to be known around the White House as the "Secretary of Shit" for his role in cleaning up messes.

Hillary was certainly not the first presidential spouse to meddle in a personnel decision. The treatment of Dale and his staff was nasty, but probably not illegal, though potentially one could make the case that Hillary was scheming to give the travel office business to her friends. Perhaps Hillary also worried that the travel office mess might lead to an inquiry into the handling of the travel expenses for the health care task force. At any rate, she was not about to take her share of the blame.

Vince Foster's notebooks show him recasting his conversations with Hillary in his own mind to minimize her role: "defend mgment decision thereby defend HRC role whatever it was in fact or might have been misperceived to be," said one jotting. In his first meeting with Podesta, Foster left Hillary out of the situation altogether, citing attorney-client privilege.

Watkins, too, was protective. "I fell on my sword," he told Foster.

Watkins's story was that he never talked to the First Lady prior to the Friday evening before the firings and knew of her interest only through Foster. And Foster backed this up, conceding only that he had "advised" Hillary about the decision to bring in Peat Marwick and may have had "a few short, incidental, non-substantive discussions subsequently to pass on my understanding from DW" about the investigation.

Podesta's report was released just before the July Fourth weekend.

Vince Foster was spared, but David Watkins and Bill Kennedy received official reprimands. Watkins did not take this well. A Clinton supporter since 1982, he had managed the finances of the 1992 campaign, averting financial disaster by arranging a $3 million line of credit with the Worthen Bank. In those days he sat at Hillary's kitchen table in Little Rock, drinking coffee and consuming Liza Ashley's famous chocolate chip cookies. Now he was being made the fall guy for her rash decision. In May 1994, he was forced to resign after a press photographer caught him hitching a ride to a nearby golf course in a presidential helicopter, an incident that may or may not have been a setup.

The problems of the travel office staff were far from over. All seven workers were investigated, and Billy Dale, in particular, had his life turned upside down. The FBI examined his finances, questioned his wife about her drinking habits (she was a teetotaler), called numbers from his address book looking for a mistress and forced his daughter to document where she got the money to pay for her wedding. To minimize his legal bills, Dale offered to plead to a misdemeanor. The Justice Department refused to deal, but word of Dale's proposal was leaked to President Clinton's personal lawyer, Robert Bennett, who mentioned it on national TV. When Dale went on trial in September 1995, Sam Donaldson of ABC and Al Hunt of the *The Wall Street Journal* were among his character witnesses, and the jury took two hours to dispose of the charges, finding Dale not guilty.

Three months later a copy of Watkins's unsent memo surfaced, but Hillary insisted that she had done no more than express "concern" about the situation: "I and others said, 'My goodness, that sounds like something that needs to be looked into,'" she recalled in a televised interview with Maria Shriver. Although Dale had been exonerated, Hillary added, "But I don't have any apologies in any way for saying I've heard there are reports of financial mismanagement."[3]

Mismanagement is a relative term. By the time Hillary made this statement, a General Accounting Office audit had found that during the first eight months of 1995, the books of the White House showed errors in the range of $200,000. Dale's accounting looked good by comparison.

As for the AAPS suit, a federal appeals court denied the group's request for a preliminary injunction, relying on a thirteen-page deposition submitted by Ira Magaziner on March 3, 1993. The counsel's office argued that all the task force participants were government employees, but later changed its strategy, trying out the novel theory that the health care proposal was the work of "an anonymous horde" who labored under conditions so "fluid" that it was impossible to say what their status was—in other words, the anarchy defense. Magaziner had many shortcomings, but the secrecy policy hadn't been his idea. Nevertheless, he took the blame, becoming the subject of an investigation by Eric Holder, the U.S. attorney for the District of Columbia. Magaziner escaped a criminal contempt indictment, but in December 1997, U.S. District Judge Royce Lamberth called his affidavit deceptive and ordered the government to pay $285,864 to cover the legal bills of the AAPS.

Judge Lamberth reserved his harshest criticism for the lawyers who helped Magaziner concoct his story. "Sloppy," "overly aggressive" and "dishonest" were some of the adjectives Lamberth used to describe the lawyers' work, adding, "It seems that some government officials never learn that the cover-up can be worse than the underlying conduct."

IT often happens that a depressed person develops an attachment to someone he sees as representing the active, fully engaged life that he feels unable to manage on his own. This was the situation of Vince Foster, who had made a heavy emotional investment in his fantasy of serving as Hillary's wise counsel, using his legal acumen to protect her from the prowling wolves in the political forest. During the campaign, Foster had worked so many late hours at headquarters that his wife, Lisa, was furious with him. Her anger continued to simmer after Foster went off to Washington, while she stayed behind in Little Rock to look after the younger of their two sons, who was finishing his junior year in high school.

During the early weeks of the administration, Foster worked shoulder to shoulder with Hillary on health care issues, and one evening he found himself on the Truman balcony of the White House, sipping cocktails with the Clintons and singer Judy Collins. But the bad days outnumbered the good. Married to his college sweetheart, Foster had never lived alone. Now, he complained to friends, he was living in a borrowed apartment, so cramped that his outstretched arms reached from wall to wall. Every morning Foster left his tiny cell and drove to the White House, where he was putting in twelve-hour days as a matter of course. Foster was researching medical malpractice law for the health care task force, and he was also looking into the escalating costs of the redecoration of the First Family's private quarters. The decorator, Kaki Hockersmith, had been the Fosters' neighbor in Little Rock, and she was distraught over the high labor costs on the project. Most of the work was being done by federal employees, whose time was billed to the foundation that supported the redecoration through private donations. Apparently, there had been a misunderstanding about federal overtime rules, but Foster had jumped to the conclusion that the usher's office was jacking up the bills.

Foster was also doing private work for Hillary, working with Brantley Buck, the Rose firm attorney who was setting up her blind trust. In addition, he was helping out Denver attorney James Lyons, who was still slogging away, doing his best to "reconstruct" the records of the Whitewater Development Company. The business of gathering information about Hillary's finances dragged on for weeks. Foster had to go through Maggie Williams to get answers to some questions, such as whether certain stocks in Hillary's name were being held in trust for Chelsea. Though handling the First Lady's personal business was not illegal, it was unusual, and it exposed Foster to potential conflicts. Foster was being paid to protect the interests of the presidency, not Bill and Hillary Clinton as individuals.

On June 5, Lisa Foster arrived in Washington, and the family soon moved into a rented Georgetown town house. Lisa found Vince stressed out. A lot of his sentences began with the words "I just can't handle . . . ," she would recall. He felt he was being criticized at home, at work and even in the pages of *The Wall Street Journal,* which had run an editorial entitled "Who Is Vince Foster?" To make matters worse, Foster's secretary had failed to answer the *Journal*'s request for a photograph, and the piece appeared illustrated by a silhouette drawing containing a question mark.

Among the books Foster kept by his nightstand were Theodore White's *The Making of a President,* Bill Clinton's campaign tract *Putting People First* and a number of books on ethics.

Early in July, Foster told his brother-in-law, former congressman Beryl Anthony, "I have spent a lifetime building my reputation and now I am in the process of having it tarnished."

Foster was always very protective of Hillary, and it is unlikely that he would have poured out his troubles to her. On July 14, however, he did meet with Susan Thomases, who was in Washington on one of her midweek working visits and staying at 2020 O Street, a private hotel known for its lush Victorian ambience. Foster discussed his fear that David Watkins would try to shift the blame for the mess now known as Travelgate onto Hillary. He also talked about his unhappy marriage to Lisa, whom he described as too dependent. Foster's main problem was that he was feeling overworked and overburdened, but despite the eighty-hour weeks he was putting in, he was not being very productive. Foster spent much of the week of June 12 writing personal thank-you notes and, apparently, handling tax and other private business matters for the Clintons.[4]

On the fifteenth, Senate Minority Leader Bob Dole sent a letter asking for congressional hearings on Travelgate. No one else seemed to take the threat of hearings very seriously, including Lisa Foster, but Vince was making a list of lawyers, deciding whom he would hire to represent him. Foster also talked of resigning, but Lisa, who had just packed up and moved the household to Washington, urged him to stick it out until the end of the year. One evening, she suggested that he calm himself by writing down all the things that were bothering him; Foster took out a legal pad and began to write.

Over the weekend, Foster and his wife got away to the eastern shore of Maryland, where they stopped by the estate of Washington attorney Michael Cardozo and his wife, Harolyn, the daughter of wealthy Democratic contributor Nate Landow. The Hubbells were guests of the Cardozos that weekend, and Lisa Foster and Harolyn Cardozo played tennis with another houseguest, tennis guru Nick Bollettieri. On Monday morning, Marsha Scott stopped by Foster's office to ask about the getaway. According to the executive assistants who worked in the outer office, Deborah Gorham and Linda Tripp, Scott closed the door and spent nearly an hour with Foster, a highly unusual occurrence. Since Scott served as a sort of surrogate aunt to the Arkansas crowd and was known to be a conduit of messages to and from the President, it is reasonable to assume that whatever Foster told Scott was of some importance.

The FBI's report of its interview with Scott says: "Scott stated that she was uncomfortable talking about some of the private thoughts Foster shared with her. She said she believed Foster had painted himself into a box with no windows. But she got the sense during the July 19th meeting that Foster had come to a decision and was, if anything, relaxed. . . ." Scott went on to mention Foster's concerns about the hostile media and about White House staff cuts that made it impossible to do thoughtful work—his exact words to her were "The staff cuts are killing us."

That evening, the President called Foster to ask him to a screening of *In the Line of Fire*. Foster passed on the invitation, and Clinton suggested a brief meeting on Wednesday to discuss "organizational" matters.

In mid-June, while Clinton was out of town, Hillary had accepted an invitation to dine out at I Matti, a fashionable Italian restaurant, with the Fosters, the Hubbells and a group of friends from Arkansas. The White House called to reserve a large table, and the Secret Service arrived early to check out the premises. But at the last minute, Hillary had to cancel and Foster sulked all evening, like a child denied a promised treat.

Hillary would later tell the FBI that she wasn't sure if she ever talked to Vince again after that evening. On July 5, she and the President left on a trip to Japan. Press reports noted that the First Lady was out of sorts in Tokyo, attributing her mood to the absence of worthwhile events on her schedule. Returning from Asia, the First Family stopped in Hawaii for a brief vacation. Clinton was back in Washington by the beginning of the week, but Hillary, Chelsea and several members of the First Lady's staff flew to Little Rock, where Hillary was to speak at a charity event, visit her recently widowed mother and sign the papers placing her assets in a blind trust. Arriving in Little Rock at about 8:20 P.M. on Tuesday, July 20, the women went directly to Dorothy Rodham's condo. They were sitting in her living room, unwinding from the long journey, when the phone rang. Hillary took the call in the kitchen, and when she returned she looked ashen, too upset to speak. Dorothy Rodham suggested that it was time for the staffers to check into their rooms at the Excelsior Hotel. Later, when Hillary was more composed, she called the hotel to say that Vince Foster had been found dead in Fort Marcy Park, an apparent suicide.

Hillary had also spoken to her chief of staff, Maggie Williams, who was in Washington. Around ten o'clock, Williams called her assistant, Evelyn Lieberman, at her home and told her the news. Lieberman jumped into her car, picked up Williams and drove to the White House, where they headed for Williams's office on the second floor of the West Wing. Lieberman later told Senate investigators that she wasn't sure what they hoped to accomplish or why they chose to head for the West Wing office, as opposed to Williams's larger office in the Old Executive Office Building: "There was no conscious thought about it. It was an automatic thing."[5]

When they arrived at the West Wing, Williams would recall, she saw a light on in Foster's office and found Patsy Thomasson, the former Dan Lasater aide who was David Watkins's deputy, seated at Foster's desk. Thomasson had been called by Watkins and Webb Hubbell, who directed her to look in Foster's office for a suicide note—something, as Thomasson put it, that "would be a comfort to

Lisa." Bernie Nussbaum arrived and he and Williams left together, leaving Thomasson alone. Sometime later, Williams had another conversation with Hillary, who asked her, "Do you know anything about why Vince would do this? What's going on? You know, is there something you can't tell me? What's happening?"[6]

Secret Service agent Henry P. O'Neill worked nights in the West Wing, escorting the cleaning crews who worked in high-security areas. Around eleven o'clock, he ran into Evelyn Lieberman and Maggie Williams. The two women were on the way out of the counsel's office, and Williams was carrying a stack of file folders. "It wasn't an enormous amount, but it was more than a little," O'Neill would tell Senate investigators. Williams denied removing anything from the office, and passed two lie detector tests to back up her word.[7]

A second Secret Service agent would add that on the following morning, he saw Craig Livingstone, the director of personnel security, carrying documents out of the White House, first in a litigator's case and on a later trip in two cardboard boxes. Livingstone, too, would deny doing anything of the kind. Foster's office wasn't sealed until ten A.M. the next day, and then only because two other White House lawyers, Steve Neuwirth and Cliff Sloan, insisted on it in rather forceful terms. Nussbaum balked at allowing the Park Police to search through Foster's papers; however, by five that afternoon, July 21, he had agreed to allow a search by FBI agents under the joint supervision of himself and Deputy Attorney General Philip Heymann.

Most people would be none too happy to learn that their attorney was dead and the police were about to rifle through his files in a search for evidence. When these situations do occur, the police typically face the burden of proving that the files are relevant. But Foster worked for the people of the United States, not the Clintons. His death was an unforeseeable tragedy that drove home a lesson about the dangers of mixing private and public business.

At any rate, there followed a round-robin of phone calls. At 7:44 the next morning, Maggie Williams called Hillary in Little Rock and spoke for seven minutes. Hillary then called Susan Thomases, who called Bernie Nussbaum. And by the time Heymann and the FBI agents arrived at ten A.M. to conduct their search, Nussbaum had changed his mind about the ground rules; he and he alone would look at the files. The agents were required to move to the other side of the room while Nussbaum hurriedly separated files having to do with Foster personally, the Clintons and White House business into three piles. At one point, when an agent stood up to stretch his legs, associate counsel Cliff Sloan accused him of trying to sneak a peek at the documents.

Later, it would develop that Nussbaum had gone through Foster's briefcase and removed the notebook containing jottings related to Hillary's role in the travel office firings. Moreover, in spite of being kept at a distance, the Park Police who were permitted inside Foster's office noticed that some of the files Nussbaum handled were related to Whitewater. As it happened, on the very afternoon Foster died, the FBI had obtained a search warrant for the offices of David Hale, in connection with an investigation of Hale's Capital Management Services by the Small Business Administration. The search of Hale's offices opened a new stage in the investigation of Whitewater, an issue that at the time appeared to have played itself out. Later, there would be much speculation that Foster somehow learned of the existence of the search warrant, a discovery that triggered his suicide. Realistically, this seems unlikely. While Foster once described Whitewater as "a can of worms," he had little personal stake in the matter, aside from his connection to Hubbell and the Rose Law Firm. By contrast, Foster had lied about Hillary's role in the travel office firings, and there was every reason to believe that he would have to choose between betraying Hillary and lying again, in public testimony before a Senate committee.

Foster had other papers in his possession likely to be at least as sensitive, including the copies of the Clintons' tax returns, a file marked "Hillary Clinton, Personal" and, for all anyone knows, documents relating to the Clintons' use of private investigators to track various women connected with Clinton in the past.

In any event, Hillary Rodham Clinton had no way of knowing exactly what might turn up in a search of Foster's office, including personal jottings made in a state of mental distress that might be embarrassing to her or his family. It seems clear from the phone records that Hillary overruled Bernie Nussbaum's plan to allow Heymann and the FBI access, relaying her wishes through Susan Thomases. Understandable as her feelings may have been, interfering with a Justice Department investigation was a serious matter.

But what may have been a rash decision made during a time of emotional stress would be compounded by the participants' lack of candor. In the summer of 1995, when the Senate special committee on Whitewater took up the question of the search of Foster's office, Susan Thomases, Maggie Williams and, to a lesser extent, Bernie Nussbaum, by then an ex-counsel, were seized by collective amnesia about Hillary's input. The sight of senior White House officials and, in Thomases's case, a close friend of the First Lady of the United States stonewalling about what could have been a simple misjudgment did a great deal to undermine confidence in the Clinton administration and spur the darkest suspicions about the contents of Foster's files.

More important in the long run, the Clintonites' lack of respect for process shocked many of the White House's permanent employees. Hillary and her aides failed to realize that there was an ethic of loyalty to the office of the presidency which the new administration had undermined to its peril.

Linda Tripp, at the time an executive assistant in the counsel's office, would recall her reaction to the search this way in testimony before the Independent Counsel's grand jury:

> There was always a sense in this White House from the beginning that you were either with them or you were against them. The notion that you could just be a civil servant supporting the institution just was not an option. I had reason to believe that the Vince Foster tragedy was not depicted accurately under oath by members of the administration. I had reason to believe that—and these are, remember, instances of national significance that included testimony

by, also, Mrs. Clinton—also in Travelgate. It became very important for them for their version of the events to be the accepted version of events.[8]

"None of the behavior following Vince Foster's suicide computed to people just mourning Vince Foster," Tripp thought. And she would go on further to recall that "the flurry of activity and the flurry of phone calls and the secrecy" began all over again two months later, when word came over the office fax machine that private eye Jerry Parks had been shot to death in Little Rock. Tripp had no information to suggest why Parks's murder, in September 1993, should have been a cause for concern in the White House. She just knew that it was. That Parks did indeed have a file reflecting surveillance of Bill Clinton comes to seem more likely.

Ironically, Linda Tripp was responsible for quelling one troublesome question about Foster's actions the day he died when she contradicted a White House security guard who recalled that Foster had been carrying a briefcase when he left the building. The guard's story raised troubling questions. Was Foster going to meet someone? What happened to the papers he was carrying? Was it possible that the briefcase concealed a gun, which he had actually carried with him into the West Wing that day, in a serious breach of security? Tripp, the career government worker, laid those worries to rest with her account, which she still stands by. But her bosses were not appreciative, in part because the Senate investigators turned up an e-mail message to a coworker in which Tripp compared Nussbaum and two associate counsels to the Three Stooges.

IT may well be that there is still more than is known about the events in Fort Marcy Park on July 20, 1993. However, the established facts were shocking enough. A presidential counselor, one who apparently had a romantic attachment to the First Lady, had worked in the White House for six months without an FBI background check or a permanent pass. Presumably, Foster had a gun in his car when he parked that morning inside the White House compound, although no weapon was detected by the gun-sniffing dogs who patrol the grounds.[9]

If it weren't for his still pending background check, moreover, Foster might have been less reluctant to consult a therapist. He had obtained referrals from his sister Sheila but never made an appointment. Instead, he had phoned his family doctor in Little Rock, who gave him a prescription for fifty-milligram tablets of Desyrel (trazadone), an antidepressant. While Desyrel was considered safe, the first doses of a new medication sometimes have unpredictable, and often agitating, effects on depressed people. The Desyrel, which Foster began taking the day before he died, may have given him just enough willpower to destroy himself.

In the wake of Foster's death, the Clintons were braced for a fusillade of questions. It happened that Foster's mentor, Phil Carroll, and Carroll's wife were guests at the White House, sleeping over in the Lincoln Bedroom, on the night Foster died. At breakfast the next morning, the President did his best to cheer them up, urging them not to let themselves be dragged down by the negative stories they were going to read in the press. Similarly, in his statement that day to the White House staff, Clinton appeared to be almost pleading for an end to speculation: "No one can ever know why this happened. Even if you had a whole set of objective reasons, that wouldn't be why it happened. . . . So what happened was a mystery, about something inside of him. And I hope all of you will always understand that."

Before Foster's funeral that Friday, at St. Andrew's Roman Catholic Cathedral in Little Rock, Clinton consoled members of the Foster family who had gathered in a room off the sanctuary. Hillary waited alone, in an alcove off the sanctuary, and was ushered directly to her pew just before the ceremony began. On the flight back to Washington on Air Force One, the President reminisced with the mourners while Hillary kept to herself, closeted in the private cabin.

While some might see the chill between Hillary and the Foster family as confirmation that she and Vince had been having an affair, there were certainly other reasons why Lisa Foster and her relatives would be cool toward her. It was no secret that Foster idealized Hillary for her assertive lawyering, her social commitment and her professional drive. Lisa Foster, the Junior League, tennis-playing stay-at-home wife and mom, could hardly avoid taking this personally. (In fact, Lisa and Hillary were not as different as they seemed superficially. Though Foster complained that his wife was too dependent, she could be outspoken and opinionated.)

During the final weeks of his life, however, Foster had begun to refer to Hillary with some bitterness as "the client." Always well prepared, and so meticulously honest that he reimbursed his firm for personal photocopies and phone calls, Foster found himself trying to cope with the fudged paperwork and indefensible secrecy of the health care task force, the bungled firing of the travel office workers and the amateurish vetting of key administration nominees—which had been left to him, Hubbell and other aides, instead of to the FBI.

Judging by the deposition Hillary Clinton gave to special prosecutor Robert Fiske's investigators, she was oblivious to Foster's distress. Nor did she see herself as being in any degree to blame for the managerial failings of the projects in which she was involved. Any frustrations that existed, she suggests, were just "general blowing off steam" over the mess the Republicans had left behind—"I think we were all amazed at some of what we found when we got here." Ironically, considering Hillary's later resentment of conspiracy theorists, it seems that she, too, had her suspicions about Vince's death. Webb Hubbell writes that Hillary asked him to check out a rumor that Foster was assassinated by a top-secret team of navy hit men. Just how one could check such a story is unclear.

On Monday afternoon, July 26, associate counsel Steve Neuwirth was collecting Foster's personal effects when he found some torn pieces of yellow paper in the bottom of Foster's well-worn black briefcase. Neuwirth and Nussbaum quickly reassembled the scraps and called Hillary into the office to see what they had found. According to Nussbaum, Hillary "began to read it but she didn't read it. She didn't appear to read it. . . . She had an emotional reaction and she said I just can't deal with this. . . . Bernie, you deal with this."

Bill Burton, the White House assistant chief of staff, who also happened to be present, had a different impression. He recalled that when Nussbaum began to read her the note, Hillary coolly interrupted him and left the room. She explained that "decisions to be made concerning privilege issues, notifying the Foster family," and so on weren't things in which she should be involved.[10]

Since Nussbaum had made a show of examining the same briefcase in front of Philip Heymann, the FBI and the Park Police a week earlier, it was difficult to imagine that it had taken a week to discover the torn-up note. More likely, continued questions in the press, including calls for a special prosecutor, had made it clear that some explanation for Foster's suicide was better than none. Even now that the note was assembled, no one knew what to do with it. Nussbaum called Susan Thomases, and also Mack McLarty, who was traveling with Clinton in Chicago. McLarty suggested holding on to the note until Lisa Foster returned from Little Rock. Eventually, the note was offered to Attorney General Reno, who was horrified and told him to give it directly to Captain Charles Hume of the Park Police.

Despite the opinions to the contrary, there can be no doubt that the note was genuine. But it wasn't a suicide note either, just a memo—or possibly a page from a memo—in which Foster exonerates himself and his coworkers: "No one in the White House, to my knowledge, violated any law or standard of conduct, including any action in the travel office," he wrote. There follows a list of charges against the FBI ("lied in their report to the AG"), the press ("covering up the illegal benefits they received from the travel staff"), the GOP ("lied and misrepresented its role and covered up a previous investigation"), the usher's office ("plotted to have excessive costs incurred, taking advantage of Kaki and HRC") and *The Wall Street Journal* ("editors lie without consequence"). Sadly, Foster felt unfairly accused and responded with a barrage of accusations against others. One suspects that this mind-set was not his alone.

Time moved quickly for the occupants of the White House. Within a few weeks of Vince Foster's funeral, Hillary, Bill and Chelsea were on vacation in Martha's Vineyard. The liberal celebrities on the Vineyard took up the Clintons in a big way. The family went sailing with Jackie Onassis and had dinner at Carly Simon's house. For the first time since the inauguration, Hillary was surrounded by smart, chic people who were excited about having a young, Democratic couple in the White House.

The vacation lifted both the Clintons' spirits and sent them back to Washington in an upbeat mood for the official launching of the health care initiative. The public relations blitz included a rally on the White House grounds and a major speech to Congress by the President, regional town hall meetings, talk show interviews and a televised health care teach-in, all leading up to Hillary's appearances before five congressional committees. As a spokesperson, Hillary could hardly have done better. Poised and articulate, she had been well briefed on the committee members' backgrounds and home districts, and she deftly mixed little bouquets of flattery into her testimony. Congressmen, Democrats and Republicans alike, fawned back. Dan Rostenkowski, chairman of the House Ways and Means Committee, actually kissed her, saying, "I think in the very near future the President will be known as your husband."

It did seem that the moment for serious action on health care had finally arrived. Polls showed overwhelming public support for reform. The HIAA, an insurance industry lobbying group, had filmed a series of ads featuring a couple named Harry and Louise, who worried aloud about how reform would affect them. On September 22, the day of Clinton's speech to a joint session of Congress, the group offered to cancel its ad campaign, provided the concerns of the insurance industry got a more respectful hearing from Magaziner and Mrs. Clinton. The White House did not bother to reply.

There was joy in Hillaryland during those days, and one member of the First Lady's loyal staff crowed to Maureen Dowd, "This is Eleanor Roosevelt time!"

The comparison, unfortunately, was all too apt. Much of the

courtly reception Hillary received on the Hill was an elaborate ritual of deference to her position as First Lady. In reality, many in Congress were furious at being left out of the process and more than skeptical of the reform's premises.

And rightly so. Asked by Senator Dan Coats to explain her contention that government could deliver health care more efficiently than the private sector, Hillary explained that "the heavy administrative costs you will find in the private sector insurance market are due to a very clear decision, which is, the more money we can spend making sure we don't insure people who will cost us money, the more money we will make . . . and in order to choose among everyone sitting in this room—who is and who is not a good risk—that takes a lot of time and a lot of manpower, a lot of personnel cost. And so I think that, if you look at the way the private sector operates, you will find an enormous amount of inefficiency. . . ." Whatever the merits of universal health care, it is hard to take seriously the cost estimates of someone who thinks that insurance companies would be more efficient if they threw out their actuarial tables.[11]

The Magaziner task force had solved all the problems of the health care delivery system—on paper. Their plan was a graduate student's dream and a legislator's nightmare. Even inside the Clinton White House, it was understood that the task force had become a rogue process. The group's final report, which was not even finished in time for the official launch date, would eventually run to 1,342 pages, which covered everything from nursing homes to the ideal racial composition of medical school classes. Magaziner projected that the reforms would cost $400 billion over six years and $130 billion a year when fully phased in. Even these figures made sense only if it were possible to drastically cut spending on patients over sixty-five—presumably by phasing in some sort of "futile care" doctrine to justify denying treatment to the very ill and very old. No one, moreover, could estimate what the impact of the plan would be on the rest of the economy.

As long as Bill Clinton continued to express faith in the health care initiative, none of his aides were about to contradict him. Male staff members in particular were leery of saying anything that would touch off a quarrel between the President and his wife. Women tended to be more outspoken, able to deal with Hillary as just another member of the President's inner circle. Donna Shalala, for one, tried to explain to her why it was a bad idea to include so many details in the plan: There was a risk of creating a negative coalition, uniting people who were opposed to one provision or another of the bill.

Hillary wasn't getting a realistic picture of the plan's chances from her public appearances, either. Her speeches were very effective, but usually made to handpicked audiences. When she did meet groups likely to be critical, she disarmed them by making reassuring promises. Addressing a meeting of the AMA House of Delegates in June 1993, Hillary assured the doctors that under the Clinton health alliance system they would never have to call an 800 number to beg for permission to treat patients as they saw fit. The line won enthusiastic applause. But how could she make such a promise? How could price caps be enforced if there was no case management?[12]

The health care reform effort began to degenerate into a passive-aggressive drama, not untypical of the way the Clintons handled personal issues. Rather than take control of the effort one way or another, Clinton simply let it slip lower on his list of priorities. The budget process had already taken precedence. Now attacks on U.S. troops in Somalia, the passage of NAFTA legislation and the crisis in Bosnia each had its turn. Hillary was reduced to bureaucratic infighting to get her husband's attention.

Frustrated, Hillary turned up the volume of her attacks on the insurance industry. In a speech to the American Academy of Pediatricians she denounced insurance companies for having "the gall" to put the "Harry and Louise" ads on TV when they were "the very industry that has brought us to the brink of bankruptcy because of the way they financed health care." An employee of the Democratic National Committee actually called up the actress who played Louise in the ads and tried to talk her into denouncing them. When the actress declined, the DNC staffer threatened to see that she never worked in the industry again.

The reply to Harry and Louise was a media campaign organized by political consultant Mandy Grunwald. It thrashed away at the villainous insurance companies but did little or nothing to answer the public's questions about what the reforms would mean to them.

Some members of the Business Roundtable, a group representing the chief executives of America's largest companies, had begun to suspect that the Clinton Health Security Act was designed to fail. The plan specified what benefits insurance companies had to offer and also the maximum amount they could charge. If the limits were unrealistic, then the companies would be forced out of business, and the administration would switch over to a 100 percent public, single-payer system. By default, Hillary and her allies would have the system they preferred from the beginning, and they could blame the greed of insurance companies and the rest of the private sector for the failure of the public-private partnership. They'd had their chance and failed.

At one Business Roundtable meeting, Hillary lost her temper when Robert Johnson, of the Johnson & Johnson Corporation, suggested that the regional alliances, the backbone of the Clinton-Magaziner plan, were really government regulatory agencies.

"I said they were purchasing cooperatives, and that's what they're going to be!" Hillary retorted, slapping the table for emphasis.[13]

Hillary and Magaziner commandeered the White House "war room," formerly the nerve center of the effort to pass the budget, renaming it the "delivery room." Months into the effort, they continued to believe they had a chance of winning over enough moderate Democrats and Republicans to pass the legislation. Meanwhile, the moment of bipartisanship had passed. Energized by the ideological tilt the issue had taken, conservative Republicans led by Newt Gingrich had begun to feel that just defeating the plan was not enough. They were thinking in terms of retaking the House.

Despite efforts by White House aides to get Hillary to tone down her rhetoric, she maintained a take-no-prisoners attitude. Tennessee Democrat Jim Cooper had proposed a compromise bill that included so many features of the Clinton plan that it was known as "Clinton Lite." Nevertheless, Hillary took Cooper's criticisms of some task force proposals as a personal assault. White House aide Jeff Eller, who was managing the effort to get congressional action on the plan, came up with the idea of planting anti-Cooper stories in the newspapers of his home state, where he was running for the Senate against Republican Fred Thompson. And, in fact, this was just what happened. In a campaign apparently stage-managed by Representative John Dingell, the chairman of the Energy and Commerce Committee, AFCSME and other unions denounced Cooper as a tool of for-profit hospitals and HMOs. Thus, the health care zealots helped to elect a Republican over a moderate Democrat.

Ken O'Donnell, an aide to New York senator Daniel Patrick Moynihan, summed up the judgment of Democratic pols when he told journalists Haynes Johnson and David Broder, "No one working for Hillary Clinton, as far as I can tell, gets politics one whit better than Ira Magaziner does. They don't get politics. That's Whitewater, that's everything. They have a War Room for everything. They don't understand it's not a fucking War Room."[14]

THIRTEEN

Lawyering Up

IN spite of all the setbacks and personal tragedies of the Clintons' first year in the White House, as Christmas 1993 approached it seemed that they were at last beginning to find their balance. The administration could boast notable successes on the budget, the family leave bill and NAFTA. Unemployment was going down. The President's approval ratings were rising.

The Clintons had planned a heavy schedule of entertaining during the Christmas season, including no fewer than four parties for members of the press. Hillary was going to do interviews for network TV in which she would discuss her continuing quest for spiritual meaning. And on the family front, the private quarters of the White House were filled with relatives, including Hillary's mother, Dorothy; Clinton's mother, Virginia; and Virginia's fourth husband, Dick Kelley.

On December 17 and 18, the Christmas spirit was blighted when the *Los Angeles Times* and *The American Spectator* released a pair of articles detailing Bill Clinton's use of members of his Arkansas state trooper security guard to arrange assignations with women. The *Spectator*'s story packed in more graphic details and gossip, including the allegation that Hillary and Vince Foster had been lovers. The *L.A. Times* article was better documented; its authors, William C. Rempel and Douglas Frantz, had used Arkansas's Freedom of Information law to obtain Clinton's phone records from 1989, documenting his frequent and lengthy phone calls to women, often in the wee hours of the morning. Together, the articles were explosive because they showed that Clinton had used state employees to procure women—and had continued to do so through the 1992 campaign, even after his supposedly soul-cleansing appearance on *60 Minutes*.

During the week after the stories appeared, the Clintons seemed to be in shock. Looking tired and drawn, they attended White House Christmas parties but were rarely observed talking to each other. Upstairs in the family quarters, the presence of two mothers-in-law, especially the ill but still feisty Virginia Kelley, did not make for a lighthearted mood. The President escaped from the pressure cooker by going out to play golf with Vernon Jordan, while Hillary was left to manage the relatives.

But people who imagined that the First Lady reacted to these bombshells by feeling that her husband had betrayed her were missing the point. First of all, it isn't clear that Hillary ever read the *Spectator* and *L.A. Times* articles. Aides tried to talk her out of it, and in fact, there would be times over the next few years when Hillary didn't even read the daily newspapers. Members of her staff, especially Melanne Verveer, did keep close watch on anti-Clinton articles in the *Spectator* and elsewhere. So one way or another, the substance of these attacks did make its way to Hillary, but often in a "Can you believe this!" context. On some level, Hillary must have been carrying a burden of unresolved anger toward her husband over his many infidelities; it is hard to imagine that she wasn't. But she also felt passionately that her husband's behavior was their business and theirs alone. She was angry at the reporters, the troopers who talked to them and, not least of all, the public for paying attention. Of course, Hillary was right that the people writing exposés of Bill Clinton's womanizing hardly had her best interests at heart. But she was also unrealistic in thinking that such matters could be consigned to a "zone of privacy"—a favorite phrase of hers. In today's world, there is probably no aspect of a United States President's life that won't be discussed in the media, sooner or later. An affair or two in Clinton's past might be of limited or no relevance to his conduct in office, but a lifelong pattern of behavior, involving the use of government employees to solicit women, was another matter.

In line with her tendency to demonize the political opposition, Hillary found it all too easy to latch onto the notion that a conspiracy of Clinton haters was working from a master plan. Her attitude, as explained by a friend who has known her for many years, was "Everyone knows that Bill has this problem. So if it's coming out now, there must be a reason." The problem was that everyone *didn't* know. Quite a few people had believed Clinton when he promised the American public the problems that led him to cause "pain" in his marriage were a thing of the past.

In her previously scheduled end-of-the-year interview with the Associated Press, Hillary hinted at her theory of a right-wing conspiracy, saying, "I find it no accident that every time we're on the verge of fulfilling our commitment to the American people" this type of story comes out.

Unfortunately for this theory, Hillary was mistaken on two counts. First, health care reform was in trouble for reasons that had nothing to do with the troopers' allegations. Second, Bill Clinton had known months in advance that the troopers were planning to go public. Warned by Raymond "Buddy" Young, his former head of security and now a $92,000-a-year regional manager for FEMA, the Federal Emergency Management Agency, Clinton had personally called trooper Danny Ferguson twice while Ferguson was on duty at the Arkansas Governor's Mansion and talked about finding jobs for him and a second trooper, Roger Perry, either with FEMA or with the U.S. Marshal's Service.

To the extent that Hillary was angry with her husband after the troopers' stories appeared, one can be sure it was over these phone calls, which amounted to an abuse of power in the interests of a cover-up. Within days, Betsey Wright had abandoned her position as a high-powered lobbyist for the Wexler Group and rushed down to Arkansas, where she got Ferguson to sign an affidavit prepared by

Little Rock attorney Stephen Engstrom, stipulating that Clinton had never explicitly connected job offers with the troopers' silence.

ON February 11, 1994, just when it seemed that the story was about to fade into the background, a woman named Paula Corbin Jones appeared at the Conservative Political Action Conference (CPAC) in Washington to announce that Bill Clinton had made an unwelcome proposition to her while she was working as a state employee at a convention held at the Excelsior Hotel in Little Rock on May 8, 1991.

The *American Spectator* article had described the incident, telling how Clinton, at the hotel for a Governor's Quality Management Conference, had pointed out a state clerk named Paula who was working at the registration desk, telling his trooper escort, the aforementioned Danny Ferguson, "She has that come hither look." The *Spectator* strongly implied that Paula had been a willing participant in a sex act and quoted Ferguson as saying that she had told him she was available to be the governor's girlfriend. Jones had given friends a very different account of her meeting with the governor at the time, and after the article appeared, one friend, Debra Ballentine, read the account to Jones over the phone. Jones had a straightforward libel case against the *American Spectator* and Danny Ferguson, but Bill Clinton was the only other person who knew what happened in that hotel room, and only he could restore her reputation. Somewhat naively, Jones decided that the President of the United States owed her an apology.

Hardly anyone took Paula Jones seriously. An article in the *Washington Post* style section sarcastically dismissed Jones's press conference as "yet another ascension of Mount Bimbo." Robert Bennett, the high-profile Washington attorney hired by Clinton to defend against the suit, dismissed her allegations as "tabloid trash." And Clinton attack viper James Carville scoffed, "Drag a 100 dollar bill through a trailer park you never know what you'll find."

It happens that plenty of people in Arkansas are every bit as offended by the label "white trash" as African Americans are by the "n" word. It defies understanding how an administration filled with people who prided themselves on their sensitivity to racial and gender stereotypes could have come up with the strategy of attacking Paula Corbin Jones as trailer trash. Nevertheless, this hate-speech strategy went almost unchallenged until November 1996, when Stuart Taylor published a lengthy article on Jones's lawsuit in *The American Lawyer*. Taylor compared Jones's claim with that of Anita Hill and concluded that, on the whole, Jones was more credible. Taylor ascribed the vastly different treatment of the two women to class bias, concluding, "It has not yet been proven that unsophisticated, big-haired, makeup-caked women from small hamlets in Arkansas—even ones who pose topless for sleazeball boyfriends—are any less likely to tell the truth than Yale-educated law professors like Anita Hill."[1]

Back in 1992, Hillary had made her decision that as a matter of principle she would maintain a defense attorney's posture, choosing to accept her husband's plea of innocence to any particular allegation that happened to arise. As it happened, this approach protected her interests quite effectively. Better to be seen as a too-trusting wife than an accomplice in a cover-up. What it did for Bill Clinton is another matter. For some reason, Paula Jones had not been on the list of problem women compiled before the election. Clinton may have found the incident at the hotel too embarrassing to own up to or, perhaps, not worth mentioning because "nothing" happened. Now, Hillary's need-not-know stance discouraged any more confessions, and Clinton couldn't tell the truth to his personal attorneys without telling his wife as well.

Repeatedly assured by the President that there was no substance to Paula Jones's allegations, Bob Bennett pursued an aggressive strategy, using private detectives to hunt down Jones's old boyfriends and hinting publicly that if the case came to trial, her reputation would be destroyed.

While the Jones case lurched its way forward through a series of court challenges, Whitewater was once again in the news. Under investigation for the bad loans his company had made to the

McDougals, David Hale was complaining that he was being set up to be the "fall person" in the case, and in September 1993 he made a deal to become a government witness. Six days after his indictment, Jean Hanson, the general counsel of the Department of the Treasury, warned Bernie Nussbaum that the Resolution Trust Corporation was issuing nine criminal referrals in connection with the investigation. The Clintons and Jim Guy Tucker were named as witnesses,

By November 1993, *The Washington Post* was badgering the White House to release Whitewater documents mentioned in the RTC referral. Senior Clinton advisers including George Stephanopoulos and David Gergen favored complying with the *Post*'s request, arguing that the facts would all come out eventually, but Hillary was adamantly against disclosure. As she saw it, her financial dealings were none of the public's business and it would be foolish to give ammunition to the administration's enemies. In late December, the Clintons' private attorney, David Kendall, arranged for the Justice Department to issue a secret subpoena for the documents. Once handed over to Justice under the subpoena, the papers were no longer subject to the Freedom of Information Act, and thus the *Post* couldn't obtain them.

Hillary was even more vocal in her opposition to calls from the press and congressional Republicans for an independent prosecutor. On this count, surely she was right. While independent prosecutors are undoubtedly necessary in some cases, it can't be good for the orderly functioning of democracy to have prosecutors with unlimited budgets and virtually unlimited power poking through elected officials' pasts, searching for some grounds for indictment. As interesting as Whitewater may have been to the press, the multiple government investigations that ensued were the products of a mania—the lawyering up of American society. No longer able to judge what's right, we rely on attorneys and judges to tell us what's legal.

Unfortunately, Bill and Hillary Clinton were ill cast to lead a stand against this madness. Lawyers themselves, they embodied the trend toward legal activism in every respect. For example, amazingly, even after Paula Jones had filed her lawsuit, Hillary supported, and President Clinton signed, the Violence Against Women Act, which gave the courts increased latitude to snoop into the private life of individuals accused of harassment and use behavior outside the workplace as evidence of their guilt. Clinton, moreover, had campaigned in favor of renewing the Watergate-era independent prosecutor law, which had been allowed to expire during the last year of the Bush administration.

Reluctantly, Hillary agreed that a special prosecutor appointed by Janet Reno would be a better bet than an independent counsel named by a three-judge panel. Robert Fiske, Jr., who was named by Reno in January 1994, appeared prepared to take a very narrowly focused look at the Clintons' role in Madison Guaranty's problems. With Fiske seemingly in control of the Whitewater investigation, the Democratic-controlled House of Representatives passed a new independent counsel law. (The Democrats favored the law by 239 to 5; the Republicans were opposed by 100 to 77.) Then, in an unexpected turn of events, the three-judge panel empowered under the new law decided to dismiss Fiske, who was already wrapping up his probe, and replace him with Kenneth Starr, a former solicitor general under the Bush administration. Stunned, the Clintons regarded this turn of events as a Republican plot against them.

A worse shock was coming. During Christmas vacation, Webb Hubbell had learned that the Rose Law Firm had begun an investigation into his use of company credit cards for personal expenses. As the inquiry expanded into charges of overbilling clients, the firm, already facing questions about its representation of Madison Guaranty, decided to take a hard line. Hillary's old friend Tyson counsel James Blair had been monitoring the situation, and in March 1994 he told the Clintons that Hubbell was in legal jeopardy and needed to resign. The Clintons' personal attorney, David Kendall, looked into the charges and agreed.

On March 13, following a high-level policy meeting at the White House, Mack McLarty took Hillary aside and told her that he and some other Clinton friends were going to try to help Webb out. Hillary was relieved. McLarty soon got Hubbell a $3,000-a-month consulting job. Vernon Jordan set up consulting work with

MacAndrews and Forbes to the tune of $60,000. The city of Los Angeles ended up paying $24,000 for Hubbell to make two five-minute phone calls. The Riadys gave $100,000 for unspecified services. Erskine Bowles and Mickey Kantor were also involved in lining up fees for Hubbell.

Were these fees, amounting to more than $400,000, an attempt to help out an old friend in trouble? Or were they hush money, as Independent Counsel Ken Starr would charge?

Undoubtedly, the Clintons did want to help Hubbell. With Vince Foster not yet a year in his grave, Hillary must have been tormented by the thought that by inviting her close colleagues to Washington, she had unwittingly engineered their destruction. Certainly no one wanted Webb to become desperate.

On the other hand, the transcripts of phone conversations made by Hubbell from federal prison suggest that not even a twenty-one-month jail sentence had persuaded him to tell all he knew. Discussing a plan to countersue the Rose firm in a dispute over the amount he owed, Hubbell's wife, Suzy, warns him that Marsha Scott had told her that the White House didn't want him to sue: "By suing it makes it look like you really don't give a shit. And that you are opening Hillary up to all this."

"Sometimes you have to fight battles alone," Hubbell assures Suzy, adding after some more back-and-forth, "I will not raise those allegations that might open it up to Hillary."

"Yes," she says. "Then I get all this back from Marsha who is ratcheting it up and making it sound like if Webb goes ahead and sues the firm then any support I have at the White House is gone. I'm hearing the squeeze play."

"So I need to roll over one more time." Hubbell sighs.

This conversation sheds at least some light on Hillary's state of mind as she faced multiple interrogations over her role in representing Madison Guaranty—from the press, the Resolution Trust Corporation, the FDIC, the General Accounting Office and the Fiske and Starr investigations. Obviously, Hillary felt that a close look at her Rose firm records might put her in jeopardy. One can only wonder whether Hillary's difficulty in explaining her work for Madison Guaranty arose because she billed more hours than she actually worked. Hillary, after all, had chosen to believe that McDougal had cheated her and Bill in the Whitewater deal. Adding a few hours to her bills would have been easy to rationalize.

In April 1994, after David Kendall revealed that the Clintons were paying $14,615 in back taxes to cover unreported profits from Hillary's commodities trades, Hillary made an attempt to put questions about her professional past behind her through an unprecedented televised press conference. Dressed in a pale pink two-piece dress, she sat alone on a raised platform in the State Dining Room and calmly answered questions for seventy-two minutes. The "pretty in pink" press conference, as it came to be called, was an opportunity for Hillary to give the Clintons' side of the Whitewater allegations. Unfortunately, it went awry, because Hillary could not bring herself to admit that she might have made some mistakes, even in the distant past. Despite the urging of Mandy Grunwald, who helped to prep her for the conference and who was hardly a soft-liner when it came to defending the Clintons, Hillary just couldn't bend.

Answering questions about her commodity accounts, Hillary not only said that she had made her own trades, she professed to have no idea why she had been so successful. In fact, in 1980 Hillary represented an Arkansas banker, Hayden McIlroy, in a fraud suit against Refco chairman Thomas Dittmer, claiming that cattle futures trades had been manipulated during the summer of 1979. Even if Hillary had no suspicion that she was receiving special treatment during the period when she was trading, she had obviously learned a great deal more about machinations in the market through her work on the McIlroy suit. Hillary's presentation of herself as a naïf, who had never given the source of her windfall another thought, was unconvincing.

Hillary's testimony under oath on various Whitewater matters followed the same disingenuous pattern. She repeatedly denied that she had ever worked on the Castle Grande real estate project. When the evidence turned up proving otherwise, she then justified her answer by claiming she knew the deal only as "IDC"—an unlikely excuse. In November 1994, moreover, she told the RTC that she knew Seth Ward as "Suzy Hubbell's father." She did not mention doing any legal work for him until investigators discovered on their own that she had. She then characterized Ward very differently, as a difficult and demanding client, and the man who represented Madison Guaranty in its dealings with the Rose firm.

In August 1995, after the RTC issued its report revealing that Hillary had indeed been one of eleven Rose lawyers who had done legal work for Madison Guaranty, Alston Jennings, who represented Seth Ward in a civil suit against Madison Guaranty, was invited to the White House. Jennings later said that he and the First Lady discussed general press inquiries about her legal work, but that Ward's name was never mentioned. It was around this time, however, that her executive assistant, Carolyn Huber, came upon the Rose firm billing records in the book room on the third floor of the White House. Huber picked up the computer printout, stored it in a box of "knick-knacks" and realized what it might be only months later, in January 1996. Hillary's fingerprints were found on two pages of the printout, one being the page detailing her work on Castle Grande.

Following this discovery, in January 1996, Hillary was summoned to testify in person before the Whitewater grand jury. Although she appeared unflappable confronting reporters outside the courthouse, friends agree that she found the experience humiliating and that it hardened her bitterness against Ken Starr and the press.

From Hillary's point of view, the Whitewater investigation was a surreal experience. In a January 1996 interview with Barbara Walters, she would compare herself to the child in the nonsense verse who said:

> As I was standing on the street,
> As quiet as could be,
> A great big ugly man came up
> And tied his horse to me.

Hillary often reacted to criticism by taking refuge in passivity, as if the things people were saying had nothing to do with her. In fact, both she and Bill Clinton took a casual approach to certain ethical matters that had a way of drawing them into deep waters. As early as 1988, Hillary had decided to remove documents dealing with the Seth Ward option from the Rose firm's files. She had no way of anticipating that she would ever be asked under oath about her Madison work, and when she was asked, she decided not to take the risk of telling the whole truth. Like her husband—and like her lawyers from Williams & Connelly, a firm well known for its hardball tactics—she seemed to see testimony under oath as a kind of word game, in which she gave answers that might be technically compliant, but that appeared to be lies to people who did not have the benefit of Ivy League law degrees. While she may have been protecting herself from legal jeopardy, politically this approach hurt her badly.

WHILE Whitewater occupied more and more of Hillary's time and exposed her to the jibes of the late-night talk show hosts, who joked incessantly about the prospect of her being indicted, the health care task force had come to a sorry end. The partisan tone of the debate, which Hillary had encouraged with her attacks on the motives of the drug and insurance companies, was coming back to haunt her. At a speech in Seattle in July 1994, she was confronted by several hundred angry demonstrators, whose boos and catcalls made it clear that they saw Clintoncare as just one symptom of a power grab by Washington that was threatening their freedom as Americans. The rage of the protesters was frightening, especially when a group of them swarmed around the First Lady's car after her speech, preventing her driver from making a quick departure.

In the Senate, a group of moderates including Democrat George Mitchell and Republican John Chafee were trying to settle on a bipartisan compromise, but the administration bill had polarized sentiment there, too. There was no pressure coming from the President and his close advisers. Indeed, they were suspected of sending covert signals that they hoped the bill would be allowed to die without coming to a vote. Not only did this happen, but the Republicans were able to use the momentum they had gained to win big in the midterm elections, sweeping into control of both houses.

But one area of the White House that was definitely flourishing was the counsel's office—the very entity whose failures led to Travelgate, violations of the advisery panel regulations by the health care task force and other snafus. Although Bernie Nussbaum was forced to resign over questions about his handling of the Foster inquiry, and William Kennedy III departed in the wake of revelations that he, too, had a nanny tax problem, the counsel's office was steadily expanding, as was its caseload. A December 1994 internal "task list" enumerated no fewer than thirty-nine areas of vulnerability, which were parceled out among fifteen staff lawyers.

The list began with "Foster document handling"—of which two subheads were "security/Livingstone issues" and "inconclusiveness re Williams removal of documents." It went on to cover such items as "Travel Office," "Use of White House resources for response efforts" and "White House operations (drugs, passes, helicopters)," before concluding with "ADFA (political favors, Larry Nichols)," "Mena Airport," "troopers" and, finally, the self-explanatory "women."

The problem described as "security/Livingstone issues" was especially interesting, because it showed that the White House even at this point recognized it had a problem with Craig Livingstone, director of the White House Office of Personnel Security. Livingstone's office was responsible for coordinating FBI background investigation files and other paperwork used by the counsel's office to issue White House security passes. A strange choice for this post, which usually went to a civil servant of unquestioned discretion, Livingstone was best known as a campaign "dirty tricks" operative. According to Dennis Casey, a consultant with the Gary Hart campaign, Livingstone had written a report detailing certain "peccadilloes and vulnerabilities" of Hart's opponents and their supporters, weaknesses that might be used against them. In 1992, he had been responsible for getting a man in a chicken suit to heckle George Bush on the campaign trail. (The operation backfired, because Bush got a kick out of the chicken and always seemed to perk up when he was around.)

According to notes taken by FBI agent Dennis Sculimbrene, Bernie Nussbaum told him that Livingstone was recommended for his post by Hillary Clinton. William Kennedy III was under the same impression. Warned by Sculimbrene that the job should go to someone who was "squeaky clean," Kennedy said, "It doesn't matter anyway. It's a done deal. Hillary wants him."

Not only did Hillary deny putting Livingstone in his job, she would later say under oath that she wasn't even aware of who Livingstone was. In 1996, when the House Committee on Government Reform and Oversight, then known as the Clinger committee, looked into Livingstone's hiring, they had no trouble coming up with a White House intern who could contradict this testimony. The intern had been present in a West Wing corridor when Hillary greeted Livingstone by name in a friendly fashion, which suggested she knew quite well who he was. More to the point, perhaps, while running the Office of Personnel Security, Livingstone continued to do advance work for the Clintons' travels. Among the trips he arranged was the Clintons' return to Arkansas for the funeral of Virginia Kelley when she succumbed to breast cancer in January 1994.

However he came to be hired, Livingstone's tenure saw the Office of Personnel Security fall into utter disarray. In previous administrations, the entire White House staff received their permanent security passes in six or, at most, nine months. Under Livingstone, only two employees had been issued permanent passes by September 20, 1993. Paperwork piled up in his office, and the FBI agents assigned to the White House seethed with frustration over their inability to get Clinton aides to fill out their SF-86 background forms.

In the meantime, Livingstone's assistant, Anthony Marceca, was assiduously requisitioning the FBI background files of hundreds of individuals who had worked in the White House during previous administrations. FBI files are filled with sensitive and potentially damaging information, including interviews with ex-spouses, psychiatric evaluations, credit information and reports of arrests and job-related problems going back to the subjects' teenage years. The usual procedure is that no one sees these records apart from the employees in the counsel's office, who make decisions about issuing security passes, and a few trusted employees in personnel security. On Tony Marceca's watch, more than 900 files that had no business being in the White House in the first place were stored on a shelf in the Personnel Security Office vault, where they were known to other employees as the "dead bin," or simply "Tony's files." Numerous employees were given access to the vault, including interns. More alarming still, the office log that kept track of who signed out files from the vault later turned out to have pages missing—a six-month gap reflecting access to the files between March and September 1994.

When the existence of the "dead bin" came to light in 1995, Marceca argued that he'd requisitioned the files in error because he was working from an outdated Secret Service list and mistook the designation "I," for "inactive," to mean "intern." To believe this, one has to imagine that Marceca thought there were hundreds upon hundreds of interns working for the White House, interns who happened to be men in their forties, fifties and sixties with names like James Baker III (Reagan's chief of staff) and Marlin Fitzwater (a former press secretary).

Even at that, Marceca's explanation fell apart when Secret Service agents John Libonati and Jeffrey Undercoffer told the Clinger committee that the Secret Service had given Marceca a more up-to-date list. In retaliation for their testimony, the Inspector General's Office opened a criminal investigation of the two agents.[2]

With hindsight, the obvious explanation for Marceca's actions is that the Clinton White House wanted the Republicans' files around as a form of insurance. Any number of Clinton aides did not meet accepted security standards, for reasons ranging from drug use to a history of sexual harassment complaints. Ironically, among the problem employees was Livingstone himself, who had been the subject of a police complaint for threatening to beat up an elderly neighbor. If problems arose, the Clintonites could always comb through the files looking for evidence that previous employees had their problems, too.

There have been several instances when critics of the Clinton administration have suspected that negative stories about them in the press were based on information gleaned from their FBI background files. Of course, this is just about impossible to prove.

During the Watergate era, the revelation that President Richard Nixon kept an "enemies list" was considered shocking. The Clintons prided themselves on their expertise at "spin," a term that often included counterattacks on administration critics. Even reporters were not immune. It was to be expected that *The Washington Post,* the newspaper that broke the Watergate story, would give a lot of attention to the unfolding saga of Whitewater. Hillary, however, had concluded that the *Post*'s Whitewater specialist, Sue Schmidt, had a vendetta against her. Sidney Blumenthal, who would eventually move from his role as Hillary's favorite journalist to her White House confidant, came up with the idea of having Chris Lehane, a junior White House attorney, write a paper that could be used to expose Schmidt as biased. The project rolled through several incarnations until Clinton press secretary Mike McCurry stuck a stake through its heart, calling it "the dumbest idea I've ever heard of in my life."[3]

More insidious has been a pattern of abuse by the IRS that became conspicuous when the agency was being run by the Clinton administration's first director, Hillary's Yale Law classmate and 1992 campaign adviser, Margaret Milner Richardson. In addition to a long list of conservative and libertarian groups and publications, targets of IRS audits in recent years have included Paula Jones, who received an audit notice five days after turning down a settlement offer from Bob Bennett; Billy Dale; Kent Masterson Brown, the lawyer who filed the AAPS suit against the health care task force; and the boyfriend of Shelly Davis, the former IRS historian and author of a book critical of the agency. UltrAir, the charter airline used by the White House travel office during Billy Dale's tenure, was audited even though it had not yet been in business long enough to file a tax return. A woman named Patricia Mendoza found herself in trouble with the IRS one month after Bill Clinton approached to shake her hand at a rally and she blurted out the words "You suck." Mendoza and her husband were also held in custody by the Secret Service for twelve hours.

The IRS has acknowledged getting ideas for audits from newspaper stories. Therefore, one could argue that no evidence directly links Hillary to any abuses by the agency—except insofar as she chose the commissioner in the first place. In at least a few cases, however, there are suggestions of a more direct White House role.

Notes from the June 1993 internal investigation of the travel office firings reflect that William Kennedy III told the FBI that if it did not move to investigate Billy Dale, he would sic the IRS on Dale. Later, according to the notes, Kennedy told two junior attorneys in the office that he had spoken to Richardson about the situation at a party and she was "on top of it."[4]

A case in which Hillary has taken a personal interest is that of radio talk show host Chuck Harder. The origins of Harder's problems may well go back to the first week of the Clinton administration when Hillary, on a trip to New York City, discovered that she couldn't get through to the White House because the switchboard was overwhelmed with calls. Later, it turned out that Clinton, announcing the formation of the health care task force, had told the public he wanted to hear their opinions. A few "right-wing radio talk show hosts," taking the President at his word, had given out the White House's publicly listed switchboard number over the air.

Harder isn't sure whether it was his show that upset the First Lady that day, though he did give out the White House number on various occasions. "We did things like that all the time," he says, proudly recalling how his suggestion that his listeners call Senator Bob Dole's office to discuss their views on GATT tied up Dole's office phones for days. Harder's three-hour show *For the People,* originally broadcast from his garage, was in 1993 the centerpiece of a nonprofit network providing programming to 300 radio stations and 83 television stations. Whether Harder's message qualifies as "right-wing" is a matter of opinion. His views are anti–free trade and pro-workingman. Ralph Nader appeared on his show weekly for eleven years, and ex–California governor Jerry Brown and attorney Gerry Spence were also repeat guests. It is true, however, that Harder did not like Bill Clinton, and during the 1992 campaign he devoted time to discussing the allegations of drug trafficking at the Mena, Arkansas, airport as well as rumors about Clinton's active sex life.

Sometime after the Clintons took office, Harder recalls, he learned that his nonprofit People's Radio Network was being audited. "A real nice lady showed up" and began looking through his books for evidence of "inurement"—using the nonprofit structure to line his own pockets. Finding nothing, the auditor departed, only to be replaced by another woman who spent every day for six months at the offices of Harder's accountant, drinking the office coffee, using office supplies and phones and taking up so much of the accountant's time that he decided to change careers.[5]

This second auditor still found no misappropriation of funds, but concluded that the network had acted improperly by selling small shortwave radios as premiums. The ruling came as a shock to Harder, who says he had been scrupulous about complying with the regulations. By this time, Harder had hired a Washington, D.C., accountant who advised him that the IRS wouldn't be satisfied until he sold out. In May 1996, Pat Choate, best known as the Reform Party running mate of Ross Perot, brokered a deal under which the United Auto Workers would become the primary investor in a for-profit company called United Broadcasting Network, buying out Harder's assets but continuing to broadcast his show.

According to a lawsuit filed by Choate and another partner, Ed Miller, against the UAW, Hillary called UAW president Steven Yokich shortly after the deal was signed and chided him for getting involved with Harder. Hillary suggested that the union could have made a "better investment" by sponsoring her brother, Hugh Rodham, Jr., who was following up a failed run for the Senate by trying to launch a career as a pro-Clinton talk show host. In August, immediately after Harder broadcast an interview with Gary Aldrich about his bestseller, *Unlimited Access,* the UAW forced him off the air. Harder also says he received only a fraction of the money he was due under his contract, and UBN later declared bankruptcy. Harder, who is still being audited by the IRS, began broadcasting for Talk America, an independent network that once ran Hugh Rodham's show. Rodham, meanwhile, has gone on to a better job as an attorney for a Miami firm involved in antitobacco lawsuits, an amazing career move for an ex–public defender with no product liability experience.

One suspects that Hillary Clinton has often been led by her combative personality and tendency to make snap judgments, setting events in motion without considering the power of her position. She seemed genuinely surprised that after taking on the controversial issue of health care she was still not protected by the same deference accorded to First Ladies in the past. And she was even more surprised to discover that the "gotcha" culture that has thrived in Washington since Watergate would be turned on her and her husband.

The President usually had at least a few aides around him who weren't afraid to tell him when he came up with a bad idea. Hillary's aides seldom thought she'd had a bad idea, and the advisers she felt most comfortable with were the dogged partisans like Sid Blumenthal and Ann Lewis, the White House communications director who could recite the day's talking points with a fixed Nurse Ratched smile. After the failure of the health care initiative and the November 1994 elections, Hillary's stock was low in the West Wing. She gave up attending staff meetings and withdrew into Hillaryland. She was also spending a lot more time with the growing cadre of White House lawyers, people who were professionally inclined to present the President and First Lady in ways that made them appear to be our defendants-in-chief.

FOURTEEN

From the White House to Your House

At the beginning of the Clinton administration, when there was much speculation about how the new First Lady would redefine her role, *Newsweek* columnist Meg Greenfield ventured what she called an "unfashionable and conceivably also felonious" suggestion: Mrs. Clinton should cultivate an appreciation of the First Lady's traditional role as hostess for the nation. In today's world, Greenfield conceded, even the wives of ambassadors could no longer be asked to serve as unsalaried social ornaments to their husbands. Nevertheless, she went on, the White House is unique. It is "a vast enterprise—a household, a symbol, a source of great potential good, a place from which important values can be reflected and in which they can be reinforced by the nature of whom the inhabitants hear, whom they recognize, whom they choose to bring together."

In fact, recent First Ladies have differed greatly in their approach to their role as hostess to the nation. Nancy Reagan was a notorious perfectionist who spent six weeks being tutored by Jackie Kennedy's former social secretary, Letitia Baldrige, before venturing to entertain in the White House. Mrs. Reagan even insisted on seeing Polaroid snapshots of the dishes White House chefs intended to serve at formal dinners. Barbara Bush had no aspirations to be a trendsetter and was content to leave menu planning and much else to the staff, yet became one of the more popular First Ladies in memory.

With health care reform and other policy matters to attend to, Hillary Rodham Clinton could hardly be expected to shine as a hostess. Yet as it turned out, she has set her own stamp on the White House, perhaps more firmly than any other recent First Lady. From her point of view, she has been a force for democratization; others might see the same changes as bringing commercialization and political correctness into the symbolic home of the American people.

In the early days of Bill Clinton's first term, it became obvious that the new administration had little interest in White House traditions and possessed even less organizing ability. Gary Aldrich tells a story about Janet Green, the aide to David Watkins. When urged to get busy finalizing plans for the annual children's Easter egg hunt, Green said breezily, "Well, I don't think Hillary or Bill will care very much about this. . . . I just don't think they will want to do this 'egg' thing."[1]

Hillary set a partisan tone by deciding to ignore the existence of Washington, D.C., society. The social scene in Washington is made up of government officials from both parties, as well as representatives of the media who often criticize them. Despite the often sharp differences that separate them in their working lives, they manage to dine together, attend charity functions and maintain long-standing friendships. The Washington scene can be intimidating—Jackie Kennedy used to get stomach cramps before having tea with the city's social lionesses—but it is impossible to negotiate at all if you start from the assumption, as Hillary did, that the Establishment is circling its wagons against you. Hillary, for example, took offense at a column by Sally Quinn, the journalist and redoubtable Washington hostess, who suggested that she was trying to present herself as "Superwoman, the role model for the '90's." Quinn and her husband, Ben Bradlee, did not become frequent guests at the White House, but then neither did many active congressmen, elder statesmen or influential media people. Young, good-looking self-described policy wonks, the Clintons should have been the toast of the town. Instead, they remained outsiders.[2]

The Clintons showed no reluctance, however, to cultivate the Hollywood elite. Screenings in the East Wing's sixty-seat theater were their favorite form of entertaining, and the stars of the films were often invited to attend. Tom Hanks spent the night in the White House after a screening of *Apollo 13;* Joanne Woodward and Paul Newman were guests after a showing of *The Hudsucker Proxy*. Jack Nicholson was invited to have dinner and watch his film *Wolf*. Director Rob Reiner was given permission to film inside the White House for his comedy *The American President,* which he frankly acknowledged was intended as a valentine to liberal Democrats. "In my perfect world Robert Dole and Newt Gingrich are the bad guys," said Reiner. "It's fine with me if *The American President* gets Clinton a few more votes."

Despite this comment, it often seemed that the White House was courting Hollywood rather than the other way around. Communications director Mark Gearan ran around with a video camera filming Barbra Streisand at a White House dinner, and the President took time away from a meeting with Russian president Boris Yeltsin to have coffee with Sharon Stone.[3]

Clinton friends suggest that it was the President, far more than Hillary, who was fascinated by the Hollywood crowd. Hillary banned Barbra Streisand from staying overnight at the White House after her frequent visits to Washington became the subject of gossip. (The official explanation was that it would have been improper for Streisand and her then fiancé to share a bedroom, though few thought this was the real reason.) On the other hand, Hillary, as much as her husband, relied heavily on the often inappropriate advice of Harry Thomason and Linda Bloodworth-Thomason.

To an extent, Hillary did succeed in democratizing the White House. The Clintons played host to some people who probably would not have been welcome during previous administrations. Among these guests was Blanche Wiesen Cook, the author of a biography that portrays Eleanor Roosevelt as a lesbian. Cook, who is currently writing a book about Hillary, also happens to be a leftist historian, known in academic circles for her intemperate anti-American views. In recent writings, she bemoans the collapse of the Soviet Union and calls the 1990s "the meanest moment in United States history—at least since the fifties."[4]

Hillary brought delegations from her favorite activist groups to the White House, including, at times, groups of children. But these occasions often had a managed, politically targeted quality that went at least marginally beyond what other administrations of either party would have considered appropriate, as when a group of children in matching red T-shirts with the logo "CAMPAIGN FOR TOBACCO FREE KIDS" were used as a photo backdrop for an appearance by President Clinton during a week when he was under fire for various campaign-funding scandals.

Similarly, when Hillary left the White House to read to sick children at the Georgetown University Medical Center, the excursion produced photographs of the First Lady with a group of suspiciously healthy-looking kids in street clothes. Hillary's aides insisted that the First Lady did not wish to have her picture taken with children who had shaved heads or unsightly breathing tubes. To the disappointment of young patients, the children of staff were substituted.

Of course, the most singular innovation of Hillary's tenure as First Lady was the exploitation of the White House for fund-raising. Like so many of the Clintons' adventures, this was not necessarily illegal; nor was it entirely unprecedented. What set the operation apart was the absence of any sense of proportion. Big donors who had never so much as been introduced to the Clintons checked into the Lincoln Bedroom and spent hours on the phone, dialing up business acquaintances and challenging them to "guess where I'm calling from." Individuals with shady pasts and even felony convictions were invited to sip coffee with the President. Everything was for sale, including trips on Air Force One, seats at White House dinners, VIP tours, visits to the White House mess and photo ops with the President. One visitor was a donor of $143,741 who had been involved in selling arms to Iran. A woman by the name of Ai Hua Qi held up a bottle of pills while having her picture taken with Bill Clinton; she was later arrested for selling similar pills, actually aspirin, as a cure for liver cancer priced at $160 a bottle.

In theory, the Clintons were billed monthly for personal guests and the Democratic National Committee was expected to pay any expenses connected with fund-raising. The Clintons, however, were not always scrupulous. When Dorothy Rodham stayed at the White House, she used an official car to go shopping, which was simply not done. And the distinction between personal and political duties often seemed unclear, even in the Clintons' own minds. Were the more than 370 Arkansans who stayed at the White House there because they were close personal friends of the Clintons, or were some of them being rewarded as donors? Was producer David Geffen in the White House because he was a "cultural figure," as the Clintons claimed, or because he had raised a million dollars for the 1996 campaign?

A prime example of the confusion of public and private was the White House's computerized database, officially known as WhoDB and familiar to some White House staffers as "Big Brother." Developed in secret by Marsha Scott, working out of a locked room in the Old Executive Office Building, the database cost the taxpayers $1.68 million and grew to contain information on more than 350,000 individuals, including birth dates, Social Security numbers, occupations, ethnic origins, religions, dietary needs, policy positions, past political contributions, reasons for meeting with the Clintons and, in some cases, sexual preferences. The argument for secrecy, made in a January 1994 memo from Scott to Hillary Rodham Clinton and Bruce Lindsey, was in part that existence of the data bank should not be made known to career employees held over from the Bush administration, a group whose "allegiance is, in my opinion, highly questionable."

The WhoDB was ostensibly created to keep track of White House correspondence, the Christmas card list and invitations to official functions. But DNC chairman Truman Arnold acknowledged in 1997 that many of the names were taken from Democratic donor lists, and the DNC in turn used information in the data bank to keep track of which donors were due to receive perks, including Kennedy Center tickets, personal notes from President Clinton and invitations to White House movies and overnight stays. Confusing matters further, the salaries of some interns and White House "volunteers" who worked on the project were actually paid by the DNC.

White House press secretary Mike McCurry called the WhoDB "elementary politics 101," arguing, "Let's face it, Bush kept lists, too." Perhaps he did, but using government computers for unofficial purposes is against the law. In 1995, the U.S. Court of Appeals, D.C. Circuit, upheld the conviction of a Defense Intelligence Agency employee named Peter Collins, who was found guilty of a felony for using his office computer to create a newsletter for his ballroom dancing club. Moreover, the federal Privacy Act (5 U.S.C., 552a) makes it illegal for a government agency to maintain a secret database or one containing personal information not required for official purposes. The sheer size of the database, not to mention the inclusion of information about personal matters like sexual preferences and the specific favors individuals were seeking from the White House, makes these concerns more than theoretical. Finally, when Representative David McIntosh's subcommittee on governmental affairs opened an investigation of the database, the White House deliberately withheld damaging memos, releasing them only in late 1997.

White House memos written by Scott describe the WhoDB as the First Lady's "no. 1 project." Nevertheless, Hillary at first denied being the one who initiated work on the database. In January 1997, she conceded to reporters that it *was* her idea—"There was no computer system and you can't run any modern enterprise without a computer system"—but she insisted she was "not aware of any specific uses" of the database.

In fact, the model for WhoDB was Bill Clinton's personal database, which had been maintained since 1982 and supported by a software program called PeopleBase, created by state representative Percy Malone. Scott went to Arkansas to study the PeopleBase system and in a June 1994 memo suggested that Clinton's personal database be "dumped into the new system and made available, when deemed necessary, to the DNC and other entities we choose to work with for political purposes." The memo carried a handwritten notation, "This sounds promising. HRC."[5]

Hillary's ability to disclaim responsibility for this enterprise would appear, once again, to be a peculiarity of her ad hoc status. People in the White House deferred to her instructions, but since she lacked an official title she did not consider herself officially responsible. This technicality may have protected the First Lady from legal repercussions, but it also created an organizational nightmare. If no one will admit to being in charge, then no one down the line can be truly responsible either. An organization run this way is unlikely ever to learn from its mistakes and correct them.

THE Clinton approach to entertaining, combined with the administration's efforts to reduce White House staff by 25 percent, led to escalating overtime bills. Domestic staff overtime alone rose from $388,472 during the last year of the Bush administration to almost $600,000 in 1995, leading Representative Jim Kolbe, whose House appropriations subcommittee was responsible for oversight of the White House budget, to call for the appointment of an inspector general to monitor spending. After the 1996 Christmas season, when the tab for political receptions and dinners reached $425,924, triple the cost under George Bush, Kolbe also called for the DNC to begin paying for these functions in advance.

Overworked White House employees received little consideration for their efforts. It was not unusual for Hillary to give the kitchen two hours' notice that she was expecting five, ten or even fifty people for dinner. And when the Clintons did give advance notice, they were frequently late. White House insiders spoke of "Clinton Standard Time," determined by adding forty-five minutes to whatever the schedule said. Unlike their predecessors, the Clintons did not often get away to Camp David for the weekends. As several friends noted, whatever Bill and Hillary's relationship might be, it did not involve spending a lot of unstructured time together in the country. But with the First Family in residence all weekend, employees—from West Wing aides to kitchen workers—were under even more pressure. "They work all the time, so we work all the time," said one staffer. "It's crazy around here."[6]

Frustrated, the kitchen staff began insisting that the Clintons follow procedure by relaying their orders through the White House usher's office. This apparently irritated the First Lady, who was perhaps already thinking about replacing chef Pierre Chambrin. Formerly of La Maison Blanche, one of Washington's most celebrated restaurants, Chambrin was known for such specialties as "composed soups" of seafood in steaming ginger-flavored broth, home-cured salmon and Virginia squab with fruits. Hillary, however, had her own ideas. On the same day that she announced a ban on smoking in the White House—though the President was spotted working in the Oval Office with an unlit cigar in his mouth—Hillary declared a preference for lighter meals featuring organic all-American ingredients. "Bill and I do a lot of fiber," as she put it.

Attacking the problem in a manner guaranteed to ruffle the feathers of any master chef, Hillary sent the kitchen thirty cookbooks, featuring the kind of low-fat nouvelle American dishes she had in mind. She also brought in Dr. Dean Ornish, who advocates keeping fat calories to 10 percent of one's diet, to tutor Chambrin in nutrition. Hors d'oeuvres disappeared from White House menus, as did the cheese course. Haddock and plain steamed vegetables were served at formal dinners. "Portion control" became a mantra in the kitchen. Chambrin had loved cooking for the Bushes, observing, "They liked a wide range of food. They had lived all over the world." Now he was miserable.

Chambrin was instructed that the President was allergic to dairy products and beef, even as the newspapers were filled with accounts of Clinton's passion for Big Macs and the gargantuan restaurant meals he enjoyed on out-of-town trips. The buzz around the White House was that Chambrin was in the middle of a struggle between the diet-conscious First Lady and a President who loved to eat. Washington gossip columnist Dian McLellan quipped that French cuisine was caught in a squeeze play between the nouvelle American and bubba food.

Chambrin was also suspicious by now that his troubles were politically inspired. A group of food writers had been lobbying in favor of having a specialist in American food in Chambrin's job, and sixty chefs had circulated a letter urging the administration to hire someone who cared about animal welfare, factory farms, organic produce and other issues of culinary correctness. Chambrin was a U.S. citizen and had lived in the country since 1968, but as he saw it, his problem was that he was fat and spoke with a French accent. "It didn't matter what I did," he said. "I didn't fit the profile of an American chef politically."

Hillary, of course, had a right to hire any chef she wanted, but her decision was complicated by a simmering quarrel that had been brewing below stairs for at least a year between assistant chef Sean T. Haddon, a young American who had been a favorite of the Bushes, and Roland Mesnier, the brilliant but excitable pastry chef. Mesnier accused Haddon of serving an orange Jell-O pie to President and Mrs. Bush; Haddon accused Mesnier of taking kickbacks on cream purchases. As with most kitchen feuds, you probably had to have been there to understand it. By late 1992, however, the personality conflict was complicated by Haddon's contention that he—a white man— was a victim of racial discrimination. Haddon had recently become engaged to an African-American woman and, he claimed, kitchen employees had told him that his hopes for promotion to sous-chef were nil, since chief usher Gary Walters did not approve of interracial marriages. On April 16, Haddon met with a counselor from the U.S. Equal Opportunity Commission, and in May he filed a formal discrimination complaint. It was the first in the history of the White House.

Regardless of right and wrong, Haddon's complaint would have been a major headache for a private employer. Chief usher Gary Walters regarded Haddon as a troublemaker who kept the kitchen in an uproar. But it is illegal to fire someone in retaliation for filing an

EEOC complaint. Moreover, there did seem to be a racial glass ceiling at the White House.

Haddon was due for a routine updating of his FBI background check. Although the FBI SPIN unit was far behind in checking out new employees, the counsel's office prodded Special Agent Dennis Sculimbrene to do something about Haddon. Sculimbrene interviewed the kitchen help, who were upset over Haddon's EEOC complaint and suddenly recalled bizarre things he had said in recent months. An assistant pastry chef claimed that in 1992, Haddon had menaced her with a knife and talked of killing the staff and storing their bodies in the freezer. Supposedly, Haddon also boasted that he could easily bring a gun into the White House, since the security guards regularly waved his bicycle past the metal detector.

Sculimbrene looked into the threats and decided they were just idle talk. That afternoon, however, he mentioned to Walters that he was looking into the possibility that Haddon might pose a threat to the Clintons. He realized that the discussion might be passed on to Cheryl Mills of the counsel's office and the First Lady, who worked as "kind of a team." Although the evidence of what happened next is under court seal, the *Washington Times* has reported that Walters consulted with both Vince Foster and Hillary Clinton before deciding on a course of action.

Within hours, two armed Secret Service agents showed up in the kitchen and escorted Haddon off the White House grounds. Haddon was interrogated for two days before the Secret Service came to the same conclusion as Sculimbrene: he was guilty of nothing but running his mouth. Really angry now, Haddon began talking to the press. Among other charges, he accused Chambrin of conning the First Lady by slipping "greasy chicken fat" into his low-fat dishes.

"Chef Haddon must go!" Walters fumed in a November 1993 memorandum to Maggie Williams. "He has slandered me and belittled the office of the chief usher, as well as undermined his fellow chefs with his allegations. I would not put up with these [*sic*] type of actions from any other employee."

Walters suggested that Pierre Chambrin be encouraged to resign. Then the entire kitchen staff could be let go and Chambrin's replacement could choose his assistants, hiring back everyone but Haddon. The memo went on to urge moving slowly to avoid scrutiny of the reasons for Chambrin's dismissal.[7]

The First Lady's office waited until March 1994, when the press was preoccupied with Whitewater, to carry out the kitchen coup. Chambrin was replaced by Walter Scheib, a young and relatively unheralded recruit from the Greenbrier Resort Hotel in West Virginia. Reviews of Scheib's cooking vary, but perhaps it's best to leave the last word to Scheib himself, who has been quoted as saying that his low-calorie dishes are "almost like normal food."

Pierre Chambrin walked out of the White House with $37,026 in exchange for his agreement not to discuss the Clintons or the circumstances of his dismissal. The severance bonus was unprecedented, but when several congressmen questioned it, Hillary Clinton denied knowing anything about the bonus or the reasons for giving it. Gary Walters, on the other hand, would testify in Haddon's EEOC suit that the First Lady's office did know. "I think they agreed with it. I mean, at some point, surely, but I don't remember—I can't characterize it. As far as I know it was, 'OK. Looks good to me.' "[8]

A week after Chambrin left, thirty-six-year-old White House usher Chris Emery learned that he was being let go because the First Lady was "uncomfortable" around him. Emery was the first usher to be fired in the 103-year history of the position. Emery's offense, it developed, was that he had spoken on the telephone four times to Barbara Bush, twice from a phone in the usher's office. Emery had taught Mrs. Bush to use her laptop computer, and the former First Lady occasionally called and left messages for him when she had technical questions. He had merely been returning her calls. "This may seem harmless, for all intents and purposes," explained Hillary's spokesman Neel Lattimore, "but it also shows an amazing lack of discretion." But, surely, the notion that spies on the staff were reporting directly to Mrs. Bush borders on paranoia. Was it possible that Sean Haddon, too, was let go because he had been close to Barbara Bush, often cooking for her and her husband at Camp David?[9]

The case of the orange Jell-O pie now became a matter for the ever growing cadre of lawyers in the White House counsel's office. The government lawyers succeeded in protecting Hillary from being deposed about her role in the affair. They also won a ruling disallowing similar employee actions in the future. Restaurant owners will be pleased to know that the White House is now exempt from the EEOC regulations that can make their lives so difficult.

Since it would not create a precedent for future actions, Haddon's complaint now cried out for settlement. After all, he had been removed from the White House under humiliating circumstances and identified in the press as a man who threatened the President. The White House preferred to litigate. Even as Bill Clinton was in Akron, Ohio, defending affirmative action as part of his "conversation" on race, an administrative judge handed down a ruling that found that the White House had used the Secret Service to punish Sean Haddon for filing a discrimination complaint.

But after four years, Haddon's case still wasn't settled. His attorney, Rodney Sweetland III, expressed amazement at the ability of the White House lawyers to transform a relatively simple case into the culinary version of *Jarndyce v. Jarndyce*. "This is the first and last [White House] case that will ever be brought under this law," observed a resigned Sweetland, "and we'll litigate this until the pyramids turn to dust."[10]

CHRISTMAS at the White House posed another challenge for the organizational abilities of the First Lady. In interviews, Mrs. Clinton repeatedly observed that the greatest surprise awaiting her in the White House was the discovery that planning for Christmas began in the spring. This was "the biggest single shot I had," as she put it, since "I've never been anyone who planned for Christmas much before mid-December, and then my husband and I would race around."[11]

This observation reveals a great deal about the mind of Hillary Clinton. She was great at thinking up new ideas for policy initiatives but naive, to say the least, about the amount of work that goes into making ideas into reality. No one who has ever planned a corporate banquet, a class reunion or even a family wedding could be surprised to learn that Christmas at the White House—an event involving 22 separate trees, 27,000 lights, 7,500 ornaments and over 100,000 holiday visitors, including tour groups and partygoers—would require months of planning.

Under Hillary Clinton's authority, even the process of getting out the White House Christmas cards was contentious. Chelsea had been the target of a number of threatening letters—as most occupants of the White House inevitably are. Such threats would disturb any parent, but Hillary was so alarmed that she decided to omit Chelsea from the White House Christmas card, in the belief, apparently, that not acknowledging her daughter as part of the family would discourage potential stalkers. Over the objections of her husband and aides, she eventually got her way, though the cards, some 200,000 of them, just barely got finished in time to address and mail before the holiday.

Inexperience, however, did not prevent Hillary from having ambitious plans. The number of holiday parties increased during the Clinton administration. During the Christmas season there were often two shifts of parties, one in the early evening and one beginning after eight. By 1996, there were twenty-five events related to DNC fundraising alone. Staff parties also became more complicated, because the First Lady's office could not agree on the definition of the word "family." At first, single staff members were permitted to bring significant others and friends. Naturally, everyone took advantage of the rule to bring as many guests as possible, so the next year, to prevent overcrowding, even spouses were banned.

In 1993, Hillary came up with the idea of using Christmas at the White House to showcase American crafts. Seventy leading American craftspeople were invited to contribute pieces toward what they were told would be a permanent White House crafts collection. Nor surprisingly, many of the contributions clashed with the period decor of the public rooms, and it soon developed that there was no space set aside to house the collection permanently. White House spokespersons suggested that the crafts collection might eventually end up in the Clinton library.

Hillary also called on crafts groups to create original Christmas ornaments, preferably angels. The request produced an abundance of Socks angels and Bill Clinton angels, including depictions of a haloed President playing a sax. Hillary loved these "funky and down-to-earth" decorations, or at least she professed to love them. Some White House visitors found them garish.

The following year, 1994, the First Lady's social staff put out the call to art students for decorations on the theme of "The Twelve Days of Christmas." In response, they received such "ornaments" as five greasy onion rings glued to a Styrofoam tray, a contribution that was prominently displayed and rated a mention in *The Washington Post.* In a much discussed passage of his book *Unlimited Access,* Gary Aldrich describes the embarrassed laughter of volunteers decorating the Blue Room Christmas tree, considered the First Lady's personal tree, as they reached into the packing boxes and came up with ornaments like a mobile of "twelve lords a-leaping," each in full erection; a variety of ornaments fashioned from crack pipes and condoms; and a gingerbread man sporting five golden rings, including a nipple ring and a cock ring. Mortified volunteers broke some of the more embarrassing items rather than hang them. Others, including the gingerbread man, found a place on the tree. Aldrich took a snapshot, which appears in the second edition of his book.

This episode ought to be a candidate for the "my most embarrassing moments" section of Hillary Clinton's memoirs. When Hillary approved the idea of soliciting ornaments from art students, she obviously didn't take into consideration their love of shocking their elders. In keeping with the last-minute mode of doing such things at the Clinton White House, the request for contributions went out late and there was no time to inspect the ornaments individually before the day to decorate the trees arrived. Social secretary Ann Stock apparently didn't notice the X-rated ornaments when she inspected and approved the tree. The result was an embarrassing gaffe, but, one might think, something to laugh about after the fact.

Mrs. Clinton is certainly well aware of what Gary Aldrich had to say about the 1994 Blue Room tree. Indeed, his description was the subject of much kidding at the going-away party for Ann Stock when she left the First Lady's staff. But the White House position is that Aldrich made it all up. In a phone interview from her office at the Kennedy Center, Stock insisted, "There were no pornographic ornaments. Nothing like that." And she completed her lengthy denial with the administration's favorite catchphrase, "That's my story and I'm sticking to it." The denial is more troubling than the actual incident, taking us into a world where the truth can be denied even to avoid embarrassment over a relatively trivial incident.

After several years of emphasizing more traditional crafts like needlepoint, Hillary reasserted her taste for the "funky" in 1997. As explained in the White House's annual Christmas message, the theme of "Santa's Workshop" was chosen "to represent the hopes and dreams of the next generation." In keeping with this theme, decorations for the eighteen-foot Fraser fir that dominates the Blue Room were solicited from leading American fashion designers, who were invited to create miniature outfits for Mr. and Mrs. Santa Claus. The commission resulted in such holiday decorations as a miniature Tommy Hilfiger ski jacket in red, white and blue, a silver lizard-skin Santa suit by Richard Tyler and a Mrs. Santa figure decked out in a red underwire bikini top with matching high-cut panties.

When an Associated Press reporter meeting with Mrs. Clinton for her year-end interview asked about these unusual ornaments, she conceded, "Some of them are a bit over the top," adding, "Mrs. Claus is going to start a diet right after Christmas."

Opinions may differ as to whether it was appropriate to turn the First Lady's Christmas tree into a billboard for clothing designers, many of them Democratic contributors. But even those who find the transformation amusing must recognize that this tree was an adult fantasy that had nothing to do with the "hopes and dreams" of children.

★　　★　　★

A similar confusion between the interests of children and those of adults was reflected in Hillary's bestseller, *It Takes a Village*. The book came into being in 1995, during a period when Hillary needed to improve her battered image, demonstrating that she was not a shrill ideologue but a wife and mother whose views were far more middle-of-the-road than her critics claimed.

In a rather unusual arrangement, Simon & Schuster advanced $120,000 to the ghostwriter Barbara Feinman, a well-connected Georgetown University journalism professor who had previously written a book with Congresswoman Marjorie Margolies-Mezvinsky as well as done research for Bob Woodward, Carl Bernstein and Ben Bradlee. In a push to get the book finished, Feinman labored on the manuscript seven days a week, sometimes staying overnight at the White House and even accompanying the Clintons on their summer vacation to Wyoming. Before the book was completed, however, there was a falling-out. When Bob Woodward's book *The Choice* appeared, revealing Hillary's efforts to commune with the spirit of Eleanor Roosevelt with the help of Jungian psychologist Jean Houston, Hillary suspected Feinman of being Woodward's source. Feinman got paid for her work eventually, but when *It Takes a Village* came out at the beginning of 1996, it lacked even a routine thank-you to Feinman. (Speechwriter Lissa Muscatine, Maggie Williams and the First Lady's editor didn't rate mentions either; in the book's brief acknowledgments Hillary thanked no one by name "for fear that I will leave someone out.") The First Lady even boasted in her weekly newspaper column of having "written a 320 page book in longhand over the last six months." No one disputed that she had done a great deal of work on the manuscript, including extensive rewriting, but the shabby treatment of her collaborator contradicted the gracious image the book was meant to project.

Overall, *It Takes a Village* was a great success. It became a best-seller, earned Hillary Rodham Clinton a Grammy Award for the audiotape version and continues to earn royalties, which she has donated to charity. The title in itself generated a cozy image of the First Lady that erased much of the bad feeling over the health care debacle. As for the text, it amounted to two books in one. On a superficial level, *It Takes a Village* was a collection of anecdotes, a few quite funny, and the sort of uncontroversial advice one expects from a First Lady: Hillary encourages parents to make time for reading aloud, to turn off the TV once in a while and to provide for their children's spiritual development. All well and good.

When it comes to the policy ideas that thread their way through the chapters, however, the picture is very confusing. Contrary to the title, Hillary's message is that the small-town values practiced by her parents' generation are passé. She paints an idyllic picture of Park Ridge in the 1950s as an almost ideal environment for children. Nevertheless, she issues a stern warning against heeding the "nostalgia merchants" who think that the "values of the 1950's" might be worth cultivating today.

In a chapter called "Kids Are an Equal Employment Opportunity," we begin to see why she feels this way: "Whenever we pose women's options as an 'either/or' choice—most commonly between work and family—we do a disservice all around," Hillary writes. And she goes on to tell how anthropologist Mary Catherine Bateson—the woman who introduced her to Jean Houston—warned her against speaking of women's efforts to balance mothering and work as a "juggling act," because "when you juggle, eventually something gets dropped." This is all very well, but most working mothers of young children do feel, at times, that their lives *are* a juggling act. And few women, no matter how well they cope with the conflicting demands of career and motherhood, would say that there aren't real trade-offs involved.

Hillary recalls no compromises in the quality time she spent with her own daughter: "We often spent hours in the backyard of the governor's mansion, stretched out on a quilt and looking at birds and clouds. We also used to lie on our backs in the front hallway when no one was around, watching the dancing rainbows the sun made as it struck a crystal chandelier." Few politically active corporate attorneys can make time in their schedules for "hours" of this sort of activity. And the image of Hillary doing so would amaze the members of the governor's trooper detachment, who roared with laughter when a *People* photographer snapped Hillary relaxing in a backyard hammock with her husband and daughter. Hillary herself writes of taking two-year-old Chelsea on the campaign trail, an exhausting routine for a toddler. And former surgeon general Joycelyn Elders recalled of her visits to the Governor's Mansion, "Hillary traveled a lot, and on three or four occasions when I was there, Chelsea would be the hostess and serve cookies. This was starting when she was just a little thing, probably no more than five or six years old. [Bill] Clinton was very involved with her life."[12] This is not to say that Hillary was a bad or uncaring mother; she wasn't. But she appears too intent on denying that working mothers, herself included, have a limited amount of time to spend with their children.

By far the most startling aspect of *It Takes a Village* is Hillary's inability to work up much excitement over what she calls the "confusion and turmoil that divorce and out of wedlock births have on children's lives." A woman of passionate convictions, with a program for every problem, Hillary is curiously passive when it comes to bemoaning the "disappearance of fathers from our children's lives." The only villains she finds in this situation are on the right, noting, "Many who protest loudly against welfare, gay rights, and other perceived threats to 'family values' have been uncharacteristically silent about divorce."

Indeed, while her book gives a nod to the "sometimes controversial" Promise Keepers, during the weekend of the group's big rally in Washington, Hillary could be found attending a book party for *On Our Own: Unmarried Motherhood in America,* written by Melissa Ludtke, a 1973 graduate of Wellesley and a friend of the First Lady, who has been a guest at the White House. Hillary had read and critiqued parts of Ludtke's manuscript, mainly those in which Ludtke describes how she grappled with her decision to become a single mother (by adoption). While Ludtke's own point of view is more thoughtful, the book is filled with interviews with educated women who became single mothers by choice, women who express bitterness over society's prejudice against unmarried mothers—which can hardly be as daunting as they claim, considering the statistics. Some of the women take personal offense at the term "family values," complaining that it is nothing more than "political shorthand for the two-parent heterosexual family." The prevailing view, it seems, is that marriage is a dying institution because it no longer answers women's needs for self-fulfillment. "Nowadays," writes Ludtke, "many women don't regard as sufficient the pat on the back that society gives them for 'domesticating' a man and bearing his children in a legally sanctioned way."[13]

No doubt Hillary Clinton has reservations about this vision. After all, she has kept her own marriage together when many other women in her situation would have bailed out. But she can't bring herself to take the plunge and come out against the trend toward single motherhood. To do so would risk offending too many friends and supporters.

The "village" Hillary advocates is one in which the state is every child's surrogate parent, ready to intervene early and often, with the ultimate responsibility for the child's welfare. Hers is a vision of an adult-friendly society in which women—and men, too—can exercise choices, maximizing their career opportunities and their personal happiness. Maybe this is the kind of world we want to live in, but surely it's hypocritical to pretend that we're doing it for the children. Many of the programs and strategies that Hillary so cheerfully advocates sound pretty grim from a child's point of view. For example, she gives a glowing account of a Massachusetts day-care center where the children play Go Fish and Concentration with cards decorated with men and women in nonstereotypical roles—"men holding babies, women pounding nails, elderly men on ladders, gray-haired women on skateboards." One suspects that the little tykes might prefer images of nurturing moms and grandmoms and strong, protective daddies. They can't have them, not because these "stereotypical" images offend parents in general, but because they threaten the brave new gender-neutral world in which certain deep thinkers in academia have made a heavy investment.

Undoubtedly, the most distressing feature of *It Takes a Village* is

Hillary's condescension toward poor and working-class parents. She tells the story of a day on the campaign trail when she approached a woman holding an infant and said to her, "I bet you're having fun playing with her and talking to her all the time." And the woman replied, "Why would I talk to her? She can't talk back!"

One would like to think that this mother was just having a bad day, but let's grant Hillary's point that poor and working-class parents spend less time talking to their children. At any rate, they don't talk to them the same way college-educated parents do. Citing a study by Betty Hart and Todd Risley, Hillary notes, "The more money and education parents had, the more they talked to their children, and the more effectively from the point of view of vocabulary development. . . . Poor parents praised their children . . . less often and criticized them even more frequently." Doubtless many such parents could improve their communication skills. On the other hand, working-class parents may do a better job of instilling self-reliance and religious commitment than, say, articulate Yale-educated parents who believe that values are relative. Whether these observations justify a massive government program to put children under the care of adults who will communicate with them in the "right" way is doubtful.

Hillary, however, believes that it does. Citing the French system of nurseries and preschools as an example to emulate, Hillary worries about the "alarming fact" that studies show that "many parents do not know what to look for when choosing child care." What she means by this is that American parents don't share her enthusiasm for government-run day-care centers and federal regulation. On the contrary, they prefer day care in settings that reflect their own cultural values and give them as much control as possible over the selection of day-care providers. These parents aren't stupid. They know, for example, that workers in government-run centers will be protected by federal employment regulations, and therefore supervisors won't be able to fire an employee just because they have a bad feeling about him. Moreover, considering that the credentialed day-care teachers Hillary extols would be educated by the same child development specialists responsible for making American public schools what they are today, it would seem that the parents have a right to be skeptical about federal standards. In Hillary's view, however, these parents aren't just wrong, they're alarmingly wrong.

IN April 1997 Hillary followed up the publication of *It Takes a Village* by hosting the first of two White House conferences on child care, known to some West Wing staffers as "Hillary's brain conference." The daylong event highlighted new research on the development of infants' brains, which shows that secure attachment to a loving, responsive caregiver has a positive effect on brain chemistry and the development of neural pathways.

One might think this research would be an argument for keeping children three years old and under at home with their mothers, or at least out of institutional settings where they receive less individual attention. But it isn't necessarily so. Child development experts have an amazing ability to slice and dice new scientific data and interpret them in a way that confirms their preconceived notions. If you start from the premise that poor mothers don't interact enough with their children, or interact in the wrong way, then the research becomes more evidence in favor of infant day care.

The conference had its rocky moments, especially when Dr. T. Berry Brazelton, the bestselling author of books on parent-child communication, used his time as a panelist to deliver a scathing critique of the 1996 welfare reform bill.

But Brazelton just didn't get it. The one reason Hillary could live with welfare reform was that putting poor mothers to work created an argument for providing subsidized day care for their children. As President Clinton predicted in his remarks to the conference, "This welfare reform effort, if focused on child care, can train lots of people to be accredited child care workers. . . . The welfare reform bill, because of the way it's structured, gives all of you who care about child care about a year or two to make strenuous efforts, state by state, to create a more comprehensive quality system than we've ever had before."[14]

Never explained is how welfare mothers who—allegedly, at least—don't talk to their own children enough or in the right way will interact more productively with other people's children once they become day-care workers.

But there was more. The number one action to come out of the conference was an executive memorandum directing the secretary of defense to take the lead in improving child care by creating partnerships and training relationships with civilian day-care centers across the country. The integration of women into the armed forces has forced the military to invest in a costly and rapidly growing day-care system. With a growing number of single mothers in uniform, not to mention dual-career marriages, the armed services now run over 800 day-care centers, some open twenty-four hours a day, a network that makes the department of defense the nation's largest child-care provider. Aside from their experience in setting up day-care facilities quickly, the military services employ quality-control techniques such as surprise inspections, a feature Hillary would incorporate into her model day-care bill—even private day-care providers, including women who care for a few neighborhood children in their own homes, would be subject to unannounced visits by federal inspectors.

DAY care is a serious problem, which makes Hillary Clinton's rosy and misleading rhetoric all the more troubling. Contrary to what she writes in It Takes a Village, the latest evidence is that children three and under do not thrive in full-time institutional day care. This is not welcome news, especially to mothers who have no other options, but nothing is gained by pretending it isn't so. There is something almost perverse about using research demonstrating a baby's need for a "secure attachment" to his mother to initiate a plan that would remake American infant care on a model created for the offspring of women who are active-duty soldiers.[15]

As the late Christopher Lasch noted, Hillary Rodham Clinton is the latest of a long line of reformers who believe that progress consists of transferring the responsibility for children from parents to the state.

"Criticism of parents' incompetence," Lasch wrote, "has long been the reformers' stock in trade."[16] Of course it may be true that some of Hillary Clinton's critics, the people she lumps together under the label "the Right," really want to go back to the era of housebound women and an all-male workplace. They won't succeed, because no one has yet figured out a way to make history run backward. The far greater danger is that Hillary and her allies will push through a "comprehensive" government program that sets their particular solution in amber.

Longer life spans, smaller families, telecommuting, better-educated young women with far different expectations—all these changes have made the stark choices that feminists thought they faced in the 1970s obsolete. The national government's role should be to create conditions that foster constructive change, not to set a standard for the "right" way to socialize little children. Most social change does not come from policy professionals. It happens in ways that are dynamic, unpredictable and often messy. There is something in the mind of Hillary Clinton that feels threatened by this. Despite lots of evidence that central planning doesn't work, she clings to the belief that just because a solution is labeled "comprehensive" or "universal," it necessarily serves everyone's needs.

FIFTEEN

The China Syndrome

A FTER a period of retreat and self-examination, Hillary was once again ready to take control of her public image. And this time, she had chosen a stage where she could concentrate on what she did best, speaking out on women's and children's issues. The United Nations' Fourth International Conference on Women was being held in Beijing in September 1995, and Hillary planned to attend as the honorary chair of the U.S. delegation.

Unlike her role in the health care task force, which had been assigned to her by the President, attending the Beijing conference was very much Hillary's idea, and one that caused a lot of nervousness throughout the executive branch. State Department and National Security Council officials argued strenuously against her making the trip. Opinion in the West Wing was said to be divided, with no one eager to speak on the record. *Time* magazine quoted an unnamed "top adviser" to Bill Clinton as saying, "He's keeping as far away from this as possible." Another senior aide predicted, hopefully, that Hillary's role would be "low key," despite talk of her "leading the women's brigade."

The immediate objection to Hillary's visit to China was the detention of Chinese-American human rights activist Harry Wu. A week before the conference opened, the Chinese gave Wu a secret trial, sentenced him to fifteen years and then deported him back to the United States. This turn of events hardly pacified critics of the First Lady's trip. On the contrary, it raised suspicions of a special deal that set a questionable precedent for the future of U.S.-Chinese relations. In an op-ed piece in *The New York Times,* Shelly H. Han of the China Strategic Institute, a group founded by Chinese dissidents in the United States, pointed out that China desperately wanted Mrs. Clinton to attend the conference because her presence would have the effect of "bolstering the ruling elite." Han vigorously protested "the use of human beings as bargaining chips . . . an inhuman way to conduct foreign policy." Dr. Charles Jones, a political science professor at the University of Wisconsin, noted that the First Lady's decision to attend the conference pointed up our lack of a clear China policy and warned that if U.S.-Chinese relations made news again soon, many Americans would suspect that Mrs. Clinton had played a behind-the-scenes role, "no matter how the White House tries to deny it."[1]

The Beijing conference proved to be a study in ironies. First of all, it wasn't in Beijing. Fearful that the conference would lead to human rights demonstrations in the capital, the Chinese had made a last-minute decision to move most of the events to suburban Huairou, an hour away. Preparations were so hasty that one American feminist described the locale as a "prettied up construction site trying to pass as a conference area." Concrete laid in haste was already buckling, and the doors to some buildings were unusable because of the danger from exposed electrical wiring.

The majority of the 36,000 women in Huairou were not members of official delegations. Rather, they were representing the so-called NGOs (nongovernmental organizations), whose side of the conference consisted of a sort of world's fair on the theme of women's issues, ranging from a workshop where women from rural villages on different continents talked about their efforts to form self-help groups to a meeting on "Consciousness Raising as a Tool for Grass Roots Organizing," where discussion leaders from NOW earnestly urged women from Germany, Cambodia and the Sudan to explore the ways in which their native languages expressed negative connotations about womanhood.

Many of the NGO events were consigned to rain-sodden tents reached by flooded walkways. NGO representatives and delegates had almost no opportunities to talk to Chinese women, and foreign journalists were segregated in special hotels, where the concierges didn't even try to hide the notes they were taking on their guests. A group of Tibetan women got into a scuffle with plainclothes police who tried to seize a copy of a tape called "Voices in Exile." However, not all conference-goers connected these conditions to the theme of women's oppression. NOW vice president Karen Johnson later said that she considered China to be ahead of the United States in gender equality, because the percentage of female scientists is higher in China than in America. (Johnson's impressions may have been drawn from a special issue of the *Beijing Review* prepared for the conference, which painted an idyllic picture of women's rights in China. Not only were women in China said to suffer less employment discrimination than in the United States, but domestic violence, rape and sexual harassment were described as rare events.) Asked about reports that China's one-child policy leads some parents to kill their female babies, Johnson added, "We tried to talk to our guide about it. That was an issue we were concerned about, but no one would address it in any real way."[2]

Hillary Rodham Clinton's speech was by far the high point of this otherwise dreary event. A drenching rain forced the event to be moved indoors, disappointing thousands who had stood in line waiting to hear her. The 2,500 or so women who managed to get seats inside the auditorium where she eventually spoke were "very excited," said Johnson. By the time the First Lady's motorcade made its way to the site from downtown Beijing, anticipation had risen to a fever pitch. In a speech repeatedly interrupted by standing ovations, Hillary avoided the touchy subject of abortion but condemned China's forced sterilizations and other human rights violations: "Human rights are women's rights and women's rights are human rights," she declared forthrightly.

After the speech, the American delegation gathered at the U.S.

Embassy in Beijing to help Hillary celebrate her triumphant appearance. "We were feeling so good that late in the evening, we got into a large circle, held hands and sang some sixties-style folk songs," recalled *Ladies' Home Journal* editor Myrna Blyth. "Guys wouldn't have done that."[3]

The First Lady's speech was indeed admirable. She had talked tough to the Chinese and delivered a ringing endorsement of human rights. Like many previous occupants of the White House, she had also discovered that it is often easier to be a hero abroad than at home. Not only was Hillary Clinton's international stature increased, but the Beijing appearance greatly burnished her image with the people who run and donate to private foundations, which often look to documents like the UN platform in setting their agendas.

Whether Hillary's appearance was an overall positive for American foreign policy is a more complicated question. The text of her speech was not reported in the Chinese press, and one suspects that the host country's leadership was quite willing to accept a few lines of criticism, which its own people didn't hear anyway, in exchange for the prestige conferred on the gathering by Hillary's presence.

KAREN Johnson explained the excitement generated by Hillary's speech this way: "She is an icon to the women of the world. She is not just the President's wife. She is a person in her own right." On the contrary, one suspects that Hillary Rodham Clinton's relationship to President Bill Clinton was of intense interest to her audience. The First Lady's presence in Beijing heightened expectations that the United States would commit itself to the conference platform and, in addition, join the 143 nations that had already ratified the UN Convention on the Elimination of Discrimination Against Women (CEDAW). In fact, there was no realistic chance that either CEDAW or the platform would make its way through the Senate Foreign Relations Committee, where Senator Jesse Helms held sway.

Like most documents produced by conferences, the 178-page

"Platform for Action" is a mixture of idealistic—and no doubt un-reachable—goals, impenetrable jargon and dubious sentiments.

Reproductive rights and homosexuality were, not surprisingly, the most contentious issues for the drafters. Many American and European feminists were bitterly disappointed that the platform included no statement on homosexual rights or a right to abortion. On the other hand, delegates from Catholic and Muslim countries found fault with the platform's preoccupation with eliminating gender stereotypes and promoting birth control. Harvard law professor Mary Ann Glendon, who attended the conference as head of the Vatican delegation, summed up the agenda as "more about the ideology of the '70's than the women's issues of the '90's." Describing the original draft document—only somewhat modified in its final form—Glendon notes, "In the few places where the drafting committee mentioned marriage, motherhood, or family life, these aspects of women's lives were men-tioned in a negative way—as sources of oppression, or as obstacles to women's progress."[4]

Prime Minister Benazir Bhutto of Pakistan echoed this criticism, calling the platform "notably weak on the role of the traditional fam-ily." In fact, the platform is filled with language that could be con-strued as mandating affirmative action programs, gender-neutral marriage laws and even censorship. An article in the June 1996 issue of *Nieman Reports,* a publication of the Nieman Foundation, expressed alarm over provisions calling on governments to establish "media watch groups" to monitor "gender bias" in the press and broadcasting.

Even the most idealistic planks of the platform pose problems when translated into national policy. The UN document comes out against female genital mutilation, for example. If the United States officially adopted the platform, would we be required to grant political asylum to the millions of women who live in countries where this rite is still practiced? Apparently so.

Although faced with opposition from Congress, the Clinton ad-ministration chose to forge ahead. Even before the Beijing meeting convened, President Clinton established the Interagency Council on

Women, with Hillary as honorary chair and Donna Shalala and Madeleine Albright as cochairs. The U.S. government was in no way legally bound to implement the provisions of the Beijing platform. Only Congress could make this commitment. The interagency council, however, seemed bent on ignoring this. It even titled its first book-length progress report *America's Commitment*. Moreover, according to Sharon Featherstone, the coordinator for the council's work at the State Department, the mission of the interagency council is to "mainstream" the provisions of the Beijing platform throughout the executive branch.

America's Commitment demonstrates why the era of big government will not end any time soon. Many of the "mainstreaming" initiatives described in its pages are good ideas, and for that matter the great majority were in existence well before the Beijing conference. Overall, however, the report treats any situation that conceivably affects the female sex as an appropriate target for a campaign of taxpayer-funded seminars, studies, conferences, pilot projects and, ultimately, legislation.

Among the initiatives outlined in mind-numbing prose in this 200-page publication are the development of teaching materials designed to "increase gender equity in mathematics, science and technology for all learners at all age levels" and the use of "cognitive research on time-use diaries" to measure the value of "unwaged work," another way of saying housework. A federal commission on the so-called glass ceiling effect finished its work in 1995; nevertheless, the executive branch is continuing its research in this area, and the EEOC is developing guidelines for pattern-and-practice suits against employers suspected of not having enough women in high-level positions.

"Society's abandonment of urban centers" has been redefined as a women's issue. Moreover, the FCC has been mandated to create rules "conducive to women business participation [*sic*]" in auctions of wireless communications licenses. (Apparently, this will help women like Hillary Clinton, who made a killing in wireless licenses in the 1980s.)

Other initiatives include the sponsorship of "over 400 women's round tables" to "insure that women's voices are heard in the White House," new rules on product labeling, a Gender Equity Expert Panel ("part of a proposed system of expert panels being developed under the Educational Research, Development, and Dissemination Act of 1994"), a "manual on school uniforms" to help schools thinking of instituting a uniforms policy, new bilingual education programs aimed at "binational women" and the creation of new management titles such as Department of Commerce federal women's program manager. There is even a "Fatherhood Initiative"—described in vague terms, suggesting that no one has yet considered what it might consist of.

One thing *America's Commitment* does not reveal is how much this "mainstreaming" effort will cost the taxpayers. Featherstone was surprised to be asked, saying, "I don't think we have a handle on that, really."

INTERESTINGLY, Hillary's trip to China coincided with the beginning of the Clintons' campaign to raise funds for the 1996 presidential elections. On her way to Beijing, the First Lady made a brief stop on Guam, where she nibbled shrimp cocktail at a buffet reception sponsored by the territory's governor, Carl T. Gutierrez. Among the civic leaders on hand to meet her was Willie Tan, who in 1992 was fined $9 million for denying overtime pay to workers in his factories in the Marianas.

Although Hillary's office denied that this event was a fund-raiser, two weeks earlier DNC chairman Don Fowler had written to Gutierrez suggesting a $250,000 contribution to the party, exactly the amount the governor raised in the weeks following the reception. Within six months, the total of contributions from Guam had grown to $510,000 in soft money, plus another $132,000 for the Clinton-Gore reelection campaign. Despite being ineligible to vote in presidential elections, the residents of Guam had become the largest per capita contributors to the 1996 Democratic campaign.

What prompted this outpouring of support?

For some years, residents of Guam had been pushing for commonwealth status. The trouble was that the proposed bill would allow local authorities to control immigration and labor policy, raising fears that businessmen like Tan would do what they had already done in the Northern Marianas, importing impoverished Asians who labor in sweatshops for substandard wages, producing clothing tagged with "Made in the U.S.A." labels. And, it seems, the donations went far to allay such fears. During 1996, Gutierrez would have two face-to-face meetings with Bill Clinton, one in March and the other in December. Also in December, Deputy Secretary of the Interior John Garamendi issued a report supporting Guam's efforts to regain control of immigration and labor matters. Labor Department officials were dismayed, though Garamendi insisted his decision simply followed the administration's "anti-colonial" policy.

A few weeks after the Guam-Beijing trip, Hillary's itinerary took her to Paraguay, a nation known as a haven for international money laundering, drug trafficking and smuggling. Hillary became the first wife of a sitting president to visit Asunción. She was officially in Latin America to attend a meeting of First Ladies and took the time to address American Peace Corps volunteers celebrating the corps's thirtieth anniversary in Paraguay. But for President Juan Carlos Wasmosy, Hillary's visit was a reassurance that the U.S. State Department did not plan to decertify his government, which would make it ineligible to receive U.S. aid.

Wasmosy had been trying to get an audience with Clinton ever since he took office in 1993, an effort that so far the State Department had thwarted in the belief that any such meeting would be an "embarrassment." Wasmosy's campaign was aided by a group of businessmen and attorneys representing the Paraguay-American Chamber of Commerce, who held a breakfast meeting in Miami in February 1995 with Hillary's younger brother, Tony Rodham. After several years on the payroll of the DNC, Rodham left to open a consulting company. He was looking for new clients, but during the course of the meeting, a Paraguayan legislator offered Rodham a bribe of more than $100,000 in exchange for his help in getting Wasmosy invited to the White House.

There are two versions of how this came about. John H. Tonelli, an attorney who was present at the meeting, later told *The Boston*

Globe that Rodham "held himself out as someone with access to the White House . . . as a semi-lobbyist, and said he could broker meetings and contacts because of his sister."[5]

Rodham's version of the story was that he would have been "more than happy" to try to arrange a meeting for nothing, but was shocked to be offered a payoff: "I was just sitting there chatting. They asked me if I could make an introduction to President Clinton. I saw nothing wrong with that. But then someone jumped in and tainted the deal by offering a lot of money."

Spurned by the First Lady's brother, the Paraguayans found a new champion in Philippine-born businessman Mark Jimenez, the owner and founder of Future Tech International, a fast-growing Miami-based company that sells computer components bound for Latin America. Future Tech is not an exporter, and thus it has no legal responsibility for what happens to its products once they leave Miami. Nevertheless, products sold by the company do seem to make their way to Paraguay, where the trade in computer parts is said to be used as a front for international money-laundering operations.

Jimenez dated his emergence as a major Democratic contributor to the 1994 Summit of the Americas in Miami, where he "fell in love" with Bill Clinton after hearing him speak. Before this love affair began, Future Tech had been under investigation by the Treasury Department. The company's situation improved after Jimenez hired the Miami law firm of Greenberg, Traurig, one of whose partners, Marvin Rosen, headed the finance operation of the Democratic National Committee. Jimenez's large donations began around this time.

In September 1995, a fund-raising dinner attended by Future Tech employees raised $20,000 for the Clinton-Gore campaign. Days later, on September 15, Arkansas attorney Mark Middleton arranged a meeting between Jimenez and Mack McLarty, who had left his post as White House chief of staff to become Clinton's leading adviser on Latin America. A month later, Hillary Clinton was on her way to Asunción, and Mark Jimenez joined the First Lady's party after her arrival, creating the impression in some Paraguayans' minds that he was responsible for her visit.

Jimenez's bonding with the First Lady marked the beginning of a remarkable series of coincidences:

On February 6, 1996, Jimenez attended a White House coffee, bringing along as his guest President Wasmosy's economic adviser, Carlos Mersan. Soon after, Jimenez gave $50,000 to the DNC.

On March 1, for the second year in a row, the State Department recommended making Paraguay ineligible for foreign aid. As in 1995, President Clinton overruled State's recommendation, signing a waiver that cited national security interests. On March 4, Jimenez traveled to Hope, Arkansas, to present a $50,000 check to the Clinton Birthplace Foundation. There followed $175,000 in donations from Future Tech to the DNC.

In April, Jimenez met at least twice with Mack McLarty to discuss his fears that a military coup against Wasmosy was imminent. NSC and State Department officials were present at one of these meetings. The coup attempt took place just as Jimenez predicted, and on April 22, the White House offered Wasmosy sanctuary in the U.S. Embassy. Jimenez donated $100,000 to the DNC the same day, and on April 23, President Clinton personally called Wasmosy at the embassy to affirm his support. The coup faltered, and Wasmosy returned to power.

In September, Jimenez was back in Washington for a meeting of the International Monetary Fund. This time, he arranged for a group of women, the wives of Paraguayan and Filipino bankers and officials, to be given a VIP tour of the White House. Mark Middleton and a member of the First Lady's staff served as guides, escorting the women through the Oval Office and other areas normally closed to the public. Afterward, the group had their pictures taken with Hillary Clinton, then boarded a chartered jet for Boca Raton, Florida, where they attended a fund-raising dinner at the home of Monty Friedkin, a major Democratic donor. There, they had a chance to chat at greater length with the First Lady, who had also flown to Florida and was the guest of honor at the event. Jimenez wrote a check to the DNC for $50,000, dated September 30, the day of the tour.[6]

When Jimenez's contributions came to light in early 1997, White

House spokesperson Lanny Davis expressed indignation that anyone would think the administration would be influenced by a mere $800,000. Senate sources familiar with U.S. policy in the region agree that it was not in our interest to allow the duly elected Wasmosy regime to fail. Still, the close coordination between Jimenez's donations and the administration's response created the appearance that American policy on Paraguay was for sale.[7]

Mark Jimenez was indicted in September 1998, for falsely representing campaign contributions as donations from Future Tech employees. He has apparently fled the United States to avoid prosecution.

A singular aspect of the Wasmosy saga was the Paraguayans' assumption that the route to influence with the White House lay through the First Lady. Tony Rodham did not even bother to report the bribe attempt, because, he said later, such offers came his way fairly often.

Doubtless this impression of the First Lady's influence in these matters had something to do with the Clintons' close relationship to the Riady family, a connection that Hillary Clinton had made through her practice at the Rose Law Firm. The Riadys sold their interest in Little Rock's Worthen Bank in the late 1980s and opened LippoBank of California in Los Angeles. But in 1990, around the time it became apparent that Bill Clinton would be in the race for the White House after all, former Rose firm partner Joe Giroir heard from his old friend James Riady, who proposed that Giroir act as his agent, negotiating joint ventures between U.S. and Indonesian companies. After being ousted from the Rose Law Firm in 1988 by a faction led by Hillary Clinton, Giroir had prospered in his independent law practice and the bitter feelings engendered by that incident had been set aside, if not entirely forgotten. Thanks to his efforts as the Riadys' American business representative, Giroir was soon earning half a million dollars a year and developing major deals with Tyson Foods and the Louisiana-based Entergy Group. Giroir arranged for Wal-Mart to open a superstore in a Riady project outside Jakarta known as Lippo

Village, and assisted Indonesian cable TV magnate Peter Gontha, who was seeking financing from the Export-Import Bank of the United States for the launching of communications satellites from Indonesia. Giroir introduced Gontha to Maria Haley, formerly Clinton's chief international trade expert in Arkansas and now one of the Ex-Im Bank's directors.

These were heady times in Little Rock. Joe Giroir told *The New York Times* that joint ventures seemed to work because Asian businessmen felt an affinity for the Arkansas style of doing business, presumably because in both places trust was based on personal relationships. Small talk at the country club revolved around investments in Asian cable and telecom franchises, and Wal-Mart millionaires were boarding planes to Jakarta to check out real estate deals. Asian businessmen, in turn, were eager to invest millions in projects that American investors had written off as white elephants. Notably, a certain Mr. Ng Lap Seng of Macao, also known as Mr. Wu, proposed a renovation of the decrepit Camelot Hotel. Ng's local agent was Charlie Trie, a friend of the Clintons and formerly the co-owner, with his sister Dailin Outlaw, of Fu Lin's restaurant, a popular lunch spot not far from the state capitol.

Early in 1994, Mr. Ng came to Little Rock to close the Camelot Hotel deal. During the course of a meeting in his hotel room, Charlie Trie mentioned that he was short of cash. Ng snapped open his briefcase, which happened to be filled with stacks of U.S. currency, and counted out $20,000. Word of this incident got around and some in Little Rock began to speculate that the rescue of the Camelot was part of a money-laundering scheme. During a tumultuous session of the Little Rock City Board of Directors, one commissioner wondered aloud if Mr. Ng had drug connections, only to be accused by Joe Giroir of anti-Asian prejudice. Deeply divided, the board refused to approve the project.

James and Aileen Riady, meanwhile, became the largest single contributors to Bill Clinton's 1992 campaign, donating $465,000. Supposedly, the Riadys' contributions were legal because they hold green cards. But according to a report of the Senate Governmental Affairs

Committee on campaign finance irregularities, also known as the Thompson committee, the Riadys hadn't lived in the United States since 1990, and at least some checks written on a Riady-owned real estate holding company represented money forwarded from foreign accounts. Therefore, the legality of these donations must be considered "doubtful."[8]

At any rate, this was only the beginning of the Riadys' generosity. Other large donations found their way to the Democrats via their family and business connections. Moreover, John Huang, an executive of LippoBank of California and the Riadys' top man in the United States, joined with Maria Tsia, an immigration counselor, to form a group called the Pacific Leadership Council, which organized Democratic fund-raising events targeting recent immigrants to California from China, Taiwan, Korea and the Philippines. Asian Americans who attended PLC functions were often startled to find that a large percentage of the guests were noncitizens. Nor did the PLC's ethnic outreach efforts stop at our borders: In 1991 it organized a tour of Asia for DNC chairman Ron Brown; the purpose was said to be to "raise Brown's profile" with the Asian business community. By 1992, the fund-raising program, now reorganized as the Asian Pacific Advisory Council, was being run by a Hawaiian couple named Eugene and Nora Lum, close associates of Brown, and Huang's role became less prominent.[9]

John Huang's friendship with the Clintons went back a long way. Huang earned an MBA from the University of Connecticut in 1972 and worked on Asian accounts for the First National Bank of Louisville and the Union Planters Bank of Tennessee. He met James Riady in 1980 at a banking seminar in Little Rock where Bill Clinton was one of the speakers, and he went to work for the Worthen Bank a few years later. After the 1992 election, Huang's name appeared on the "must consider" list for a job with the Clinton administration. In addition to his services as a major fund-raiser, his application file was replete with impressive recommendations. One letter, from Maely Tom, a prominent Asian-American fund-raiser, noted that the Riadys "invested heavily in the Clinton campaign. John is the top priority for placement because he is like one of their own. The family knows the Clintons on a first-name basis." Huang, Tom adds, "advises the Riady family on issues and where to make investments."

Despite these impressive qualifications, Huang did not get his desired appointment to the Department of Commerce. Exactly why is unknown; however, there were good reasons for officials responsible for moving his application forward to be cautious. During the mid-1980s, the Riady family had begun shifting its interests from Indonesia to mainland China. The family soon had more than $8 billion in various projects, including a $1.9 billion tourist complex with a harbor and industrial park in Fujian Province, across the straits from Taiwan. Shortly thereafter, a company called China Resources, a wholly owned subsidiary of China's Ministry of Trade, began to buy shares in the Riadys' Hong Kong Bank, ultimately controlling 65 percent of the operation. When LippoLand, the Riadys' ambitious real estate project on Java, ran into trouble, China Resources also came through with financing. Whether they liked it or not, the Riadys were now in business with a company believed by many analysts to be a front for the Chinese intelligence services. Thus, the Clinton-Gore campaign of 1992 may have been supported, even if indirectly, by the Chinese government.

There were also reasons to be suspicious of John Huang. Although he was probably born in Fujian Province in 1945, he has also listed his birth date as 1941 on official documents. In addition, it seems unlikely that Huang really intended to sever his longtime connection with the Riadys, raising the question of why he was eager for a government job in the first place. "I was uncomfortable with Huang," Commerce Undersecretary Jeffrey Garten told the Thompson committee. "My instinct, as someone who had lived in Asia, was that he wasn't the kind of person who ought to represent the American government."[10]

In January 1994, Commerce Secretary Ron Brown granted John Huang an interim top security clearance, but Huang's desired appointment as a deputy assistant secretary of commerce for international economic policy still did not come through.

That spring, Webb Hubbell resigned his position with the Justice Department under suspicion of embezzling from the Rose firm and began "networking," looking for ways to raise cash and stave off personal bankruptcy. Although the Clintons had been warned by their Arkansas friend James Blair that Hubbell was in legal jeopardy, Webb and his wife were still welcome at the White House, attending a number of movie screenings and parties, including a June reception for the emperor of Japan.

On Monday, June 20, according to White House WAVES records, which track individuals entering and leaving the building, Hubbell had a meeting with his old friend Hillary Clinton. There followed a busy week during which John Huang made at least five visits to the White House, usually in the company of James Riady. Interestingly enough, this was the same week that Mr. Ng Lap Seng reached the United States with his bulging suitcase of cash. Riady, Huang, Ng and Charlie Trie all had consultations during this period with White House aide Mark Middleton. Other highlights of the week included breakfast and lunch meetings on Thursday, June 23, between James Riady and Webb Hubbell, followed by Riady's and Huang's attendance at the President's Saturday morning radio broadcast. White House videotapes show that after the broadcast, when the other guests departed, Riady and Huang remained behind for a private meeting with Clinton.

On the following Monday, June 27, Hong Kong China, Ltd., a Riady-owned company, issued a check to Webb Hubbell for $100,000. That same day, John Huang received a letter of congratulations from the Lippo Group in Jakarta on his appointment to the U.S. Commerce Department.[11]

During his tenure at Commerce, John Huang had a "reader profile," which meant that classified diplomatic cable traffic relating to commercial issues was routinely routed to his office. He also received thirty-seven briefings from the CIA, during which he was supplied with hundreds of classified CIA documents on subjects such as U.S.–China technology transfers, the Asian nuclear power industry, famine in North Korea and a variety of commercial topics. At least three of the documents Huang saw were coded to indicate that they contained information that could imperil the safety of U.S. intelligence assets.[12]

Huang frequently left his office and walked across the street to the Washington branch office of the Stephens, Inc., brokerage house in the old Willard Hotel, where the staff permitted him to use an empty office that was maintained as a courtesy to Stephens executives visiting from out of town. Huang used the Stephens company fax machine, sent and received overnight packages and made phone calls. And from his own office at Commerce, he also placed hundreds of phone calls to the California headquarters of LippoBank. The bank's president, James Per Lee, was amazed to learn from press reports that most of these calls were to his own secretary, who told him she was just "relaying messages." The secretary has since left the country.

Huang also enjoyed unusual access to the White House, visiting at least sixty-seven more times, and had dozens of phone contacts and/or personal meetings with officials of the embassies of the People's Republic of China, Korea and Singapore. When these contacts came to light during the course of the Thompson committee investigation in 1997, Huang's superior, Jeffrey Garten, said he was "taken aback" to learn of them, since they had no obvious connection to Huang's duties.

"At bottom, Huang's stint at Commerce is difficult to understand," the Thompson committee concluded in its final report. This statement qualifies as wry humor. Huang's activities could be explained in only one way: he was a spy. Whether he was helping out the Riadys or the Chinese government remains an open question. It is one that Huang so far refuses to answer; however, a representative of the CIA told the Thompson committee that the agency was pursuing an "unverified" lead that Huang had a financial relationship with the Chinese government.[13]

DID Hillary Clinton play a key role in overcoming resistance of career employees to Huang's employment at Commerce? The circumstantial evidence that she did has been confirmed by Nolanda Hill, a Texas

TV station owner who was the lover and confidante of the late Ron Brown. In a June 1997 interview on ABC's *Prime Time Live,* Hill recalled Brown saying that Huang got his job at Commerce because of the First Lady: "He believed that she was the person that made the call."

If so, this wouldn't have been Hillary's first intervention in Commerce Department business, at least according to Hill's sworn statements. Brown, Hill says, credited Hillary Clinton with thinking up the idea of using trade missions as a fund-raising tool. (The Clintons, indeed, had pioneered the idea of "commercial diplomacy"—using trips abroad to help make business contacts for key contributors—during their years in Little Rock.) Ron Brown started out as a great admirer of Hillary, and one of the things he liked about her, according to Hill, was her philosophy. She understood that "everything is politics and politics is driven by money."

After the 1994 midterm elections, however, Brown's opinion of Hillary began to change. The White House was in a panic over the Republican takeover of Congress, and Hillary was desperate to raise more cash to ensure her husband's reelection in 1996. Hillary, says Hill, began pressuring Brown to auction off places on trade missions for $50,000 apiece, and her demands that he cater to the whims of big-money contributors became intolerable. Brown began to refer to the trade missions as "doing Hillary's chores," and on one occasion he complained, "I'm a motherfucking tour guide for Hillary."

Brown's objections to trading favors for cash had little to do with morality, it seems; rather, Hillary was making trade-offs between campaign donations and inclusion in the trade missions too obvious, putting him in jeopardy. Brown's business activities were being scrutinized by the Justice Department, and documents on the trade missions in particular had been subpoenaed in connection with a lawsuit filed by Larry Klayman of Judicial Watch, a dogged litigator intent on exposing the Clintons as corrupt. According to Hill, Brown told her that Leon Panetta, then Clinton's chief of staff, and Panetta's assistant, John Podesta, were urging him to delay producing the documents and Brown was "very worried."

Hill's last conversation with Brown about the documents came in early 1996. Brown opened up his ostrich-skin portfolio and produced a packet of documents about one inch thick. Hill scanned the top five or six pages and saw that they were letters from Melissa Moss of the Commerce Department's Office of Business Liaison, and were addressed to various trade mission participants. Each of the letters, Hill recalled, "specifically referenced a substantial financial contribution to the Democratic National Committee."[14]

Nolanda Hill never saw Ron Brown alive again. In April 1996, he perished along with thirty-four other officials and business executives on a trade mission to Croatia when their plane crashed while preparing to land in Dubrovnik. Also killed in this accident was Charles F. Meissner, who was John Huang's immediate supervisor at Commerce. Melissa Moss denied under oath that she had written the letters Nolanda Hill claims to have seen. Left with unresolved financial and legal problems after Brown's death, Nolanda Hill had reasons to feel bitter toward Brown and his family. However, it is also difficult to imagine what she would stand to gain by concocting accusations against Hillary Clinton. At the time Hill outlined Brown's comments about Mrs. Clinton in a sworn affidavit, she said that she feared retaliation. A week after she repeated her story in a federal district court, she was indicted on charges of fraud and tax evasion.

A new phase of the Clinton fund-raising odyssey began on Sunday evening, September 10, 1995, immediately after Hillary's return from Beijing. During a meeting with key aides, including Al Gore, Maggie Williams, Harold Ickes and DNC chairman Don Fowler, the Clintons approved Dick Morris's plan to jump-start the 1996 campaign by spending a million dollars a week on ten weeks of TV ads, paid for with "soft money," DNC funds that by law could be used for issue-oriented appeals not directly related to an appeal for votes. The ads ran mainly outside the major markets of Washington, New York and L.A. and managed to boost Clinton's poll numbers while attracting remarkably little media scrutiny. But paying for the media blitz was no easy matter. Harold Ickes, the veteran Democratic operative and onetime boyfriend of Susan Thomases who had been brought to the White House initially to deal with Whitewater issues, virtually took over the DNC fund-raising machinery, doing away with any semblance of vetting donors.

As part of this effort, John Huang decided to leave the Commerce Department and resume his role as a fund-raiser. DNC officials weren't eager to give Huang a job, however. Their reasons aren't clear, but by this time an investigative article on the Clinton-Riady-Giroir connection had appeared in the conservative *American Spectator*, and DNC officials had to be concerned that there would be questions raised about the sources of any money Huang brought in. After they failed to respond to Joe Giroir's suggestions that they give Huang a job, President Clinton himself intervened to get him hired.

Longtime observers of Hillary Clinton have no doubt that she played a key role in approving, and perhaps even in directing, the strategy of targeting foreign donors. It must be said, however, that she managed to keep a low profile. It isn't clear that she attended the weekly sessions at the White House where the soft-money ad campaign was discussed, although Dick Morris mentions the participation of Maggie Williams as well as Evelyn Lieberman and Ann Lewis, the White House director of communications, who were both close to the First Lady. Even Hillary's visits to Guam and Asunción attracted relatively little attention. In contrast to Al Gore's visit to the Hsi Lai temple, where he accepted checks laundered through Buddhist nuns, there were no embarrassing photos and colorful witnesses to keep the story alive.

One notable exception was the First Lady's entanglement with Johnny Chung, a man who has been described as a "Hillary Clinton groupie."

A pudgy Torrance, California, businessman, Chung first met Hillary in the Governor's Mansion in Little Rock in 1992. Ostensibly, he was in town trying to get state business for AISI, his California-based fax distribution company. The business was a dud—Chung was eventually sued by one investor who claimed to have been bilked out of $900,000—but Johnny Chung did have a talent for getting his picture taken with important people, including Republican governors Christine Whitman of New Jersey and Pete Wilson of California. Hillary, however, was his favorite pinup. AISI brochures featured a dozen pictures of Chung with the Clintons, primarily Hillary.

After the death of Hugh Rodham, Sr., in the spring of 1993, Chung wrote Hillary a letter of condolence, and he was soon a fixture in Room 100, the reception area of the First Lady's office suite in the Old Executive Office Building. On as many as fifty-one occasions, Chung dropped by, apparently just to admire the display of photos from Hillary's recent travels and the life-size Hillary cutout, next to which many a visitor has posed for snapshots.

Some people might be surprised to learn that a private citizen with no particular business to conduct would be able to enter the White House complex at will and hang around the First Lady's office. Maggie Williams would explain to surprised investigators from the House Committee on Government Reform and Oversight that the welcome offered to Chung was in accord with the standard of "graciousness" set by Mrs. Clinton. Said Williams, "Now that sounds a little crazy, a little offbeat, but that was a model of graciousness that we had in the White House. . . . Although I do know that certain people in the White House, people outside of the White House, pay special attention to people who give the money, there are also people in the White House who pay special attention to people just because they are people."

Williams went on to describe Chung as her personal affirmative action project. "As an African American, I can tell you, I know what it is to be different in politics in America, and be on the outside of things and struggle mightily for insider status and recognition. And so I, perhaps, had an especially high tolerance for Mr. Chung."[15]

A good deal of tolerance was needed in dealing with Chung, because he had a way of causing problems. After he brought a group of businessmen to a White House Christmas party, a picture of one of them taken with the Clintons showed up in a Chinese beer ad, to the embarrassment of the White House.

In February 1995, after Chung showed up with another delegation including Zheng Hongye, an executive of COSCO, the shipping company allied with the People's Liberation Army, DNC finance chairman Richard Sullivan questioned whether donations in the form of checks drawn on Chung's company were in fact laundered foreign funds.

In early March, Chung found himself with another VIP delegation of Chinese businessmen who were hoping to visit the White House. No longer welcomed by the DNC, he wasn't sure he could deliver. As he put it, "I was on the limb."

On the morning of March 9, Chung dropped by the First Lady's office and asked Williams's assistant, Evan Ryan, if he could arrange for his friends to eat lunch in the White House mess and meet Mrs. Clinton. Ryan went to speak with her boss and returned fifteen minutes later. "The First Lady has some debts from the DNC," she told Chung, mentioning a figure in the neighborhood of $80,000.

"Then a lightbulb goes off in my mind," Chung would recall, "I said, 'I will help for $50,000.' "

The next morning, Chung returned with a check and a note in an unsealed envelope. The note read, "To Maggie—I do my best to help." The group went off to lunch, and on their return Williams told Chung that the First Lady would have a few minutes to see them before she met with a delegation of teachers. Chung asked if Mrs. Clinton was aware of his contribution, and Williams said, "Yes, she definitely knows." Chung and his party were taken to the Map Room, where Hillary greeted him warmly, saying, "Welcome, my good friend."

But the businessmen whom Chung was escorting still weren't satisfied. They wanted to meet the President. Back in Williams's office, Chung called Betty Currie and tried to arrange for the visitors to attend Clinton's Saturday radio address. Currie gave Chung a number at the DNC, and Chung placed the call from Williams's office.

On Saturday morning, when Clinton saw Chung at the radio address with men who appeared to be high-level officials of the People's Republic of China, he was disturbed, commenting that the delegation probably shouldn't have been there. His concern led to a debate over whether to release the photos that are routinely taken of the President and his guests at these events. National Security Council analyst Robert Suettinger okayed the release of the pictures but commented in his memo that Chung was a "hustler."

Maggie Williams claims to have been unaware of this warning, though she wouldn't have been impressed if she had known. Asked about Suettinger's memo during hearings conducted by Dan Burton's House Oversight Committee, she exclaimed, "My God, if we kept hustlers out of the White House and the halls of Congress, who would there be?"[16]

Chung still wasn't finished causing problems, however. During his visits to the First Lady's office suite, he became friendly with Gina Ratliffe, a college senior who was doing an internship at the White House for academic credit. In early April, around the time her term came to an end, Chung told Ratliffe that he was "impressed by her professionalism" and offered her a "dream job." Chung talked of his plans to move his business affairs into a $3 million house in Virginia. Ratliffe could live in the house rent free, help with the redecoration and earn $28,000 a year handling arrangements for the Chinese visitors he squired around Washington. Robert Suettinger counseled Ratliffe that she should view the offer with a "touch of suspicion," but she accepted anyway and departed immediately on a trip to China with Chung. On their return, Ratliffe continued to work as an unpaid volunteer in the First Lady's office, even though she was on Chung's payroll. The situation continued for four months until Ratliffe, who had yet to see the house in Virginia, quit her job and returned home to Michigan.

Thus, during the summer of 1995, when Hillary was planning her trip to Beijing, Chung had an employee who was also working with the First Lady's staff. Moreover, word later got back to the DNC that while he was in China that spring, Chung had been heard boasting that the Clintons had entrusted to him the task of negotiating the release of Harry Wu. Presumably, this was another of Chung's fantasies. But who can be sure?

At any rate, Chung's claims to access made an impression on Chinese military intelligence. Between July and September 1996, it passed $300,000 to Chung via Lieutenant Colonel Liu Chaoying. Colonel Liu is a vice president of China Aerospace, a company that deals in satellite and supercomputer technology, as well as the daughter of General Liu Huaqing, long one of the most powerful members of the politburo of the Chinese Communist Party. At least $100,000 of Colonel Liu's money found its way to the Democratic National Committee.

In May 1997, Hillary Clinton took part in the wedding of Maggie Williams to Bill Barrett, an employee of the Council for National Service. Hillary's role in the somewhat unconventional wedding ceremony was to read a love letter that Charles Drew, the blood bank pioneer and father of District of Columbia councilwoman Charlene Drew Jarvis, wrote to his fiancée in 1939. At the reception, held at the home of Evelyn Lieberman, some of the guests took turns singing love songs. The President obliged with an Elvis-like rendition of "Love Me Tender."

Williams's marriage at forty-two was a happy ending to a stressful stint of public service. Between investigations of her role in the search of Vince Foster's office and allegations that she violated the Hatch Act by accepting Johnny Chung's $50,000 check, Williams had accumulated some of the heftiest legal bills of any Clinton aide, and this in a time when even aides called as incidental witnesses can wind up owing their attorneys $100,000.

As it happens, however, the most alarming sections of Williams's House and Senate testimony have little to do with concrete charges of wrongdoing. Whether or not Williams violated the Hatch Act by accepting $50,000 from Johnny Chung is surely less important than the total lack of concern on the part of Williams or, it seems, anyone in the First Lady's office about what he might be up to. Chung's impersonation of a wealthy businessman was laughable. His "guests" were all too obviously representatives of a foreign power—one that happens to be a totalitarian state and the only serious threat to U.S. military security in today's world. No one, it seems, wondered why

Chung's friends were so eager to get close to the First Lady of the United States and what they hoped to gain by it. What others would recognize as a potential security problem, Williams apparently saw through the lens of political correctness and personal loyalty, the two scales of value that really mattered to the First Lady and her staff.

Said Williams, "I realize I may have pushed the limits, but my experience had been at the White House that people of color and others in my view were not given overall the kind of respect that white males were, and I decided that I'm the boss of this office. This is one office where I can run it the way I want to run it, and the guy is genuinely, whether right or wrong, grateful to Mrs. Clinton. . . ."

SIXTEEN

The Making of an Icon

S NORKELING, sightseeing and a presidential round of golf with
Greg Norman were on the First Family's itinerary when they
visited Australia during the course of a quick trip to Asia after the
1996 election. Bill Clinton had just won a second term in office, the
first Democrat since Franklin Roosevelt to do so, and the Australian
newspapers were filled with upbeat articles about the youthful, energetic American President and his impressive wife. In one of her few
official appearances, Hillary delivered a speech at the Sydney Opera
House on "Women of the Twenty-first Century" and found herself
addressing such a warmly supportive audience that she decided to
follow up the policy-wonkish address by taking questions.

Responding to one woman who wanted to know how she ever
found time to write a book, Hillary told of working on her manuscript
while crouching next to a campfire in Wyoming. Another questioner,
however, hit a nerve by asking how she felt about the criticisms that
had forced her to keep such a low profile during the 1996 election
campaign. For a First Lady, Hillary said ruefully, "the best way to
escape the politics of one's time is to totally withdraw. Perhaps put a
bag over your head and somehow make it clear that you have no
opinions and no ideas about anything."

"But we all knew there wasn't much chance of that with Hillary

Rodham Clinton," the *Sydney Morning Herald* would observe the next day. And, indeed, Hillary's behavior during the official reception immediately after her talk dispelled any suggestion that the world would have a chance to get used to a kinder, gentler Hillary Rodham Clinton.

Janette Howard, the wife of Australia's Liberal prime minister, had made the mistake of assuming that the brief gathering after the speech would be a traditional First Lady's tea. The guest list she compiled included ranking Liberal Party women as well as her own best friend, a prominent charity fund-raiser, her daughter and her daughter's best friend. Hillary had seen the list in advance and was not pleased. Her staff sent word to Mrs. Howard that Mrs. Clinton wanted to talk with women who shared her policy interests, women like Catherine Harris, director of Australia's Affirmative Action Agency, and Sue Walpole, the Federal Sex Discrimination Commissioner, both members of the opposition Labor Party.

Despite some last-minute changes by Mrs. Howard, the guest list still wasn't to Hillary's liking, and minutes before her speech members of her staff circulated through the crowd at the Opera House, informing the female stars of the Labor Party that Hillary hoped to see them at the reception. When Hillary finished her talk, Clinton security people handed out hastily scrawled name cards to the last-minute invitees and spirited them into the reception area through a service corridor. Hillary spent a few minutes in stilted conversation with the humiliated Mrs. Howard and her guests, then dropped them cold as she moved on to chat with the Laborites about health care and affirmative action.[1]

A few hours later, while waiting for her husband to return from a golf game with Greg Norman, Hillary sat in the backseat of her official car with *Time* correspondent Ann Blackman discussing the past four years, saying, "I'm sure there are lots of lessons, things we did that could have been done better." Looking forward to the next four years, she added that she expected to play a "formal role" in monitoring the progress of welfare reform. "I want to travel around and talk to people about what is happening on the ground. I'm interested in how the actual implementation of welfare reform will work at the local and state level."

One could safely assume that many of the nation's governors, having just been given authority to redesign their states' welfare programs, would not be eager to have the First Lady show up, uninvited, to monitor their efforts. When Hillary's comments became public, followed closely by a protest from Wisconsin's Republican governor Tommy Thompson, White House spokesman Mike McCurry, who was traveling with the President in Manila, curtly assured reporters that it was all a misunderstanding: "I am not aware that there is any formal role that is planned for the First Lady."

The President was said to be annoyed at McCurry for undercutting his wife by denying the "formal role" story a bit too forcefully. And well he might have been. In an interview with Barbara Walters the previous October, Clinton himself had been the first to suggest that Hillary might monitor the progress of welfare reform, but he had realized in midsentence that the suggestion was controversial and added the qualifier, "It's not a formal role." Whether Hillary had intended to one-up her husband by deliberately using the words "formal role" is unclear. Perhaps she just misspoke. But now that the phrase had been the cause of embarrassment, Blackman found herself accused by the First Lady's communications director, Marsha Berry, of putting words in Mrs. Clinton's mouth. Even after she produced a tape recording to back her quotes, Berry continued to blame Blackman for taking a passing comment out of context.[2]

THE Australia trip served notice that the "New Hillary" who had emerged during the 1996 campaign would not be staying on permanently. From February through November, the First Lady had worked tirelessly for her husband's reelection, and done so without generating the kind of controversy that marked her efforts in 1992. She managed this by concentrating on her strengths, appearing before women's organizations—especially feminist groups, on campuses and at Democratic benefits. And she had raised $11.3 million, largely from women, who traditionally contribute at a much lower rate than men.

The Clinton campaign had gone to great lengths to keep the First Lady away from reporters who might prod her into making the kind of brittle, off-the-cuff remarks that had caused problems in the past. In early July, when her Republican counterpart, Elizabeth Dole, was basking in the spotlight, Hillary found herself in Revolutionary Square in Bucharest, Romania, during an eleven-day tour of Eastern Europe, which campaign aides argued would appeal to ethnic voters in places like Illinois. There followed her carefully scripted appearance at the Democratic convention, where her speech was heavily salted with references to "my husband," its structure organized around her hopes for her daughter Chelsea's future.

The popularity of Elizabeth Dole, a career woman who married late and never had children, raised the question of why Democratic strategists worked so hard to stress Hillary's domestic side. No one questioned that Hillary was a devoted mother, if somewhat over-protective. Chelsea Clinton was perhaps the most carefully chaperoned teenager in America, and White House staffers had been warned that even the most innocuous public comment about Chelsea could be grounds for dismissal. Still, the media could be so cruel that Hillary's zeal in shielding her daughter evoked admiration. She had even managed to embarrass into silence the TV comedians who made fun of Chelsea's adolescent gawkiness.

Ironically, considering Hillary's reputation as a shrewd professional woman and keen student of public policy, her public record was questionable. Her too-evident ambition made her an easy target for her critics. And even her friends would have to admit that she would be judged more successful if she did not try to do quite so much.

With the beginning of a new term, Hillary had an opportunity for a fresh start. Somewhat surprisingly, after years of favoring comprehensive federal programs, she now let it be known that she was emphasizing cooperation with state and local governments and private business because "we are living in a more entrepreneurial time." And instead of taking on one big assignment, she would divide her time among many smaller projects. Indeed, Hillary had so many events on her schedule that she needed to hire a second speechwriter, and so many pet projects that the general public often had no clear image of what she stood for.

In early 1997, Hillary talked of "doing a lot of work" in the District of Columbia. She visited schools and hospitals, and announced her support for the Washington Interfaith Network, an affiliate of the Industrial Areas Foundation, a group founded by radical organizer Saul Alinsky. She also read storybooks aloud to groups of children in connection with her "reading initiative." If parents in the District of Columbia spent more time reading to their children, she said at one event, "it could make a significant difference in this city."

After a few months, however, little was heard of Hillary's work for the District. In the spring, the Clintons and Gores headed to Philadelphia to celebrate Colin Powell's Volunteerism Summit. Spring also saw the first of Hillary's "brain conferences," inspiring a flurry of magazine stories on new theories of child development and setting the stage for the administration's $22 billion day-care bill. By autumn, Hillary was kicking off the Millennium Project, whose goals ran the gamut from preserving the original Star-Spangled Banner and robotship missions to Mars to solving government agencies' Y2K problems and producing a series of "Millennium Moment" television spots.

The somewhat confused image of the Millennium Project was typified by the first White House Internet "cybercast," a lecture by Harvard historian Bernard Bailyn, followed by an on-line question-and-answer session hosted by Bailyn and Mrs. Clinton. It was no secret that Hillary and the spin doctors of the White House counsel's office were hardly fans of the Internet. In 1995, the White House had issued a strange 300-page document called "Stream of Conspiracy Commerce," which proposed to instruct reporters on how conspiracy theories about the death of Vince Foster and related subjects made their way from the Internet into the mainstream media. And it is certainly true that if the Internet represents the future, the Clintons and their political heirs are in trouble. While anti-Clinton websites like the Free Republic forum bristled with up-to-the-minute news, discussion and media links, the sites sponsored by Clinton defenders, such as James Carville's Information and Education Project, were invariably as stale as steam-table meat loaf.

The chat with Bernard Bailyn, far from celebrating the power of the Internet, soon devolved into a cautionary lecture, with Hillary expressing alarm over the power of a medium "without any editing or gate keeping function. . . . I mean it's just beyond imagination what can be disseminated. . . . I'm a big pro-balance person. That's why I love the founders—checks and balances; accountable power. Anytime an individual or an institution leaps so far out ahead of that balance and throws a system, whatever it might be—political, economic, technological—out of balance, you've got a problem, because then it can lead to the oppression [of] people's rights, it can lead to the manipulation of information, it can lead to all kinds of bad outcomes which we have seen historically." The idea of the Internet as a tool of "oppression" is questionable. And of course, conspiracy theories flourished long before it came into being; the notion that Lyndon Johnson and/or the CIA killed Kennedy and the "October Surprise" both received more mainstream media attention than theories about the murder of Vince Foster.

By now, Hillary had become so wary of the press that only a small group of reporters was invited to the White House Map Room to hear her describe what she considered her number-one project of the second term: micro-enterprise. Also known as micro-credit, the movement was begun by Dr. Mohammed Yunus in Bangladesh. Yunus founded nontraditional banks that gave small loans to women who otherwise would not qualify for credit. Borrowers were organized into groups that served as mutual-aid societies and used social pressure to ensure that their members worked hard and paid back their loans on time. In recent years, micro-enterprise has become *the* hot-button topic in the nonprofit sector. Doubtless this is a good thing. It makes sense for foundations to put money directly into the hands of poor people to help them start businesses as opposed to, say, funding multimillion-dollar studies.

On the other hand, some micro-enterprise enthusiasts have come to see the program as a global panacea. In Bangladesh and other Muslim countries, extending credit to women has been seen as a method of undermining religious fundamentalism, and Alan Jolis, Dr. Yunus's biographer, writes that micro-credit holds the key to resolving "a host of intractable, long-term social ills related to poverty: In Norway's arctic circle it is helping to repopulate the Lofoten Islands. In Oklahoma, thanks to Chief Wilma Mankiller, micro-credit is helping to reduce alcoholism. In Chicago, it is helping to get unwed mothers off welfare."[3] The World Bank, meanwhile, has embraced the principle of economic independence for women as the key to eradicating poverty in less developed nations, and it sent a team of "gender experts" to the Beijing women's conference to tout micro-enterprise programs.[4]

Hillary's enthusiasm for the idea goes back to the mid-1980s, when she brought Dr. Yunus to Arkansas to set up a micro-credit program known as the Good Faith Fund. After earlier visits to micro-enterprise programs in India and South Africa, in 1995 she toured sites in Chile and then in Nicaragua, where a group of tortilla bakers and mosquito-net sellers renamed their community bank in her honor, Madres Unidos Hillary. The Nicaraguan women, Hillary wrote of this trip, "asked me about women in India who were doing the same thing. Apparently, they had seen some television coverage of my trip. . . . At that moment, I felt I had witnessed a profound connection among women around our world."[5]

Whether funding micro-enterprise with tax dollars makes sense is another matter. When a government-funded program decides to give Jane a $500 loan to start a hair-braiding business but denies a loan to Mary, who makes decorative candles, there are bound to be questions of fairness. In developing countries, borrowers are closely monitored and come under heavy social pressure from the group to repay their loans. In the United States, many loans are never repaid, and funding programs that serve women almost exclusively may be discriminatory in itself.

But the most curious aspect of the micro-credit initiative was the price tag, which was projected to reach $1 billion within five years. At this rate, every woman of working age in the country could expect to get a $500 loan at some point in her life. Needless to say, this was not the plan. Programs doling out tiny loans were just the warm, fuzzy side of a more sweeping program to channel federal tax dollars to nontraditional banks known as community development financial institutions, or CDFIs. The definition of a CDFI is highly technical, but the best-known example is Chicago's Shorebank, where Hillary's Wellesley roommate, Jan Piercy, was once a senior vice president. Another example is the Southern Development Bancorporation of Arkadelphia, Arkansas, which Hillary helped found in the mid-1980s.

Community development projects like Shorebank's rehabilitation of Bronzeville, the South Side neighborhood that was the birthplace of the Chicago blues sound, can be a good social investment, and the CDFI concept had supporters in both political parties. But it is also obvious that without strict controls, nontraditional credit has the potential to become a boondoggle for the well connected. And this is exactly what happened.

First-stage funding for the so-called CDFI Fund was approved by Congress with little fanfare in 1994, but the fund did not get organized and hire a director, Kirsten Moy, until the latter part of 1995. The White House, meanwhile, wanted to announce the first round of grants during the 1996 State of the Union message. This proved impossible, but continual pressure to get the money flowing undoubtedly had a lot to do with the problems that ensued.

Most government agencies charged with handing out grant money, including even the National Endowment for the Arts, have established program guidelines and a point system for evaluating applications. The CDFI staff, however, argued that such procedures would be too inflexible for a "venture capital" operation. Instead, project evaluations were farmed out to freelancers. One consultant, Nancy Andrews, was paid up to $133 an hour and a total of $228,065 in nineteen months, not counting over $31,000 for travel to Washington and stays at the Mayflower and other luxury hotels. The fund had no controller or chief financial officer, and it avoided Treasury Department limits on consulting fees by funneling some payments through set-asides for minority and disabled contractors.

In June 1996, the fund's deputy director, Steve Rohde, found himself in a "contentious" meeting where he was told that the Treasury planned to "roll out" the awards during a "Presidential event" the following month. Although Rohde protested that the paperwork for the grants was still incomplete and even threatened to resign in protest, this time the White House would not wait, and the first $37 million in grants was announced on the last day of July.

The largest grant by far, $4.5 million, went to Shorebank, which had been described in a 1993 article in the *ABA Banking Journal* as "[Bill] Clinton's favorite bank." Shorebank also received technical grants—up to $336,500 in one instance—for advising three loosely affiliated institutions that between them won an additional $6.5 million. These banks, often described as part of the "Shorebank system," were the Louisville Development Bank, the Douglass Bank of Kansas City and the Southern Development Bancorporation. Southern, for which Hillary was formerly head counsel, passed muster even though its subsidiary, Southern Ventures, had its license revoked in 1994 amid allegations of sweetheart loans and unpaid consultants.

In addition, a $2 million grant went to the Enterprise Corporation of the Delta, headquartered in Jackson, Mississippi, despite its having been characterized as "risky" during the initial screening process. A number of prominent Arkansans have been connected with Enterprise and its parent corporation. Hillary once sat on the board, as did Bob Nash, the White House director of personnel; Donald Munro, a personal friend of Bill Clinton; and Marlin Jackson, a former head of the Arkansas Department of Finance and Administration (ADFA). Enterprise had also presented impressive supporting documents, including a letter of recommendation from President Clinton himself.

Almost as soon as the CDFI grants were announced, murmurings spread through the banking community. Aside from their connections to friends of Hillary, the four banks that captured one-third of the money happened to be the only for-profit institutions among the applicants. In the spring of 1997, the investigations subcommittee of the House Banking Committee decided to review the grant process, and Chairman Spencer Bachus sent word to Director Kirsten Moy that investigators from his committee would be coming to the fund's offices on April 18, to review its files. Moy happened to be in Europe attending a conference on women entrepreneurs, and Steve Rohde was in a panic. The paperwork for Shorebank, Southern, Louisville and Douglass was still incomplete. Although they were the four big winners, there were no evaluation reports in their files.

Rohde stayed up all night creating the missing paperwork. He left the evaluations undated but worded them as if they had been written a year earlier, prior to the decision to award the grants. This ruse failed to fool Bachus's subcommittee, which noticed that of the 268 applications, Rohde had handled only those of the four major grantees. Forced to admit that the paperwork was generated after the fact, Rohde was undaunted, explaining that he had always intended to tell the truth if caught. Therefore, by his lights, he had done nothing wrong.

"If one set out in advance to create a government grant program that operates as a political slush fund, it certainly would look something like the CDFI Fund during the first round [of grant awards]," Representative Bachus concluded. "It is hard to imagine how the Fund could have been more poorly managed during the four years of its existence." His report also noted that the quality of applications had begun to fall off sharply after the initial awards, raising the question of whether the program was overfunded.

The House subcommittee was particularly scathing about the staff's attempt to deceive Congress through misleading answers and faked documentation: "Senior CDFI officials view Congressional oversight as a kind of 'cat and mouse' game, in which executive branch officials pursue a strategy of concealment and obfuscation."[6]

Steve Rohde, who cooperated with congressional investigators, and Kirsten Moy, who didn't, were forced to resign after the revelation of the doctored files. Mrs. Clinton's spokesperson, Marsha Berry, denied that the First Lady dictated who got what grants, but this was a charge no one had made. The staff of the CDFI Fund didn't need to be told which programs were Hillary's favorites, and given the lack of procedural safeguards and the White House's demands for quick decisions, no one could be surprised by the money that flowed in their direction.

BY the time the CDFI Fund ran into trouble with the House Banking Committee, Hillary had shifted her focus to the international scene. In March 1997, she and Chelsea left on a two-week tour of five African nations. As noted by Ruth Rosen in a highly positive article in the *Los Angeles Times,* when meeting with women leaders abroad Hillary was given to using terms like "male democracies," which would have been controversial, to say the least, had she used them at home. "Prevented from promoting a progressive agenda here," writes Rosen, "she has turned herself into an international activist who challenges patriarchal customs and sows subversive ideas abroad."

Rosen used the word "subversive" as a compliment, but it raises a legitimate question. One can admire Hillary's humanitarian intentions while still wondering if her conduct on these journeys, which were paid for by the taxpayers, necessarily served the best interests of U.S. foreign policy.

Hillary's foreign trips enjoyed the strong support of Secretary of State Madeleine Albright and undoubtedly were designed to further certain official aims, but her outspoken ways sometimes did not please host governments. Following the Clintons' visit to Australia, in November 1996, Hillary went on to Thailand, where she led reporters and news cameramen on a tour of a clinic for AIDS-infected prostitutes, much to the embarrassment of the Thai government. In Argentina in October 1997, her ringing endorsement of birth control drew a standing ovation from an audience of activist women but offended Roman Catholics.

In November 1997, Hillary toured Russia and Central Asia, where she visited hospitals and neonatal facilities, urged the Uzbekis to develop a "civil society" and attempted to explain to a puzzled student what a "policy wonk" was. As on her other journeys, Hillary was a vigorous advocate of micro-credit programs. To Americans, the merits of such programs may be obvious, but in Muslim countries they are regarded by traditionalists as antireligious and antifamily. In Bangladesh, micro-credit banks have been torched by mobs, and borrowers physically attacked. And, indeed, the writings of micro-credit enthusiasts make it clear that one of their goals is to undercut the growth of fundamentalist movements by getting women out of the home and making them economically independent. Even if one agrees with Hillary Clinton that this would be a good thing, it isn't clear whether the First Lady of the United States should be endorsing such programs. If an emissary from a foreign country came to the United States to promote social programs that many view as antireligious, doubtless Americans would resent her as a meddler.

In May 1998, Hillary traveled to Geneva, Switzerland, to accept a $40,000 prize for her work on behalf of women and children. It was donated by the United Arab Emirates and awarded through the World Health Organization. The money went to a charity that was dedicated to reducing infant mortality, but the prize was presented at the same WHO session where Fidel Castro made a speech accusing the United States of "attempting to carry out a genocide against our people." For an official representative of the United States to sit quietly through such a speech by the Cuban dictator would have been unthinkable. Hillary, apparently, was accepting the prize as a private individual, although she traveled to Geneva with a considerable entourage. Perhaps the most surprising aspect of the occasion was that it went virtually unreported in America.

Later in the month, during an interview sponsored by the U.S. Information Agency and broadcast to Palestinian students, she declared her support for a Palestinian state. The comment violated a long-standing policy under which the White House has refrained from taking a position on this question. Suspicions that Hillary was sending a signal to Palestinian leaders that President Clinton supported their demands outraged Israelis and threatened to derail the delicate peace negotiations, much to the dismay of Middle Eastern experts in our own State Department.

Hillary's higher profile was a part of a gradual shift in her thinking that began, aides said, when she flew to London to attend the funeral of Princess Diana. The outpouring of grief unleashed by Diana's death crystallized Hillary's appreciation of the power of celebrity. It was Diana's face on the cover of a thousand magazines, not her résumé, that had been the source of her power, enabling her to wield enormous influence.[7]

The death of Diana left her favorite charities without a high-profile patron and prompted intense speculation about who might replace her. Hillary was obviously a candidate, though her position as the wife of a sitting U.S. president placed her under certain limitations. The beautiful, articulate Queen Noor of Jordan was even more sought after. The American-born queen, formerly Lisa Halaby, attended Princeton University in the early 1970s, where she took part in demonstrations against the Vietnam War. Since marrying King Hussein in 1978, she has compiled an impressive résumé. Her charity work includes sponsorship of a micro-credit program inside Jordan, and she has been active in Princess Diana's favorite causes, the Landmine Survivors' Network and the International Campaign to Ban Landmines.

Hillary and Queen Noor soon discovered that they had much in common, including an uncertain future. It was no secret inside Jordan that Queen Noor would like to see her older son, Hamzah, succeed to the throne, but Noor was King Hussein's fourth wife and there were other children who would take precedence. So far, Hussein had finessed the issue by naming his younger brother as his heir. Queen Noor was far from universally popular in Jordan, and if her husband should die without changing his will, protocol would dictate that she leave the country.

Hillary Clinton visited Jordan in the spring of 1998, and her friendship with Queen Noor deepened over the summer, after King Hussein was diagnosed with cancer of the lymph nodes and he and Noor took up residence in their home in suburban Maryland, so that the king could undergo chemotherapy. The queen's ability to combine a glamorous image with an ironclad sense of privacy seemed to provide Hillary with another, more enticing model of international activism, very different from Diana's. Perhaps it wasn't necessary to be a policy wonk or a celebrity. She could aspire to be an icon of international do-goodism, raise a lot of money for worthy causes and still maintain her privacy.

Hillary's fiftieth birthday, on October 26, 1997, marked her coming out as a cover girl. *Time, U.S. News & World Report,* an A&E *Biography* television special and countless newspaper stories hailed her as the symbol of baby boomer womanhood. Once defensive in her rare conversations with reporters, Hillary now entertained her interviewers, doing a raucous imitation of a tobacco-chewing client back in Arkansas and, according to *Time,* telling a funny story about "the migrations of screw worms." She talked of missing Chelsea, who had overruled her mother's desire to keep her within an hour or so of Washington, D.C., and bravely gone off to attend Stanford.

Hillary's aides and friends, meanwhile, let it be known that the First Lady had been privately lobbying her husband in favor of the ban on land mines, and also favored some sort of settlement in the Paula Jones case. Whether she was prepared to see her husband deliver the apology that Jones was demanding is unclear, but Hillary was said to be prepared to withstand some degree of "embarrassment" as the price of putting the matter behind them. Her flexibility on this issue was a sharp departure from her usual stance of refusing to concede an inch to her enemies, but Clinton continued to balk at admitting any misconduct and settlement talks reached an impasse.

The First Lady's birthday celebrations, meanwhile, took on the aspect of a royal tour. They began with a surprise party thrown by her staff, to which everyone wore "Hillary masks" constructed from that week's *Time* magazine cover photo, and continued with a bash at the Fairfax Hotel, co-sponsored by Ann Jordan and Buffy Cafritz. The following day, Sunday, there was another birthday party at the White House, this one based on a "state fair" theme, with booths set up to celebrate each decade of the First Lady's life. On Monday, Hillary flew to Chicago, where high school bands turned out to serenade her on her arrival at O'Hare Airport and Mayor Daley had named a park in her honor. Then it was off to Park Ridge to celebrate Hillary Rodham Clinton Day.

Opinion in Hillary's hometown was far from unanimous in her favor. Not only is Park Ridge a center of anti-abortion activism, but some residents were not pleased by Hillary's suggestion in *It Takes a Village* that their way of life was a throwback to some earlier, unenlightened era. During Bill Clinton's first term, when the public library proposed displaying the First Lady's portrait, there were so many protests that the idea was abandoned. This time, local pride had won out. Only a few anti-abortion protesters showed up to mar the this-is-your-life discussion group at the elementary school where old friends gathered to reminisce about childhood games and the lessons in social awareness they had learned attending the Reverend Don Jones's University of Life. One story that came out at this meeting had to do with a sixth-grade classmate of Hillary's whose father forbade her to wear nylons to a dance. A children's advocate even then, Hillary circulated a petition to get the father to change his mind. He was not impressed.

The advantage of being an icon was that appearances come to be accepted as reality. No matter how many of Hillary's projects fail to live up to their billing, her reputation as a very smart woman is unlikely to suffer. *Time*'s cover story, anticipating the day-care bill that was to be Hillary's next major domestic initiative, noted, "She promises to lay out the problem's complexities with her customary intellectual rigor." The theme of the *Time* story was that Hillary was now satisfied to act as an advocate, leaving the execution of policy to others. As far as day care was concerned, this may have been true. The Clinton legislative proposal, unveiled in January 1998, alarmed conservatives by accepting the premise of federal regulation, but it was still far from the big-government solution Hillary had once favored, including tax credits for parents as well as grants to state and local governments.

Hillary, however, still had lofty intellectual ambitions. Guided by Sidney Blumenthal, who had become her closest confidant in the West Wing, she saw herself securing her husband's legacy by promoting a new era in international politics, the triumph of the so-called Third Way. In early November 1997, following a visit to Belfast, Hillary flew to a private meeting at Chequers, the country home of the British prime minister, Tony Blair. Also present at the meeting were DLC president Al From, Deputy Treasury Secretary Lawrence Summers and Housing Secretary Andrew Cuomo, as well as a number of Labour Party theoreticians, gathered for a discussion of how the United States and Britain could promote an international alliance of Third Way parties.

The Third Way, which exists somewhere between socialism and free enterprise, is an old idea. Thomas Edsall, writing in *The Washington Post,* gives a fairly neutral definition, calling it "a version of liberal-left politics that calls for competitive, free-market strategies while using government to prevent the market from devastating those least prepared to live without the protection of the welfare state." In practice, the Third Way can mean almost anything, from a "pragmatic" approach to governing—a favorite concept of Tony Blair's—to a technocratic nanny state, based on an alliance between big corporations and big government, similar to the "integrated labor market" envisioned by Marc Tucker and the NCEE.[8]

Ensconced in a windowless office in the West Wing, Sid Blumenthal had become a part of the White House spin machine, to the dismay of Clinton aides like Mike McCurry. Blumenthal, who was so enamored of conspiracy theories that his nickname in the White House was G.K., for Grassy Knoll, had been involved along with Carl Oglesby, the radical polemicist Hillary had read while at Wellesley, in the so-called Assassination Information Bureau, which proposed alternative scenarios about the death of John Kennedy. Blumenthal talked to Hillary every day, and his influence seemed to lead in two sharply divergent directions. He encouraged the idea that Clinton could secure his historical legacy as the leader of a global Third Way movement, and at the same time he fed her sense of being besieged by ideological enemies at home.[9]

Both tendencies were reflected in a speech Blumenthal gave at the John F. Kennedy School of Government at Harvard in the spring of 1998. He began by extolling the Blair-Clinton discussions, describing them as "the emergence of a trans-Atlantic one-nation politics." (Exactly what Blumenthal meant by "one-nation politics" is impossible to pin down. He had to understand that this was a loaded phrase, alarming to some who cared about old-fashioned concepts like national sovereignty, but he appeared to glory in the implications.) But if the nature of the movement the Clintons were trying to create was a bit nebulous, Blumenthal had no doubt who its enemies were, and his speech segued into a vitriolic attack on Independent Counsel Ken Starr. In a lengthy tirade, he assailed Starr as, among other things, an ignorant and vindictive man who had appointed himself "Grand Inquisitor for Life." For good measure, he described Starr's deputy, Hickman Ewing, who is a member of the Fellowship Evangelical Church, as a "religious fanatic."

In fact, if Ken Starr were as rabid as Blumenthal claimed, he could long ago have indicted Hillary Clinton for perjury, a course some in his office favored. Whether Starr had the evidence to convict is another matter, but weak evidence has not stopped other zealous prosecutors from bringing unwinnable cases in the past.

The Clintons' real peril as 1997 drew to a close lay with the Paula Jones case. Back in 1994, when Jones filed her lawsuit, James Carville had implied that she was bought and paid for, a pawn of conservatives whose charges were a lie from beginning to end. This had turned out not to be true. Jones stood fast in her demand for an apology, turning down substantial financial offers and breaking with her lawyers, Joseph Cammarata and Gil Davis, who were eager to arrange an out-of-court settlement. Only then did the conservative Rutherford Institute agree to back Jones's suit. It had taken three years to bring Paula Jones and conservative money together. Now, the attorneys retained by the Rutherford Institute were hiring private investigators and even using billboard advertising to call on women harassed by William Jefferson Clinton to come forward.

The Jones attorneys managed to locate Juanita Broaddrick, the woman allegedly raped by Clinton at the Camelot Hotel in the late 1970s. Broaddrick gave a statement acknowledging that something traumatic had happened to her on the night in question, but on January 2, 1998, she submitted an affidavit denying the incident ever occurred. On September 26, the *New York Post* reported that Broaddrick had repudiated her affidavit in the presence of investigators from the Office of the Independent Counsel, and quoted her attorney as saying, "We're not admitting anything; we're not denying anything." Information about Broaddrick's situation was part of a file of confidential evidence submitted by Ken Starr that was being held under seal by the House of Representatives, available only to members of Congress.

Dolly Kyle Browning, served with a subpoena demanding that she turn over her correspondence with Clinton, had her sister, Little Rock attorney Dorcy Corbin, phone Bruce Lindsey, who helped negotiate a compromise under which only correspondence Clinton had sent on official stationery was covered. (Since the Jones suit involved Clinton's actions in work situations, Browning's private relationship with the governor, if any, was irrelevant.) Browning thought she had an understanding with Lindsey that "Clinton won't tell lies about me and I won't tell the truth about him."

However, when it developed that Browning was giving an affidavit acknowledging a sexual relationship with Clinton, there was trouble. White House attorneys let it be known that the President had notes of a conversation he had with Browning at their 1994 high school reunion in Hot Springs, during which—allegedly—she berated him for having sex with Gennifer Flowers when he had never had sex with her. Supposedly, Browning also said that she needed the money and planned to write a novel based on an affair between the two of them that never happened. Marsha Scott, who had been Clinton's escort at the reunion—Hillary was not in Arkansas on this trip— claimed to have been present and to have overheard Browning's wild threats. *Newsweek* reporter Michael Isikoff interviewed witnesses who had been at the reunion, where the President's long conversation with Browning at a corner table had obviously been a matter of great interest. Their accounts conflicted with the Clinton/Scott version of events that evening on key points. Browning, indignant over being called a liar, would eventually file a lawsuit charging the President with defamation of character.

Another Clinton accuser, Kathleen Willey, originally came to the attention of Jones case attorney Joseph Cammarata in January 1997, through a still mysterious phone tip. The caller, who claimed to be Willey herself but perhaps was not, said that Willey had been groped by the President in a room off the Oval Office when she went to him to ask for a job. The date of the encounter was memorable because later that same day, November 29, 1993, Willey had learned that her husband had committed suicide. An ardent Clinton supporter, Willey didn't get the ambassadorship she wanted, but she was eventually named by the President as a member of a delegation to an international conference on biodiversity in Indonesia. She also got a presidential appointment as a member of the United Service Organization's Board of Governors. Willey did not want to become a witness against the President, and there is evidence that she was the target of a concerted campaign to get her to deny that the incident ever happened. For example, Nathan Landow, the former chairman of the Democratic Party in Maryland and father-in-law of Webb Hubbell's friend Michael Cardozo, chartered a jet to bring Willey to his Maryland estate for the weekend and offered to pay for her to go Christmas shopping in New York.

Meanwhile, several "Jane Does" approached by the Jones attorneys denied any "improper relationship" with Clinton. Their use of the identical language the President would eventually use in a sworn deposition suggested that they had been taking advice from the same attorneys. Trooper Danny Ferguson, no longer a target of the Jones suit but a likely witness, was being advised by attorney Bill Bristow, who would become the Democratic candidate for governor of Arkansas in 1998. Betsey Wright, the "bimbo eruptions" specialist of the 1992 campaign, worked closely with Bristow, and later served as a consultant during his run for office.

But unlike those of 1992, the "bimbo eruptions" appeared to be reaching epidemic proportions. Michael Isikoff had been pursuing Linda Tripp, the former secretary to Bernie Nussbaum, who happened to be present when Kathleen Willey emerged from her meeting with the President on November 29, 1993. Tripp said that Willey had appeared disheveled but also "flustered, happy, and joyful" after her encounter with Bill Clinton. She added that she was speaking to *Newsweek* because she wanted to make it clear that whatever happened in the President's office, "this was not a case of sexual harassment."[10]

However clumsily, Tripp's statement put forth a version of what happened between Willey and Clinton that was far less damaging than the story eventually told by Willey, who broke down and gave the Jones lawyers their deposition on January 10, 1998. In return for her loyalty, Tripp was insulted by Clinton attorney Bob Bennett, who told Isikoff that Linda Tripp was "not to be believed." Tripp, by this time working at the Pentagon, was a presidential appointee; like Billy Dale, she could lose her job anytime the President decided it no longer served his interests to have her in the government. She had two children in college, and with the attorney of the President of the United States publicly calling her a liar, her career prospects in or out of government did not appear to be bright. It is little wonder that Tripp decided to purchase a tape recorder and begin taping her conversations with another Pentagon employee, Monica Lewinsky, who was giving her an earful about her stint as the "Special Presidential Assistant for Blow Jobs."

By the fall of 1997, Kathleen Willey's encounter with the President was already being discussed in *Newsweek,* and the press was also raising questions about Clinton's relationship with Shelia Lawrence, the widow of ambassador to Switzerland Larry Lawrence and owner of the Coronado Hotel, where Clinton sometimes stayed and played golf on trips to California. (The story died abruptly after Shelia Lawrence filed suit against columnist Arianna Huffington.) Resolving the Jones matter was more urgent than ever.

Another complication arose on December 17, when the President learned that Monica Lewinsky had become the Jones attorneys' Jane Doe number 6 and was listed as a potential witness in their case. With the date scheduled for him to give his deposition to the Jones lawyers fast approaching, Clinton still might have found a way to settle. And if he couldn't settle the case, he could have defaulted, justifying his decision on the grounds that it was inappropriate for a President of the United States to discuss his sexual conduct under oath. Clinton's critics would have scoffed at this explanation, of course, but at least he could have claimed with some plausibility that he was acting on principle, for the good of the country. Clinton took neither action. Clinton's effort to find Lewinsky a job in New York was reinvigorated. Lewinsky was sent to see Vernon Jordan, and on the advice of an attorney recommended by Jordan, she signed an affidavit denying that she'd had sexual relations with the President.

What Clinton didn't know was that on January 9, eight days before his scheduled deposition, a nervous Linda Tripp had played tapes of her conversations with Lewinsky for her attorneys. They advised her to hand the tapes over to Bob Bennett. Tripp, who had no reason to trust Bennett, walked out of the meeting and contacted Ken Starr.

JANUARY 7, 1998, may be recalled as Hillary Rodham Clinton's last truly happy day in the White House. Escorted into the East Room by a delegation of children, the Clintons announced what the President called "the largest single commitment to child care in the history of the United States." To applause and laughter, Clinton said, "I thank my wife, who has been talking to me about all these things for twenty-five years now, and is sitting here thinking that I have finally got around to doing what she has been telling me to do. I was thinking it would be nice to have something new to talk about for the next twenty-five years."

This was a "be careful what you wish for" moment. Ten days later, Clinton gave his deposition to the Paula Jones lawyers, ensuring that he and Hillary would have a great deal to talk about from that day on, assuming that they were still on speaking terms. The day after the deposition, Sunday, Clinton was concerned enough to call in his secretary, Betty Currie, and quiz her about what she recalled of Monica Lewinsky's visits to him. "We were never alone, right?" he asked.

By Wednesday, Lewinsky's name was in *The Washington Post.* Ironically, that very evening, Hillary played hostess to Richard Mellon Scaife, a Pittsburgh multimillionaire and the financier behind numerous efforts to dig up dirt on Bill Clinton, including Christopher Ruddy's investigations into the death of Vince Foster and the so-called Arkansas Project. It turned out that Scaife, in addition to contributing to right-wing causes, also gave to Stanford University, where Chelsea Clinton was a student, and to the White House Endowment Fund, a charity that supports the renovation of the White House and the restoration of national treasures such as the original Star-Spangled Banner. Scaife had contributed before but never received an invitation to the White House. How he came to be included among the 100 guests at that evening's black-tie dinner is an enigma. Though the White House press office confirmed that the First Lady personally approved the guest list, one has to wonder whether this was a sign of open-mindedness or just another slipup. "It's more than the Bushes ever did for me," Scaife told his own paper, the *Pittsburgh Tribune-Review.*

There can be no doubt, of course, that Hillary knew that something had gone on between Lewinsky and her husband. In addition to a copy of *Leaves of Grass,* one of the gifts he was said to have given Monica was a brooch—and Clinton's interest in collecting antique brooches, some of which ended up as gifts to various women, was well known in Arkansas. Monica, in turn, knew Clinton well enough to be aware of his penchant for frog-related knickknacks. These gifts told the story of a relationship that had been affectionate at times, even if one-sided. For a woman who still cared about her husband, this had to be far more hurtful than any allegation of a quick grope or sexual encounter.

But what could Hillary do? As much as some might have taken vicarious satisfaction in seeing the First Lady dump her husband's clothes on the White House lawn, any show of spite on her part would have embarrassed the nation and ultimately made her look weak and out of control. The very thing she had dreaded—becoming an object of pity as Lee Hart had been in 1988—had now come to pass.

Over a period of days and weeks, there were signs that Hillary was closing her mind to the truth. According to Sidney Blumenthal's later testimony before the independent counsel's grand jury, Hillary told him "that she was distressed that the President was being attacked, in her view, for political motives, for his ministry of a troubled person. She said that the President ministers to troubled people all the time, that he has ministered to—that he does so out of religious conviction. She said to me on that occasion, 'If you knew his mother, you'd understand it.' " The definition of "ministering" in this context is as elusive as Bill Clinton's definition of words like "alone" and "is," which he would struggle to pin down in his grand jury testimony. But that last sentence rings true.

It was hardly a secret that Clinton supporters had engaged in a campaign to silence inconvenient women, sometimes by buying them off, sometimes by harassing them and occasionally by soliciting affidavits. Somehow, the press had been slow to put the story together and suggest that this amounted to a pattern of illegal behavior. Oddly enough, the columnist now saying that this effort had been a criminal conspiracy all along was Dick Morris, himself exiled from politics after being caught in a sex scandal on the eve of the 1996 Democratic convention. In an appearance on KABC radio in Los Angeles, Morris also volunteered his opinion that Hillary was a lesbian, or at least frigid—a suggestion that blamed her for a pattern of behavior that Clinton had followed even before his marriage.

Hillary, meanwhile, had decided to accept her husband's denial, a situation that confined her to a shrinking universe of people who could claim, with a straight face, to believe the President's version of events. Long reluctant to speak candidly about her marriage to friends, she now retreated into a self-imposed silence, knowing that anything she said might end up as raw material for the grand jury.

Focusing their anger outward on a common political enemy had long been a technique that helped keep the Clintons' relationship going in times of stress. In Arkansas, they had turned their emotional heat on Jim Guy Tucker, Frank White, Sheffield Nelson or whoever happened to be the competition of the moment. During the 1992 campaign, they had trashed the bimbos. Now it was Ken Starr. "The 'enemies' thing is a very, very powerful unifying force," said one friend of the couple. But it was Hillary who seemed to supply most of the energy. When Hillary was in residence at the White House, she would talk to Bill in the evening and get him pumped up for the next day's fight. When she was away, his attention tended to wander.[11]

As spring and summer wore on, the White House spin machine went into overdrive. Freelance Clinton defender James Carville complained bitterly that Ken Starr and his lieutenants were leaking to the press, and Clinton attorney David Kendall filed a motion of protest with the court. However, it would seem that most of the leaks were coming from outside lawyers, witnesses and, above all, the White House. Terry W. Good, the White House records director, admitted in a deposition in connection with a Judicial Watch lawsuit that he had been ordered to search for "anything and everything that we might have in our files on Linda Tripp." Defense Department spokesman Ken Bacon leaked information about Tripp's arrest in connection with a teenage party prank. The *Nation* magazine, about as far from being right-wing as it is possible to get, reported that Hillary's confidant, Sid Blumenthal, was calling reporters and urging them to investigate the alleged homosexuality of a member of Starr's staff. The White House was forced to acknowledge that I.G.I., Terry Lenzner's detective agency, had been retained by Bob Bennett to investigate Starr and his prosecutors.[12]

George Stephanopoulos, who had left the administration and gone to work for ABC News, noted that there had been talk in the White House of an "Ellen Rometsch strategy"—a reference to a Kennedy-era attempt to blackmail congressmen into keeping quiet about JFK's affair with an East German spy.

On July 27, Monica Lewinsky at last struck an immunity deal with the independent counsel. As a sort of present, she turned over the long-rumored blue dress, stained with what was either "spinach dip" or the semen of the President of the United States. Clinton's legal team received a request for a presidential blood sample from Starr's office on August 10, and they drew the obvious conclusion that laboratory tests had turned up DNA evidence. With Clinton scheduled to appear before Starr's grand jury the following week, the attorneys counseled him that he would have no choice but to admit to sexual activity.

Although Hillary was said to be working closely with her husband's legal team, she was among the last to know of the change in strategy. As late as August 13, while rumors circulated through the White House, Hillary assured her staff that the President was standing by his denial of an affair. Late that evening, hours before the early edition of the next day's *New York Times* broke the story that Clinton was ready to capitulate, he told Hillary the bad news.

For the second time in eight months, Hillary was faced with having to confront the painful truth. So, too, was Chelsea Clinton, who had soldiered through a semester at Stanford, ignoring the occasional remarks and the awkward sympathy of classmates, bolstering her confidence with her father's assurance that the charges were just more political gamesmanship. A friend of the Clintons, who had talked to them in January, relayed the observation that "half the kids at Stanford come from broken homes anyway," a factoid that may have comforted Chelsea's parents but didn't seem particularly relevant. Now, Chelsea, too, had to endure a second round of disillusionment.

Amazingly enough, considering how vigilant Hillary normally was about defending her daughter's privacy, she called on the Reverend Jesse Jackson, who rushed to the White House to comfort Chelsea, then wrote about his ministerial visit in *Newsweek,* revealing that he had hugged Chelsea, prayed with her and counseled her on the need to forgive "men in their weakness." Chelsea, Jackson assured us, had told him, "I can cope." One hopes this is true, though more believable accounts say that while accepting the situation, Chelsea suffered stomach pains and broke up with her boyfriend, who couldn't be the rock of emotional stability she demanded and needed.[13]

On Monday, August 17, Clinton appeared before the grand jury. Afterward, at ten P.M., he went on national television to deliver a speech he had drafted the previous evening, conceding that he'd had an "inappropriate" relationship with Lewinsky and that he had "misled" the public. Most of the brief speech, however, consisted of an angry attack on Ken Starr. More conciliatory language, drafted by Paul Begala and other aides, was dropped, in part for fear of offending Hillary.

Both Clintons were angry that evening, furious that a President of the United States should be required to answer questions about his most intimate sexual activities in front of a grand jury. Almost anyone would have felt the same way. And yet, the President could hardly be surprised to find himself in this situation. Anyone could foresee that his statements under oath, even in a civil deposition, were bound to be dissected by the press and, very likely, end up being the subject of further legal investigation by someone, even if not Ken Starr. Even Monica Lewinsky, hardly a great legal strategist, had pleaded with Clinton to avoid giving a deposition under oath to the Jones lawyers, reporting back to Linda Tripp that the man she called "the big he" was "in denial."

Hillary was said to be bitterly angry with her husband, though it was hard to say whether she was more upset about his infidelity or his sheer ineptitude about concealing it. It was no secret among her staff and friends that she saw her time as First Lady as a chance to build a "platform" for her career after she left the White House. The failure of health care reform and her Whitewater troubles had been setbacks. She had regrouped and was building on her international reputation as an advocate for women and children, only to face humiliation and political ignominy.

The atmosphere was chilly on the morning after the President's speech as the Clintons, hand in hand, embarked for their summer vacation on Martha's Vineyard. At the airport was a "spontaneous" gathering of friends and supporters—actually summoned to the airfield by a round-robin of phone calls. Bill and Hillary spent most of their stay in seclusion, avoiding the usual rounds of golf, boating and dinner parties. The President, however, would soon be venturing out to speak before friendly audiences. He seemed little changed. According to one witness who declined to be identified, he even made a pass at an attractive woman during one of his first public appearances, unfazed by the reactions of witnesses, who were gasping in amazement.

Meanwhile, in Washington, the First Lady's spokesperson, Marsha Berry, issued an unusual statement, assuring the press that Mrs. Clinton, like the rest of the country, had been "misled" by her husband and, therefore, had never herself intentionally misled the American public. "She learned the nature of his testimony over the weekend," insisted Berry. "Like I said, she was misled. The President said that, and that's true."[14]

This disclaimer could not alter the fact that Hillary Rodham Clinton's most significant public accomplishment by far had been saving Bill Clinton from political ruin and making him a viable candidate. She had stood by him when he was under attack, supported him in his lies and told a few lies of her own, not to mention taking the fight to a new level of paranoia by denouncing her husband's enemies as right-wing conspirators who must be not only defeated but punished. Hillary was not the first politician's wife to put a false face on her marriage. One thinks of Jacqueline Kennedy, who no doubt put up with a great deal, yet never lied to the voters or blamed the opposition for her husband's behavior. Even though Hillary may have been misled on certain points of information, she was, as she reminded us during her January 1998 *Today* show appearance, the person who knew Bill Clinton better than anyone else in the world. Therefore, she must be held accountable for her bad judgment in thinking that his private flaws would not interfere with his performance of his public role as President of the United States. She had taken her problem and made it our problem.

And yet, as Bill Clinton's stock fell, Hillary's rose. Hillary had always been popular with the leftish activists who still predominate in state Democratic organizations. Many who had been tepid in their support of Bill Clinton from the beginning found Hillary's politics more attractive and were happy to rally to her side. Always an effective speaker, she was now also a victim twice over—first of a white male and then of the party's right-wing enemies. For this constituency, Hillary represented the seductive proposition that true morality lies in having the "correct" positions on issues such as day care, gun control and tobacco advertising, not in personal behavior.

But even voters who were deeply troubled by Bill Clinton's lies could feel a certain kinship with Hillary Clinton. Her judgment about her husband and other matters may have been deeply flawed, but just as she had considered and rejected divorce, many of us, too, still found good qualities in Bill Clinton and wondered if the trauma of getting rid of him would really be worthwhile.

As the November elections approached, Hillary was once again out on the hustings. She helped tip the balance in favor of Democratic senatorial candidate Chuck Schumer in New York and restored the flagging campaign of California senator Barbara Boxer, her relative through the union between Tony Rodham and Boxer's daughter. And as had happened so many times in the past, Bill Clinton redeemed himself on Election Day. Inside the White House, the Democrats' unexpectedly strong showing, not to mention the President's consistently high approval ratings, was taken as evidence that the public had forgiven Bill Clinton.

Throughout the worst of her husband's troubles, Hillary had steadfastly kept to her agenda. At the end of August, she flew to Belfast to attend a conference called "Vital Voices: Women in Democracy," the first of a series of convocations designed to encourage greater participation by women in public life. The series was a favorite project of Hillary's, as well as of Secretary of State Madeleine Albright and Teresa Loar of the State Department. A second conference, in Montevideo, Uruguay, was the highlight of the First Lady's ten-day tour of Latin America the following month. One sponsor of the conference, Muni Figueres of the Inter-American Development Bank, noted that the trip would serve as a respite for her, since Latin Americans in general "have a clearer sense of the separation of personal life from public performance than Americans do." Joanna Mendelson, a Latin American specialist from American University, put it more bluntly, observing that the wives of Latin American politicians would be sympathetic, since they have learned to accept their husbands' infidelity as "a given of public service."[15]

Back home, not long after the Starr referral went to Congress, Paula Jones let it be known that she had decided that the words of apology she had long sought were now superfluous. Although her case had already been dismissed by Judge Susan Webber Wright, Clinton had reason to fear that it would be reopened, this time with an emphasis on the circumstances that had led Monica Lewinsky and Jane Doe number five to give their affidavits. Rather than allow that to happen, the President agreed to an $850,000 cash settlement, more than it would have cost him to dispose of the case with an apology at any time during the previous four years. Since Clinton had no savings of his own, Hillary would write a check for the $375,000 portion of the judgment not covered by his insurance.

The neat separation between public and private realms that seemed so clear to the First Lady's Latin American friends was harder to discern when viewed close up. The President repeatedly denounced the politics of personal destruction but failed to dissociate himself from the efforts of pornographer Larry Flynt and investigator Dan Moldea to dig up dirt on his Republican opponents, raising suspicions that the Clintonites were now tacitly, and perhaps even actively, pursuing the so-called Ellen Rometsch strategy of competitive character assassination. Hillary continued to stand by her husband, joining him on the White House lawn to greet Democratic members of Congress who were bused over the Hill for a pep rally immediately after the House voted articles of impeachment in a party-line vote.

And yet, there were signs that the First Lady was finding it hard to regroup emotionally, as she had done so many times in the past. Back in September, as she was presiding over an announcement about advances in colon cancer treatment, a reporter in the back of the room had shouted out a question about whether the President had any thoughts of resigning, and Hillary had briefly lost her composure.

Referring to a demonstration of rectal examination techniques going on in the next room, she said, "I invite the press to go for their tests in the Green Room." As the weeks went by, White House employees leaked stories about bitter fights between the President and the First Lady, even more bitter than those reported in 1993. There were times, during the Clintons' trip to Israel in December, for example, when she looked exhausted and shrank from the President's touch, as he tried to take her hand in public.

And little wonder. She, after all, was not just another wife of a straying politician. She had put her personal credibility on the line to defend Bill Clinton and might yet pay a price for it. Hillary had planned her years in the White House carefully, believing that she would be building a portfolio she could translate into a role as an international advocate for children. There was talk of establishing a foundation, which she would head, and of finding a post for her in international affairs—with the United Nations, with the World Bank or, perhaps, as an ambassador. No one could say for sure what impact Bill Clinton's problems might have on these plans. A Washington attorney was already looking into the possibility of Hillary's selling her memoirs and taking on lucrative speaking engagements after her husband's second term. Ironically, the Lewinsky scandal ensured that the First Lady's story would command a huge price—money she would need to cover the Clintons' legal bills, which had topped $5 million even before the impeachment proceedings began.

Ironically, the one alternative for Hillary's future that appeared un-affected by the President's problems was the possibility that she might run for office herself, perhaps aiming at the New York Senate seat about to be vacated by the retiring Daniel Patrick Moynihan. It was difficult to imagine that Hillary, who'd found the constant scrutiny of life in the White House so difficult, would care to subject herself to a Senate campaign. On the other hand, after all her troubles, the ardent courtship of leading New York Democrats had to be im-mensely flattering. The First Lady's aides persisted for months in say-ing that their boss was not seriously considering a run for office, but friends who knew Hillary well enough to be familiar with the reach

of her ambition saw it differently. A woman who continued to believe, after twenty-five years, that but for her marriage she could have
been a senator—or, for that matter, president—would be sorely
tempted to prove she was right. If she didn't seize the golden opportunity being offered to her by Judith Hope, the chairman of the New
York Democratic Party, there would be other offers in the years to
come.

A more difficult question about Hillary Rodham Clinton's future
was whether she was prepared to focus her energies more effectively
than she had in the past. Did she really wish to be a full-time legislator—in the year 2000, or ever? Would she demonstrate a more
lasting commitment to the needs of her constituents than she had to
the people of her adopted state of Arkansas, where many voters felt
that both Clintons had conspicuously neglected their interests after
going to Washington? Would a Senate seat become just one more
Hillary project, to be juggled along with her favorite causes, her international travels, and the pressing need to rehabilitate her personal
finances?

In the short run at least, the prospect of yet another political campaign—this time with Hillary as the candidate—appeared to have a
rejuvenating effect on the Clinton marriage. The President, friends
said, was among those urging Hillary to run. Looking forward to
acquittal by the Senate, combined with a future electoral bid by his
wife, Clinton could begin to see that the troubles of 1998 had been
just one more cycle in the pattern that had defined his and Hillary's
marriage from the beginning: His mercurial moods and lack of self-
discipline provoked a crisis, from which he then redeemed himself by
working his political magic. That magic, it now seemed, might extend
to Hillary. Perhaps Bill Clinton's real legacy was that by embarrassing
his wife he had given her the touch of vulnerability necessary to make
her a more sympathetic candidate.

On January 20, 1999, Clinton once again demonstrated his remarkable resiliency when he went to the Hill to deliver his State of
the Union address, just hours after White House counsel Charles Ruff
had argued in his defense before the Senate impeachment trial. Hillary

was in the gallery, the leader of a support group that included civil rights pioneer Rosa Parks and Dominican baseball hero Sammy Sosa, singled out for his charity work in his hurricane-battered homeland. Saluting his wife for her campaign to preserve the Star-Spangled Banner, Clinton paid her a misty-eyed tribute: "I'd like to take just a minute to honor her, for leading our millennium project, for all she has done for our children. . . . I honor her. Thank you. And here she is." Hillary, acknowledging the flood of applause, wore an expression that suggested that she, at least, was painfully aware of the irony of this use of the word "honor."

But the public psychodrama that was the First Couple's marriage had not yet played itself out. On February 24, NBC's *Dateline* aired an interview with Juanita Broaddrick, filmed the previous month. Hillary, of course, had to have been aware of Broaddrick's charge of sexual assault, which had been well known to Clinton campaign workers as early as 1990. Now, however, she could no longer dismiss Broaddrick as just another disgruntled groupie or a tool of her husband's right-wing enemies. A statement by Patricia Ireland of NOW put the White House on notice that the feminists would no longer tolerate such tactics, and Hillary couldn't defend Bill Clinton without alienating the very friends whose support she was counting on in her future endeavors, whatever they might be. The sixties saying that "the personal is political" had been turned inside out. For Hillary Rodham Clinton, the political was now personal. All the private pain and convoluted rationalizations that had allowed her to keep her marriage together had become part of her public portfolio. She and Bill were once again discussing divorce, and as their political interests diverged, it became more difficult to imagine how they could find a common ground to repair their battered partnership.

The Half-Life of a First Lady

I N 1992 nothing had infuriated Hillary Clinton's admirers so much as the charge that she was using her husband to further her own ambitions and would dump him as soon as his term was over. Seven years later it was Hillary's friends who predicted that she would soon be dissociating herself from Bill Clinton and moving out of the White House in order to seize the opportunity to become the junior senator from New York, while conservatives strenuously doubted that such a thing could happen. Granted that Mrs. Clinton could not have foreseen, and certainly would not have wished for, some of the circumstances that brought this situation about. Still, the reversal was stunning.

This time it was the conservatives who were indulging in wishful thinking. The woman who had urged Bill Clinton to run in 1992, knowing full well that she and her daughter would have to tough out the inevitable allegations about other women, was not about to be intimidated by the prospect of a battle with a bunch of New York Republicans. Hillary was serious about running.

On the other hand there was a question about her mental and emotional commitment to going her own way. The pattern of the Clintons' marriage over many years had been that every time this couple were on the point of separating, they would manage to reunite

THE FIRST PARTNER

in order to win the next election. Campaigning had provided the jolt of adrenaline that revived their fading marriage time and again in the past, and one was entitled to wonder whether Hillary was simply resorting to a time-tested remedy. And indeed, reports that the Clintons were one step away from divorce appeared to be premature. By the time Hillary's fifty-second birthday rolled around in October 1999, her Senate bid was looking more and more like a Bill-and-Hillary production. Onstage at the First Lady's birthday fund-raiser, the President hung on to his wife's hand like a man still in the throes of puppy love and serenaded her on the saxophone with his rendition of "My Funny Valentine." Hillary's spokesperson Marsha Berry had even begun to hint to skeptical reporters that physical passion had made its way back into the Clinton partnership.

The most candid explanation of why Hillary Rodham Clinton was looking toward a U.S. Senate seat from New York was provided by Harold Ickes, her unofficial adviser and mentor. "She understands how short the half-life of a 'former' is," Ickes said in a June 20 Associated Press interview. "Many of her friends were saying to her, 'Oh, the world is your oyster, Hillary, there are so many options.' " According to Ickes, Hillary's reaction to such comments could be summed up: "One, she thinks being a senator is a big deal, and it's a very attractive alternative given what she thinks are her other alternatives. Two, she's not interested in making money for the sake of making money."

So much for the notion that Hillary might be content as head of the World Bank or UN high commissioner for refugees, two posts often mentioned as suitable for her. Hillary and her friends were much given to talking about "platforms." The First Ladyship had been a platform; the Senate would be another. The specific duties of the job were less important than having the visibility to go about her self-directed mission, which Hillary was given to describing as "fighting for the causes I care about." In an age of celebrity candidates who connect with voters via the TV screen, this approach was not entirely unrealistic. Nonetheless, it also gave cause for apprehension that Mrs.

Clinton might not have a deep inner drive to get elected. If circumstances turned against her, there were always other platforms.

In the revised version of Clintonian history, it was New York congressman Charles Rangel who first suggested—during a meeting at a Harlem church in October 1998—that Hillary might run for the seat of Senator Daniel Patrick Moynihan, who was due to retire in 2000. This overlooked earlier published reports that described how Hillary and Democratic state chair Judith Hope had discussed the possibility of a Senate bid in late 1997. If the truth be known, Hillary must have thought about this option years earlier. The First Lady—who once hoped to serve as attorney general in her own husband's cabinet as Bobby Kennedy had served in his brother Jack's—could hardly have failed at least to daydream about following Bobby's path to the U.S. Senate and from there to a presidential bid of her own. Hillary's earliest political mentors had been RFK acolytes from a segment of the Democratic Party that still subscribed to the legend that Bobby, had he lived, would have rescued Great Society liberalism and changed the course of American political history. The presidency of Bill Clinton had turned out to be a mixed blessing for the party's liberal wing. But if Hillary could get herself elected to the Senate, she would be symbolically rewriting history, reversing the course of the last three decades.

Even a woman who did not think in such ambitious terms would have found it hard to pass up the opportunity presented by Moynihan's retirement. Leaving the impeachment-beleaguered White House to campaign for Democrat Charles Schumer in New York, the First Lady had been cheered by adoring crowds, and the state Democratic Party had done everything but lay down a red carpet stretching from 1600 Pennsylvania Avenue to the Empire State. As Bill Clinton had learned early in his career, the chance to run for a vacant, traditionally Democratic Senate seat didn't come along very often, and it was characteristic of Hillary to seize the moment, counting on her ability to do two jobs at once.

Under the circumstances it was easy to forget that the comparisons with RFK did not necessarily run in her favor. Bobby Kennedy had the advantage of running at a time when his assassinated brother was revered as a martyr. He had grown up in New York and had a large family, which he planted in Glen Cove, Long Island, creating at least the impression that the state was truly his home. Even at that, Bobby Kennedy had just squeaked by in the 1964 elections, drawing two million votes less than the presidential candidate, Lyndon Baines Johnson. Nor had Kennedy been especially happy once he did get to the Senate, complaining to aides that so much of its work was carried out by "five or six old men sitting around."[1] Hillary's popularity was based in part on sympathy for a betrayed wife, and her connections to New York were so slight that she was forced to drag out vague memories of having driven through the state as a child during one of the Rodham family's summer vacations.

Hillary's New York adventure would begin with what her campaign chose to call a "listening tour"—a clever if somewhat cynical effort to defuse the carpetbagger issue and introduce the candidate-to-be to party activists. Senator Moynihan offered his farm in Pindars Corners, New York, as the site for the tour's kickoff. Never an admirer of the Clintons, Moynihan did enjoy playing the role of elder statesman, however, showing off the converted one-room schoolhouse that serves as his office and puckishly observing that in the mid-nineteenth century (before federal education initiatives, though he didn't say this) the students had enjoyed such reading matter as *Paradise Lost* and *The Pilgrim's Progress*. "My God, I almost forgot," Moynihan said, interrupting his musings. "I'm here to say that I hope she will go all the way. I mean to go all the way with her." It was left to Mrs. Clinton to remind the press of her many years as "a tireless advocate for what I believe in."

In spite of this less than fervent endorsement, Hillary's precampaign campaign got off to a brisk start. The First Lady had never looked more fit and sported a becoming new haircut that she could manage without help as well as improved makeup, said to include a machine like a super–Water Pik that sprays on a water-based foundation. The pastel suits of a few years earlier were gone, replaced by sensible black or brown designer pants ensembles. According to a press release issued by the Bruno Magli company of Palermo, Sicily, Hillary's wardrobe also included eighteen pairs of designer shoes, purchased for a total of about five thousand dollars. Persistent rumors that Hillary had been spotted entering the office of Park Avenue plastic surgeon Dr. Cap Lesesne, her face half hidden by a bandanna, were heatedly denied by Marsha Berry, though not by Dr. Lesesne himself, who noted coyly that he had a busy practice and many prominent clients.[2]

With seasoned pros like Harold Ickes and Mandy Grunwald advising her, Hillary had the kind of organization rarely available to first-time candidates, not to mention the ability to charge a goodly percentage of her travel expenses to the public tab. (Hillary's people insisted that she was not doing anything other incumbents did not do routinely, and they were correct up to a point. The difference was that the First Ladyship was a courtesy office. Hillary's frequent trips to New York before her announcement, not to mention later jaunts to Ireland, Italy, and Israel, the homelands of many New York voters, served no necessary purpose. She herself made this point when she announced that she would be drastically scaling back her official schedule in the year 2000.) Unlike past candidates for statewide office, who have had trouble distinguishing between Oneonta and Onondaga, Hillary was impeccably briefed as she progressed in her blue van from one upstate hamlet to another, holding "conversations" with hand-picked average citizens who turned out to be mainly Democratic activists. The scenario of these meetings seldom varied: Citizens talked about their personal trials—one woman had been stepped on by a bull—and Hillary empathized. They expressed their concerns about such issues as HMOs, education, and joblessness, and Hillary agreed that she was concerned that they were concerned. In a foray to New York City, meanwhile, she was named "Principal for a Day" at a Queens high school. Teachers' union officials responsible for issuing the invitation argued that they weren't violating a rule against using school events for political purposes, since Mrs. Clinton was present in her capacity as First Lady.

The press fretted about the artificiality of it all. One reporter wrote of watching as campaign aides set up a lemonade stand, complete with a deliberately misspelled sign, and then brought in a pair of children to run it so that Hillary could stop by for a photo op. Not for the first time there was talk of showing off the lighter side of Hillary, possibly not a good idea considering her predilection for jocular phrases like "Okeydokey, artichokey." Those who have observed many Hillary Clinton speeches gradually come to recognize a certain programmed quality in her mannerisms: She nods sagely when another speaker says something she agrees with, and does a double take when she spots a familiar face in the crowd, her jaw dropping open in feigned surprise. Still, it would be a mistake to think that the majority of voters are so cynical. It isn't every day that a First Lady visits Cooperstown, Rome, or Glens Falls, not to mention smaller hamlets in between. Many women, in particular, were genuinely excited and touched that a celebrity of Hillary's magnitude would take the time to find out about their problems.[3]

Even the ongoing soap opera of the Clinton marriage worked to Hillary's advantage. Women who had been two-timed by their husbands empathized. Men wanted to protect her. The question of how to exploit these emotions politically was a delicate one, and from the time she began to consider running, Hillary was bombarded with advice, solicited and otherwise. Some of her aides, women in particular, were adamant that she needed to express moral disapproval of her husband's behavior, especially in light of the public airing of Juanita Broaddrick's allegations of rape. The chosen vehicle was to be an interview with Lucinda Franks, which appeared in the premier issue of Tina Brown's magazine, *Talk*. This interview, or rather series of interviews, gave us the heretofore unimaginable spectacle of a First Lady of the United States describing her husband, the President, as undisciplined and prone to "sins of weakness." Hillary went on to describe Bill Clinton as a victim, a man who at the tender age of three had been a pawn in a "terrible conflict" between his mother and grandmother, commenting: "A psychologist once told me that being in the middle of a conflict between two women is the worst possible situation."[4]

Reaction to the interview was overwhelmingly negative. One way or another most Americans had made up their minds about Bill Clinton's behavior. They didn't need to hear any more excuses. The question everyone was asking was: Why did Hillary put up with it? In fairness, Mrs. Clinton was uncomfortable with the whole idea of talking about so deeply personal a subject, and she undoubtedly meant what she said. She was married to a very successful man whom other women found attractive, a man who would never leave her because no one else would ever take care of him the way she had. Every time Bill Clinton strayed and Hillary forgave him and helped cover up for him, she reinforced her sense of herself as the grown-up in the marriage, the responsible one who bore the burden of taking care of a great man who was really just a charming bad boy deep down inside. This pattern was so deeply ingrained that perhaps it didn't occur to Hillary that she was just another strong woman contending for the soul of Bill Clinton—one who was getting a measure of sweet revenge by publicly branding the most powerful man in the world an emotional cripple.

Of course there was another question that many people would have liked to see answered in *Talk*. Why had Hillary helped Bill Clinton lie to us? Why had she denied her husband's affair with Gennifer Flowers when she now acknowledged in passing that Flowers had been telling the truth? Why had she dismissed the allegations about Monica Lewinsky as the work of a "vast right-wing conspiracy"? Finally, since she knew all too well her husband's history of "undisciplined" behavior, why had she allowed his accusers to be portrayed as sluts and liars and, in Monica Lewinsky's case, a stalker? Since wifely loyalty is always a sympathetic excuse, even a tepid expression of regret would have gone a long way. The amazing aspect of the *Talk* interview was that Hillary showed not a glimmer of awareness that her conduct, either public or private, might be less than noble and self-sacrificing. Explaining her ability to forgive her husband, she even compared herself to Jesus, who forgave Saint Peter three times.

The *Talk* piece, not to mention the publication of Gail Sheehy's book *Hillary's Choice,* revived the interpretation of Hillary Rodham Clinton as a classic enabler. In Sheehy's formulation Hillary just can't help herself because Bill Clinton is her fatal flaw—she is "addicted to love." But the reality is that Hillary is no addict. She went into her marriage with her eyes open, and over the years there were a number of crisis points when she made considered decisions to keep her marriage going rather than split up the family and lose all that she had invested in her and Bill's political partnership.

Hillary was actually correct to think that she should not have to apologize for her failure to seek a divorce. That was, indeed, her choice to make. The public's interest in the matter stemmed from the way in which the Clinton marriage has always turned on the axis of a shared ambition. There never seems to have been a time when Hillary considered the possibility that she and Bill should take a break from politics to work out their personal problems. Instead she became caught up in the task of presenting an acceptable facade to the public. It speaks volumes that Hillary fiercely defended her husband even as his denials of various allegations became ever more improbable, only to be furious with him when he finally told the truth. Bill Clinton's sin was not his infidelity but that he abandoned his pact with her to lie about it.

In the *Talk* piece Hillary acknowledges knowing about her husband's dalliance with Gennifer Flowers—contrary to what she said in 1992—but insists that as far as she knew, Bill had sworn off womanizing in about 1988. To believe this, one has to imagine that Hillary drew no conclusions even from on-the-record evidence, such as the 1990 release of the governor's phone records, showing lengthy late-night calls to Marilyn Jo Jenkins. Whether the subject is her marriage, Whitewater, her commodities trades, or her controversial tenure as a board member of the New World Foundation, which indirectly funded Palestinian organizations, Hillary often resorts to pleading naïveté rather than responding to criticism in a straightforward manner. Her tendency to see any criticism as a purely vindictive personal attack is a characteristic of the narcissistic personality. Despite many predictions that the New York press would give Hillary the grilling of her life, to an amazing degree they were willing to go along with the pretense that—as Hillary's press director, Howard Wolfson, put it—"the message is listening." Any other politician who proclaimed the existence of a "vast conspiracy" would never hear the end of it, but when a representative of the *National Review* managed to ask the First Lady if she regretted threatening on national TV that "when all this is put into context . . . some folks are going to have a lot to answer for," Hillary went rigid with anger and dismissed the query as not worth an answer. "I'm not going back. I'm going forward," she said frostily.

An insatiable hunger for attention is another characteristic of the narcissist. Hillary's need to occupy center stage even when doing so is clearly ill advised led to the first serious blunder of her (non)campaign. The debacle started innocently enough during an appearance on the *Today* show, when Katie Couric asked her, "Are you a big Knicks fan?"

"I'm *becoming* a big Knicks fan," Hillary joked.

"More and more every day, huh?" parried Couric.

The subject then shifted to the Yankees, due at the White House later that same day for a ceremony belatedly recognizing their victory in the 1998 World Series, which had occurred while Bill Clinton was preoccupied with trying to keep his job. Hillary, who had been doing rather well in this conversation, defusing the carpetbagger issue by poking fun at herself, abruptly volunteered the information that she had "always been a Yankees fan."

A surprised Couric pointed out that Hillary had always claimed to be a Cubs fan in the past.

"I am a Cubs fan," Hillary agreed. "But I needed an American League team to cheer. When you're from Chicago, you can't root for both the Cubs and the Sox. So as a young girl I became very interested [in] and enamored of the Yankees."

And sure enough, when the Yanks showed up in the Rose Garden that afternoon and manager Joe Torre presented both Clintons with Yankee caps, Hillary clapped hers on her head and kept it on while photographers snapped away.

It may actually be true that the young Hillary Rodham took a passing interest in the Yankees. Even as a girl Hillary always liked to associate herself with winners. She doesn't fit the profile of a loyal follower of the Cubs, whose frustrated hopes for a championship season are the stuff of colorful local legends. Hillary should have realized, however, that a recovered memory from childhood couldn't compete with the passionate fandom of her likely opponent in the Senate race, New York City mayor Rudolph Giuliani, who plausibly claims to have attended "maybe a thousand" Yankee games over the course of a lifetime and often wears his team jacket in his office at City Hall. Giuliani knows the Yankees, and a month after their White House visit, some team members were still seething over being used to promote the First Lady's political ambitions. During a City Hall ceremony honoring him for pitching a perfect game, Dave Cone took the opportunity to tell the mayor, "I'd like to say on behalf of all the Yankee players that your sincerity as a Yankee fan really comes across. I mean there's a lot of politicians who *say* they are baseball fans and put on the cap—" Cheers and laughter from the crowd drowned out the rest of Cone's remarks.

Hillary compounded her mistake when she failed to back up her talk of being a fan by attending a game at Yankee Stadium. Sports fans never forget, and the longer she stayed away, the more certain it became that her appearance would provoke a demonstration. As a result, when the Yankees got to the 1999 World Series, Hillary was conspicuously missing from the luminaries who made the obligatory trek to the Bronx to see and be seen.

On a more serious note, Hillary was dogged by a comment she made during a 1998 chat via satellite with Israeli and Palestinian youngsters who were attending a conference in Switzerland. Answering a student's question, the First Lady said quite pointedly that she believed a Palestinian state would "be in the long-term interests of the Middle East." As a matter of long-standing policy, the U.S. government has avoided taking a position on Palestinian statehood, believing it to be an issue for Israel and the Palestinian Authority to work out. Hillary's comment caused an uproar in Tel Aviv, and White House press secretary Mike McCurry pleaded that the remark expressed the First Lady's personal opinion, which was "not the view of the President." Few believed this, however, assuming that Hillary's remarks were intended to put pressure on Israeli prime minister Netanyahu, even as they encouraged Yasir Arafat to think that he would not have to make significant concessions.

In July 1999, hoping to still criticism that she had tilted U.S. policy in Arafat's favor, Hillary sent a carefully worded letter to an Orthodox Jewish organization affirming her support for moving the capital of Israel to Jerusalem. Meanwhile a member of her staff suggested to Seth Gitell of the Jewish weekly the *Forward* that an inquiry into Mrs. Clinton's past might turn up evidence of Jewish roots. The *Forward* hired a Chicago-based genealogist who duly discovered that Hillary's maternal grandmother, Della Murphy Howell, had made a second marriage to a Chicago businessman named Max Rosenberg.

The *Forward*'s write-up of its findings struck a friendly note, observing that this information "opens a window into Mrs. Clinton's warm relations with the Jewish community." It also quoted Democratic consultant Hank Sheinkopf, who predicted hopefully that "Jews will now feel she is almost one of their own," though he cautioned that "professional Hillary-haters will say, '[W]hy didn't she tell us sooner?'"

Actually it is obvious why Hillary didn't mention her Jewish relatives sooner. During the 1992 campaign Hillary's mother, Dorothy Howell Rodham, had made it abundantly clear to reporters that her childhood was a painful subject she would rather not delve into. Now, thanks to the prodding of Hillary's press secretary, the *Forward*'s researcher had turned up Cook County Superior Court records dating back to 1926, which showed that Della Howell lost custody of her two daughters after her husband accused her of an act of domestic violence. According to Hillary's grandfather Edwin Howell, his young wife liked to go out in the evenings and nagged him constantly to take her places: "I couldn't go. I worked most of the time. She became abusive and angry, and scratched and bit me, flared up at me." Edwin Howell's testimony was confirmed by two witnesses, one of them Della's own sister.

Della lost custody of her two daughters, who were shipped off to California to be raised by their father's parents. Della's fortunes soon improved, however, and in 1933 she met and married Max Rosenberg. Three years later, by which time Dorothy was sixteen and working in California as a live-in mother's helper, Della petitioned to regain custody. Once again the court ruled against her, but Dorothy soon moved to the Chicago area to work and renewed contact with her mother. Hillary had occasional contact with Della, Max, and their daughter, Adeline, until her thirteenth year, when the Rosenbergs moved to California.

Those who find accounts of Hillary Clinton's own temper difficult to credit may now ponder the disclosure that some seventy-five years ago a judge heard similar allegations about her grandmother. Whether the story sheds any light on Hillary's views about the Middle East is another matter.[5]

If there was one segment of the population whose support Hillary could count on, it was African-American voters, whose allegiance to the Clintons remained solid even during the height of the Lewinsky scandal. The first sign that her campaign could complicate even the assignment of holding on to the allegiance of black voters came when the Reverend Al Sharpton turned up at the White House for, of all things, the Rose Garden ceremony honoring Joe Torre and the New York Yankees. Best known as the adviser of Tawana Brawley, the Dutchess County teenager who in 1987 invented a lurid story about being gang-raped by white policemen, Sharpton has been a fixture on the New York scene for almost two decades, converting every racial incident into an occasion for self-aggrandizement and inflammatory rhetoric. In 1995, for example, he inserted himself into a Harlem real estate dispute, branding a Jewish discount-store owner a "white interloper" in the neighborhood. This comment inspired a crazed follower to shoot up the store and set it on fire, trapping seven people inside.

Despite this incident, Sharpton did well enough in the 1996 mayoralty race to force a runoff with Manhattan activist Ruth Messinger and make himself a force to be reckoned with in New York Democratic politics. His rise was due not so much to his own strength as to the weakness of the rest of the party. David Dinkins, New York's first African-American mayor, had presided over a disorganized administration, and Messinger was a seventies-era activist who spouted policy-wonkish jargon while insulting large segments of the electorate, intentionally or otherwise. (Messinger, for example, called her Republican opponent, Rudy Giuliani, the candidate of people who frequent catering halls in the outer boroughs, presumably a slam at Orthodox Jews, Italians, and white ethnics in general.) Instead of building a coalition of working people of all races and backgrounds, New York Democrats appeared to see their party as a coalition of minorities who have nothing in common but their differentness. As a result they found it necessary to appease Sharpton, lest he run against them and siphon off much-needed minority votes. They allowed him to play point man in attacking Giuliani and joined his campaign to protest the police shooting of a West African, Amadou Diallo, taking no notice when the leader of a group representing African immigrants in the United States complained that his efforts to open a dialogue with the mayor had led to death threats against himself and his five children.

Dealing with Al Sharpton had become a problem for all New York Democratic candidates for state and national office, but few seemed so eager for his approval as Hillary Clinton. And Sharpton, it must be said, played Hillary like a violin, alternately damning her with faint praise (calling her "good on health care") and berating her as a wimp. In late November he said that if Mrs. Clinton was "too scared and too intimidated and too much of a lackey to challenge Giuliani, then [she should] step out of the way and let me take him on." Sharpton then boycotted a meeting that Mrs. Clinton had set up with African-American ministers, insisting that the First Lady of the United States come to his office for a face-to-face talk.

This dance culminated in Hillary's agreement to appear at a Martin Luther King Day celebration at Sharpton's Harlem headquarters. Speaking to a rapt audience, she reminisced about hearing Dr. King speak when she was a teenager and declared her intention to be a unity candidate: "I want to be defined by who I lift up, not who I push down." But she also referred to "the murder of Amadou Diallo," following Sharpton's line that the shooting of the unarmed civilian had been deliberate homicide, and at the prompting of her press spokesman, Howard Wolfson, she inserted a rebuke to a previous speaker who made anti-Semitic remarks. (A month later, while being interviewed by WNBC reporter Gabe Pressman, Hillary agreed that her reference to the Diallo case was a "misstatement." Asked if she was sorry for saying that Diallo was murdered, she replied, "Absolutely, yes. I mean, that's what a jury should decide.")

Hillary appeared to be placing her faith in the doubtful premise that it is possible to achieve unity by playing to the most militant factions of each ethnic group in turn. In the case of her relations with New York's Hispanic voters, she was put in a tricky situation by her husband's decision to offer clemency to sixteen convicts who were members of the Puerto Rican terrorist group known as the FALN, provided they agreed to make statements renouncing violence. During the late 1970s and early 1980s the FALN was responsible for at least 130 bombings, resulting in the deaths of six people and many more injuries. Unremorseful after serving almost a third of their seventy-two- and seventy-four-year sentences, the prisoners had issued a joint statement as recently as 1997, declaring that their violent tactics were justified under international law because they were part of the struggle against colonialism, itself a "crime against humanity."

While it may be a truism that one person's terrorist is the next person's freedom fighter, the people of Puerto Rico have had opportunities to express their views on the political status of the island at the polls, and the independence option consistently comes in last, earning only a small percentage of the total votes cast. The FALN was a small Marxist sect that sought to accomplish through terrorism what it couldn't do through persuasion, and while the sixteen prisoners who were offered clemency were not directly tied to any of the acts that led to death and serious injury, they had taken part in crimes aimed at supporting terrorist goals, including armed robberies and the storage of explosives in safe houses. Nevertheless a campaign to win their freedom had garnered the support of such luminaries as

Coretta Scott King, Cardinal John O'Connor, and former New York City mayor David Dinkins, as well as the National Lawyers Guild.

The United Methodist Church signed on to the clemency effort in 1996. A statement issued by the church's General Conference called for the unconditional release of the FALN sixteen and advised all United Methodists to work for "justice and freedom for the Puerto Rican political prisoners." According to Eliezer Valentín-Castañón of the General Board of Church and Society, the church's policy arm, the Methodist position was based purely on the argument that the sentences were excessive. But the language used to describe the FALN in both official and unofficial church statements suggests otherwise. Speaking for himself, Valentín-Castañón described the prisoners as "community activists" who were "committed to the armed struggle." He rejected the suggestion that the long sentences were justified by the prisoners' role in furthering a terrorist conspiracy.

Hillary Clinton was first lobbied on the church's position during a White House meeting with a group of Methodist women a few months after the General Conference's statement. She was noncommittal at the time, and there appeared to be little hope that Bill Clinton would act on the matter since he had previously commuted prison sentences in only three minor cases out of thousands of requests. However, the lobbying of Hillary as well as other administration officials continued, and departing White House counsel Charles Ruff recommended the clemency offer as one of his last official acts. A subsequent investigation by Representative Dan Burton's House Committee on Government Reform later turned up an E-mail from presidential adviser Jeffrey Farrow that showed that the political ramifications of taking action were considered inside the White House. Farrow argued that clemency would please New York's Hispanic members of Congress and "help the VP's Puerto Rican position." The Farrow memo did not specifically mention Hillary.[6]

The President went along with Ruff's recommendation, and his decision was made public on August 11, a few days before he and Hillary arrived in New York State, where they planned to spend part of their summer vacation. The timing of the announcement encouraged suspicions that Clinton hoped to boost his wife's Senate bid by currying favor with New York's large Puerto Rican community. Meanwhile the FALN's sympathizers were far from grateful: They wanted complete vindication. The prisoners spurned the demand that they renounce violence, and their attorneys issued a statement protesting certain conditions placed on their clients' parole. As the stand-off continued, the information was leaked to the press that the FBI, Janet Reno, and local law enforcement agencies had all opposed the pardons. Victims of FALN terrorism came forward to protest the deal, among them ex–New York City police detective Richard Pastorella, blinded by an FALN bomb, and Joseph and Thomas Connor, who had been about to celebrate their ninth and eleventh birthdays on the day their father died in an explosion at Fraunces' Tavern, a New York City landmark, site of George Washington's farewell to his officers in 1783. Influential Democrats, among them Senator Moynihan, also condemned the clemency deal.

Hillary insisted that she had known nothing about the clemency decision until after the fact. Given her church's interest in the matter and its relevance to New York, this was difficult to credit, especially in light of her recent boast in *Talk* that she and the President talked policy over breakfast and while watching TV at night. At any rate Hillary initially supported the clemency offer. Then, on Saturday, September 4, she called for it to be withdrawn, saying: "It's been three weeks and their silence speaks volumes." Hillary had broken the first law of politics by changing her position without warning her allies, New York's Hispanic politicians, who felt blindsided when they learned over the weekend of her change of heart.

Now Hillary's stance angered both sides. Although the newspapers described her statement as a reversal, she had never actually said that the FALN sixteen were not political prisoners. Indeed, it now also seemed that President Clinton did not intend to leave his offer open indefinitely. Sensing that their window of opportunity would not stay open forever, twelve of the sixteen prisoners signed a statement renouncing violence—without, however, accepting any responsibility for the bombings or apologizing to the victims.

Looking back, the strangest feature of this saga was both Clintons' emphasis on the need for the prisoners to promise that they would abstain from violence in the future. Surely Marxist revolutionaries who consider the United States to be an oppressive colonial power and view the bombing of innocent civilians as a legitimate form of political expression could easily rationalize lying about their future intentions, especially if the lie was extracted as a condition of gaining their freedom. The chief value of the prisoners' renunciation would seem to be that it shifted responsibility for the consequences of their release away from the Clintons. Should any of them get into trouble in the future, Bill and Hillary could claim to be victims of a broken promise.

WITH most of the FALN convicts out of jail, Hillary faced pressure from other groups with agendas of their own. Supporters of Jonathan Pollard, an American serving a life sentence for spying for Israel, had long hoped that Clinton would commute Pollard's sentence and allow him to emigrate. Now their hopes were revived.

The revelation that Hillary had a Jewish step-grandfather seemed to have done little to resolve her problems with Orthodox Jewish organizations. Either the delicate choreography of New York ethnic politics had eluded Hillary's staff or she simply had not paid enough attention to their advice. Her overtures to the Palestinians abroad and the Reverend Al Sharpton at home aroused apprehensions. The Crown Heights Community Council, a group dominated by the Lubavitcher Hasidim sect in a neighborhood riven by a race riot in the early nineties, complained when Hillary came to the area in August to speak at a black church but never scheduled a meeting with them. Under fire for canceling a visit to Israel earlier in the year, when the trip did take place in November, she was criticized for failing to include a symbolic stop at the Western Wall. Her itinerary was revised to include the Wall, but she also found time to inspect a clinic in the Palestinian stronghold of Ramallah. Her hostess was her Palestinian counterpart, the outspoken and unpredictable Suha Arafat. The theme of several of the speakers that day was that the Israeli government was conducting a war against Palestinian children. Mrs. Arafat herself spoke of "the daily and intensive use of poison gas" against children, as well as the contamination of "80 percent of water sources used by Palestinians" and the "smuggling [of] spoiled food-stuffs" into Palestinian towns, all part of "a clear policy of annihilating our people." Though Mrs. Clinton did not appear to be pleased by this turn of events, she made no protest, acknowledging her hostess's remarks with an embrace and the ritual double air-kiss.

When she belatedly issued a tepid statement deploring Mrs. Arafat's inflammatory language, Mrs. Clinton offered two reasons for her fail-ure to reply: First, she had been a victim of bad translation. Second, acting in her capacity as First Lady, she hadn't wanted to spark an international incident. Of course, the problem was that she had al-ready done just that when she endorsed Palestinian statehood in 1998. Suha Arafat had reason to believe that the American First Lady was a sympathizer and might be similarly indiscreet under the right circum-stances.

The only unusual feature of this incident was the coverage it re-ceived in the American media. Hillary's foreign travels have often involved her in activities that wouldn't have played well at home. One recent example was her appearance as the keynote speaker at the Dawn of the New Millennium women's conference in Reykjavík, Iceland, an event Hillary's staff discouraged American reporters from covering. Perhaps they worried that the press would pick up on some of the delegates' fatuous antiman, antifamily rhetoric or feared that some would see a certain irony in Hillary's promise that she and her husband were working to end the international trafficking in women. Conservative columnist Joseph Farah, one of the few American jour-nalists to take note, later pointed out that Ng Lap Seng, a major figure in the campaign finance scandal, had allegedly made millions running prostitution rings in Macao.

★   ★   ★

IF Hillary had been slow to control the damage from her position on clemency for the FALN prisoners, it may have been because she was preoccupied by house hunting. The search for a suitable New York residence had turned out to be more complicated than she could have anticipated. Upscale buildings in New York City tend to be co-ops, whose boards are notoriously averse to celebrities, especially ones likely to station Secret Service agents in the lobby and on the roof. Luxury rental properties in the suburbs are scarce and expensive, and the Clintons soon found themselves in the market to buy. In August Hillary's scouts discovered a five-bedroom Dutch colonial on 1.1 acres with a swimming pool in the Westchester County hamlet of Chappaqua. Despite Hillary's warning in *It Takes a Village* that the clock can't be turned back to the 1950s, it seemed that for her it *was* possible. In an area where substantial minority populations are the rule, Chappaqua is an exception—"as white as Antarctica," in the words of Pope Brock of the *Westchester County Times*. Moreover, the house, sometimes known as Little Brook Farm, was situated on a cul-de-sac protected from through traffic and gawking sightseers.

The President and Chelsea joined Hillary in inspecting the house on their way to the Hamptons for the first leg of their New York vacation, but almost immediately there was a hitch in their plans. With legal bills totaling as much as two and a half times their assets, Bill and Hillary were not good candidates for a mortgage on a $1.7 million property. The Clintons were under the impression that former White House chief of staff Erskine Bowles would guarantee the financing, but on September 1 they learned that Bowles was having second thoughts. There was another bidder prepared to match their offer on the house, and Hillary was frantic.

The Clintons, meanwhile, had moved on to the Finger Lakes resort of Skaneateles. Devoid of media stars and short on registered Democrats, Skaneateles was not the kind of place where either Bill or Hillary was likely to spend five days of unstructured time in the absence of a political motive. (One village luncheonette was offering a

Hillary Special: bologna on white bread.) Democratic fund-raiser Terry McAuliffe had arranged for the Clintons to borrow a picturesque lakeside house, but reporters who descended on the town expecting abundant photo opportunities were frustrated. Hillary, it seems, was preoccupied by the threat of losing the house in Chappaqua. Reportedly ex–treasury secretary Robert Rubin was approached through an intermediary and declined to help out. More surprisingly, Bill's childhood friend Mack McLarty, who had opened his wallet for the Clintons often in the past, also turned them down. Eventually McAuliffe offered to deposit $1.35 million in cash with the financing institution, Bankers Trust, as collateral on the mortgage.

The hyperactive McAuliffe is known for mixing fund-raising, lobbying, and private business deals in ways that sometimes attract negative attention. His friendship with the Clintons dated from 1994, when he submitted a memo to Harold Ickes, suggesting overnights in the Lincoln Bedroom as a reward for major donors. More recently he had been mentioned in connection with a plan to funnel Democratic soft money through the Teamsters Union political fund. But McAuliffe had the magic touch when it came to investing. According to The New York Times, he had recently made eighteen million dollars from a one-hundred-thousand-dollar stake in the start-up fiber-optics network company Global Crossing. He could afford to put up $1.35 million to help his friends, though a businessman involved in a booming company would normally be reluctant to park such a large chunk of cash when he could be getting much higher returns in the stock market.[7]

When the press questioned the deal, Hillary assured reporters that it had passed muster with the Office of Government Ethics. In fact the ethics office had been asked to rule only on the narrow question of whether the President would be required to report the guarantee as a gift. Stephen D. Potts, the director of the ethics office, protested the misrepresentation in an October 8 letter to the White House. Within the week the Clintons announced that they had refinanced their mortgage through a subsidiary of the PNC Bank Corporation. PNC did not ask for a guarantor, and its terms were described by

Kenneth Harney, who writes a widely syndicated column on real estate, as "an extraordinarily good deal": 7.5 percent interest, no points, and a down payment of $375,000, less than usually required for homes in a similar price range. One is hardly surprised that the President of the United States would get favorable treatment in the mortgage market. Still the arrangement evoked a disquieting echo of Whitewater, which began with the Clintons getting involved in a mortgage that was a little beyond their means even as Bill was running for governor. For the time being the Clintons were living on the President's salary of $200,000 a year plus the income from their blind trust, recently depleted by the Paula Jones settlement as well as the amount of the down payment on their new house. They were carrying $109,000 a year in mortgage payments plus about $26,000 in taxes, not to mention incidentals like insurance and utilities, and the considerable cost of decorating and furnishing their first home since 1983.

After inspecting the Chappaqua property on November 4, a cheerful Hillary pronounced her new house "in great, great shape," but when asked whether Bill Clinton would be changing his legal residence from Arkansas to New York, she said, "That's a good question. I haven't really talked to him about that." Such statements helped fuel speculation that Hillary was not fully committed to the Senate race. By mid-November, following her return from the Middle East, Democratic circles in New York and Washington were abuzz with rumors that she was dispirited by the reaction to her embrace of Suha Arafat as well as by her steadily sagging poll numbers. Lending some support to the talk, Hillary's exploratory committee showed few signs of gearing up for a major campaign effort. Out-of-town and foreign journalists complained that their voice mail messages went unanswered, and even major New York dailies and veteran TV correspondents were unable to land interviews. Even the First Lady's all-star birthday gala was less than a complete success. The tickets, originally priced from $250 to $1,000 each, failed to sell out, and organizers had been forced to paper the house, handing out tickets to passersby on the street.

New York Democrats had already spent $338,000 in soft money to run pro-Hillary ads, and now it appeared that she might not even run. City Councilwoman Ronnie Eldridge, an RFK supporter back in the 1960s, called on Hillary to withdraw, saying, "I just think there are too many problems in the campaign." "Maybe she needs to give up her day job," suggested Judith Hope.

On November 24 Hillary moved to stifle the rumors. Although word of her impending announcement went out from the office of Andrew Cuomo, and the talk shows were already rounding up talking heads for comment, Hillary's reassurance took the form of a staged dialogue between herself and teachers' union president Randi Weingarten—a ploy that allowed her to avoid answering questions from the press. Hillary hired Bill de Blasio, formerly a deputy to HUD Secretary Andrew Cuomo, as her campaign manager, but signs remained that her organization was less than stellar. The hillary2000.org campaign website, for example, was as dead as a boiled mackerel. On November 29 the site was still decrying Republican obstructionism on the federal budget even though Bill Clinton and Congress had reached an amicable settlement twelve days earlier.

Nevertheless the outlines of Hillary's positions on the issues were beginning to emerge. She told the members of District Council 37, the city workers' union, "I don't think Reaganomics and supply-side economics [were] the right direction for this country," adding that anyone who favored them "must not have a good memory for what was happening in New York in the 1980s." It would be interesting to know what she meant by this. After all, Reaganomics conquered runaway inflation, and some of the Reagan-era policies that hit New York hardest, such as cutbacks in support for low-income housing, have been continued by the Clinton administration. As for the economic decline of upper New York State, it is a matter of opinion whether this trend was caused by Reaganomics or the high-tax policies of Governor Mario Cuomo. Hillary claimed to have learned from the failure of the health care task force, though the only lesson she mentioned was tactical—the need to put the program through piece by piece, rather than in an omnibus bill whose problems were all too obvious. She came out for extending the family-leave law to businesses with fewer than fifty employees—great for workers, but a frightening prospect for the owners of mom-and-pop stores and shoe-string start-up companies. She even appeared before representatives of New York's biotechnology firms, advocating government/private sector research partnerships but offering no assurance that she wouldn't once again decimate the industry by attacking pharmaceutical companies or advocating price controls.

Hillary Clinton has never accepted the commonly held notion that she is far to the left of her husband. Her spokesman Howard Wolfson used the *m* word—moderate—to describe her, and former White House scandal manager Lanny Davis argued heatedly on the TV talk show circuit that Hillary's (reluctant) acceptance of welfare reform proved that she was hardly an old-fashioned liberal. Davis had a point. Even in her student days Hillary had never put much faith in "people power," nor has she always been a passionate defender of entitlement programs that put money directly into the pockets of the poor. Her true political ancestors are the Progressives of the early twentieth century, who looked to the government to dispense services, offered on the condition that the poor conform to certain social norms. Traditionally this meant a tough stance on crime as well as the promotion of self-improvement—education, hard work, sobriety, and middle-class manners. Ironically enough, Hillary's nemesis, Rudy Giuliani, was almost a throwback to old-fashioned Progressivism, while she favored an updated version centered on self-esteem, multiculturalism, and addressing the grievances of assorted classes of victims. Though at times she expressed impatience with this politically correct agenda, she rarely seemed able to reach beyond it.

But Mrs. Clinton's number one campaign issue lay elsewhere: She had come out firmly against "meanness." As her *Talk* interview made plain, she saw herself and her husband as prime victims. "People are so mean," she told Lucinda Franks while recalling criticism of Bill Clinton's behavior. "People need love and support."

This theme appeared again in Hillary's direct-mail fund-raising appeals. One envelope that appeared in the mailboxes of potential Democratic contributors featured an alarming warning from the First Lady: "They can never stop me from speaking out for my beliefs. They can never deter me from working to help build the kind of future our families need and our children deserve." Checking inside to discover who "they" might be, one found that Mrs. Clinton was referring to the people who "launch anti-Hillary websites . . . [and] publish mean-spirited books and press releases." Flattering as this might be to Mrs. Clinton's critics, anti-Hillary websites and books hardly had the power to stifle one of the most celebrated women in the world. Nor was there ever a campaign to stop Hillary Rodham Clinton from speaking out. It was she who declined to give interviews to anyone inclined to ask substantive questions about her record.

Hillary's first close encounter with the meanness on the campaign trail came on January 19, when Tom Bauerle, host of a popular morning drive-time show on Buffalo radio station WGR, asked her an impertinent question: "Mrs. Clinton, you're going to hate me. You were on television last night talking about your relationship with the president, Bill Clinton. Have you ever been sexually unfaithful to him, and, specifically, the stories about you and Vince Foster—any truth in those?"

"Yes, Tom, I do hate you for that," Hillary replied, "because those questions, I think are really out of bounds."

Two points must be made about this exchange. First, in the land of the First Amendment, anybody can ask anyone anything. Second, a campaign is not a confessional, and no candidate is required to answer questions she considers offensive. The instant verdict of the media was that Hillary had handled herself beautifully if only because she kept her temper and even squeezed out a giggle when asked a follow-up question about whether she had ever used pot or cocaine. With hindsight, however, it seemed that she had let herself be rattled. As an attorney who has taken (and given) many depositions, Hillary knew that a refusal to answer requires no elaboration. Instead, in replying to Bauerle she rambled on, assuring him that "everybody who, you know, knows me knows the answers to those questions." In the end, with only a minimum of prodding, she gave Bauerle a flat "No, of course not." She also denied ever having used illegal drugs.

Since most voters undoubtedly consider Bauerle's question both rude and irrelevant, it is safe to say that Hillary won this exchange. On the other hand, one recalls that Mrs. Clinton did not consider marital infidelity out of bounds in 1992 when she urged Gail Sheehy to raise the issue of George Bush's alleged affair with the "other Jennifer." Her sweeping and quite unnecessary denial was, moreover, exactly how she had gotten herself in trouble over Whitewater, and how her husband flat-footed his way into impeachment, when he lied under oath in response to questions he now argues he never should have been asked in the first place.

By allowing herself to be backed into answering the infidelity question, however dismissively, Hillary had failed to close the door on the issue once and for all. Within a week a supermarket tabloid was reporting an unsourced story about the First Lady's supposed dalliance with a New York restaurateur. Tom Bauerle, meanwhile, retreated into seclusion. In what was surely a first for a local talk show host who had managed to make national news, he refused all offers to disclose the interview on the network talk show circuit, apparently under orders from the station's management.

OF course, the number one "meanie" in Hillary's universe was Rudy Giuliani. Democratic activists wasted no opportunity to suggest that Giuliani was temperamentally unsuited to the Senate, and Mrs. Clinton repeatedly chided him for being perpetually angry. "It is certainly clear that there have been so many controversies and difficulties," she told reporters during a campaign stop in Rochester. "And you know, at the end of the day, that's a lot of wasted energy. People ought to be working together to solve problems, particularly when it comes to the children." This was a startling observation coming from the First Lady who fired a White House usher because he spoke to her predecessor on the telephone, instigated the firing of Billy Dale, signed off on campaigns to discredit Paula Jones as a slut and Monica Lewinsky as a stalker, and, we're told by Gail Sheehy, refused to speak to her own husband for months at a stretch.

Even Giuliani's admirers would acknowledge that he has a confrontational personality. For the most part, however, he had used his anger to shake up the bureaucracy and get the city moving forward. (Surely it was the Clintons who frittered away their political capital on domestic discord, stonewalling, and contentious relations with Congress.) Still, Hillary's tactics might appeal to a public that just wanted to be left alone. Giuliani didn't just talk about personal responsibility; he made demands. He tried to hold the director of a public museum accountable for how he spent taxpayers' money and ordered taxi drivers to pick up African-American fares. He even offended a lot of New Yorkers, who regard pedestrian rights as sacred, by trying to get them to use crosswalks designed to prevent traffic gridlock. Hillary, by contrast, represents the triumph of hope over experience—the utopian belief that government programs can solve social problems at no cost or inconvenience to any of us.

THE clash of philosophies was most apparent when it came to the issue of homelessness. During the antiauthoritarian 1970s, social scientists argued that forced hospitalization of the mentally ill violated their civil rights. As a result untreated psychotics came to form a large percentage of New York City's hard-core homeless. In 1999, voluntary outreach programs having failed miserably, the mayor ordered police to give street people a choice between jail and transport to a shelter. This hardly solved the problem, but it did give shelter employees a chance to evaluate the mental state of homeless individuals, and it put pressure on the state to fulfill its traditional responsibility to provide treatment centers, apparently to some effect. Hillary Clinton's response was to denounce the mayor for "criminalizing the homeless with mass arrests"—though in fact, the only people who ended up in jail were those with outstanding warrants. She also described the baby Jesus as a homeless child, implying that the mayor was lacking in Christian charity. Mrs. Clinton was apparently unaware that New York City maintains the largest network of shelters in the country, and she had forgotten the Book of Luke, which tells us that Mary and Joseph were away from home because "there went out a decree from Caesar Augustus that all the world should be taxed."

Meanwhile James Carville called the mayor a "thug," and Housing and Urban Development Secretary Andrew Cuomo announced that he was seizing control of sixty million dollars in federal housing grants, on the grounds that Mayor Giuliani had unfairly slighted one AIDS advocacy group out of the many that received moneys.

In Hillary Clinton's ideal world psychiatric care would be available to everyone, but nothing in her approach tells us how to deliver care to the people who need it most—and are often too sick to realize it. Even pro-Democratic journals like the *New Republic* have begun to discuss the tendency of federal programs to spend heavily on high-ticket extras while failing to solve the problems that brought them into being in the first place. A recent article by Matthew Miller, a fellow of the Annenberg Public Policy Center, singles out the effort to reduce the size of school classes—a pet cause of Hillary's—as a prime example of this phenomenon of "goal displacement." Miller notes that shrinking classes is a "pricey" program that appeals to middle-class voters even though there is little or no evidence that it leads to higher student achievement. What it does accomplish is to lead experienced teachers to desert inner-city schools for better-paying jobs elsewhere, creating a new crisis—a shortage of qualified teachers—which in turn requires still more federal subsidies.

The moral is that one-size-fits-all federal programs don't solve the schools' problems, and sometimes even do more harm than good. On the other hand, as Miller also points out, the momentum in their favor is powerful. Even though many voters are at least mildly cynical, they may vote for the candidate who favors increased federal spending in the hope that this time things will be different. As Miller puts it, the operant philosophy is: "When in doubt, throw money."[8] This dynamic may explain why Hillary's record—the disappointing results of educational reform in Arkansas and her support for the schemes of the very people who failed to improve the schools in Rochester, New York—may not tarnish her credentials as the education candidate.

THE progress of Hillary Clinton's campaign against "meanness" was being watched with interest in the state of Arkansas, where the *Democrat-Gazette* ran a cartoon labeling the First Lady's upstate New York listening tour as "the Hillary Witch Project," and columnist John Brummett suggested that New York's unionized teachers ought to consider the history of the Clintons' relations with their counterparts in Arkansas.[9] Hillary has not been seen much in Arkansas in recent years. She did not find time to visit the state at all in 1999, even missing the dedication of the Clinton Birthplace Memorial in Hope. Nor was she along in mid-December when her husband received an award from the Little Rock Chamber of Commerce. Bill Clinton's speech on that occasion was in part an effort to rekindle enthusiasm for his presidential library, once projected as the centerpiece of a downtown Little Rock renaissance. The original target of a January 2001 opening is now but a memory. Little Rock voters rejected a tax intended to pay for the purchase of the land the library will occupy, and the city has been diverting funds from admission fees to the zoo.

In his remarks Clinton said of the library, "I want it to be a place of touch and involvement and learning."[10] He also referred repeatedly to the conspicuously absent Hillary, telling the audience that he had spoken to her as recently as one A.M. the previous night. Some reassurance was necessary because Arkansas Democrats were feeling a little miffed with the Clintons. While they may blame Ken Starr for conducting an investigation that turned a lot of lives upside down, there is a feeling that Bill and Hillary failed to appreciate what so many old friends went through on their behalf. This feeling could only intensify when, two weeks after his speech, the White House announced that in the year 2000 Bill Clinton would be registering to vote as a New Yorker.

It seems unlikely that Hillary will take any further interest in the project she pointedly refers to as "his library." Nor is there much chance that the Hillary Rodham Clinton papers will end up in Little Rock. The First Lady had also made it clear that she wouldn't be spending much of her time in Washington, and family friends were reported as saying that Chelsea, never known for her love of politics or the routines that go along with it, was now to fill in for her mother in the role of hostess at official functions. Commuter marriages are a fact of contemporary life and probably more common in Washington than anywhere else. Still, there was something poignant about a two-term president left alone in the White House to contemplate his legacy. Nor was it without irony that Hillary Clinton, protector of all the children in the world, might allow her college-student daughter to act as a substitute for her.

One is tempted to interpret these developments in light of a familiar pattern: A woman in midlife, stung by revelations of her husband's dalliance with a younger woman, is striking out on her own, her marriage over in all but name. It is always dangerous, however, to presume that the Clintons' partnership is in any way typical. For another perspective on the future one needs only to recall one of the less noticed passages in Lucinda Franks's *Talk* interview. Even as Hillary chattered on about what a "relief" it was to make her own decisions and vowed that Bill Clinton would not have a "front and center" role in her Senate campaign, she told Franks in the next breath, "He's already said he wants to run my constituency. Meet the people and all. But he'll stay behind the scenes for a while." Although Hillary reportedly laughed when she said this, she didn't sound like a woman who had declared her independence quite yet. On the contrary, her words raised the possibility that the New York Senate race would be another "two for the price of one" candidacy.

Lending support to this interpretation, *The New York Times* reported persistent rumors, denied by Hillary's spokesman, that the President was actively soliciting funds for his wife's campaign. *The Washington Post,* meanwhile, commented that Clinton gave the keynote speech at a luncheon hosted by two Connecticut women, Sandra Wagenfeld and Francine Goldstein, thanking them for "being so wonderful to Hillary." Two days after this event, Wagenfeld and Goldstein, who happened to be backing Bill Bradley over Al Gore, donated $150,000 to the Democratic National Committee. Officially, such donations can't be earmarked for any particular candidate, but the presumption was that they would translate into soft money support for Hillary's campaign.[11]

While outsiders may see Hillary's marriage as stressful, it has undoubtedly been a source of emotional security for her. She has, for a long time, been the grown-up in her marriage, the one with the real power. Where else would Bill Clinton find a woman who forgives him when he strays, defends him to the world, and knows how to manage his sometimes volatile moods? She, in turn, has the companionship and support of an undeniably brilliant man who is also a supreme political tactician.

If infidelities or geographical inconvenience could destroy the Clintons' marriage, they would have done so a long time ago. One issue that does have the potential to land Bill and Hillary in divorce court someday is money. Both Clintons have great potential earning power, but neither is highly motivated to earn. Whether Hillary remains active in domestic politics or makes an early exit to become a human rights activist on the world scene, one can be sure that her activities will require large infusions of cash. Bill Clinton stands to earn millions from book royalties and lecture fees, and he will also have his government pensions and, if he wishes, a partnership in a prestigious New York law firm. But in the past, during the rare intervals when he was not pointing toward the next election, he tended to be restless, unfocused, and even mildly depressed. This usually meant trouble between him and Hillary, and all signs point to another difficult adjustment. Facing the end of his days in the White House, the President sounds very much like a star athlete who isn't quite mentally prepared for retirement. He talks about playing golf and spending time in Arkansas, where his presidential library is projected to include an apartment for his use. Vague hints that he might consider following in the footsteps of John Quincy Adams, who became a congressman after leaving the White House, probably don't reflect a fully formed intention, but it would be hard to bet against his taking another shot at electoral office.

Of course, all politicians need money, the Clintons just seem to need it a little more than most. Whether Bill or Hillary is the candidate, they play to win, which means a long and expensive campaign. On the personal level they are a couple already in their mid-fifties with no home equity and all the expenses entailed by settling into a new house. Their longtime Arkansas friends are preoccupied with raising funds for the Clinton library and birthplace memorial. As far as their legal bills go, a suggestion that Bill and Hillary would try to win government reimbursement met with a decidedly negative reaction. As Hillary's comment to Harold Ickes about the short half-life of a First Lady demonstrates, she understands very well that the Clinton name will become less potent once a new administration takes over. In order to raise money for any purpose, she needs to project the image of a woman whose future is at least as interesting as her past.

Despite many missteps, Hillary's chances of winning election are still reasonably good. The geography of New York State favors her effort: Residents of New York City are overwhelmingly registered Democrats, and upstaters tend to be suspicious of politicians from the big city, regardless of party. In the era of celebrity candidates, never having held public office may actually be an advantage. Voters can indeed cast their ballots "for themselves," projecting their hopes and expectations, however unrealistic, onto the candidate's glossy image.

If elected to office Hillary would remain a media favorite and be mentioned as a possible candidate for national office. Whether she would commit herself to the long and tedious process of turning ideas into legislation is another question. Always a hard worker, Hillary is also impatient for quick results. Temperamentally she is an advocate, better at energizing audiences with a vision of progress than hammering out deals, in which the realities of the economic bottom line inevitably come into play.

In terms of her personal finances, election to office would create a new set of complications. Strict ethics rules prohibit senators from earning outside income from speaking engagements and service on corporate boards. Hillary could write a memoir, and doubtless will, but unless she is able to be more candid about her inner life than she has been in the past, she is unlikely to write the kind of book that would solve her financial problems for the long run. Bill Clinton would still be free to pursue the opportunities that flow to an ex-president, but any unusually lucrative offers would inevitably raise questions as to whether the money was intended to influence his wife's votes in the Senate.

WITH the impeachment crises over and Hillary taking center stage, the public was eager to believe that we've seen the last act of the Clintons' soap opera marriage. In fact, what occurred was that the Clintons' political partnership had moved its base from Arkansas to New York.

On February 6, 2000, Bill Clinton was at his wife's side, smiling in silent approval as she officially opened this new phase of her career, making her candidacy for the Senate official at last. The announcement came during a rally held in the gymnasium of the State University of New York at Purchase, an event whose scale was more presidential than senatorial. Speaking under an oversize banner that shouted HILLARY! the newly minted candidate reached back eight years, borrowing her theme from her husband's 1992 campaign. Hillary proclaimed herself a New Democrat and insisted that she did not see the government as "the source of all our problems or the solution to them." But "when people live up to their responsibilities, we ought to live up to ours."

On the theory that some New Yorkers had formed a misleading impression of Hillary from the press and the administration's enemies, the rally also included an eighteen-minute documentary by Linda Bloodworth-Thomason, coproducer of *The Man from Hope*. Reintroducing us to Hillary (no longer Rodham) Clinton, the film featured her in soft pink, revealing her domestic side for the first time since the chocolate chip cookie wars of '92: "I make a mean tossed salad and a great omelet," she now said.

The overproduced event at Purchase highlighted the excessiveness that had become Hillary's defining characteristic during her years in the spotlight. One couldn't help thinking that if only Hillary had done one-tenth as much as First Lady and taken herself a little less seriously, her appeal would be much greater and she wouldn't need the most expensive campaign effort in history to power her within reach of the Senate. But then, of course, she wouldn't be Hillary, a woman who has always tried not merely to have it all but to have it all at the same time.

Acknowledgments

A number of individuals were kind enough to send me documents from their personal files. Alan Houseman and William F. Harvey were especially generous. Ronald Radosh, Steven Schwartz and Martin L. Gross shared the fruits of their research. Rodney Sweetland III provided me with court papers from the Sean T. Haddon discrimination case, Michael Chapman directed me to information about the Arkansas Governor's School.

Thanks also to Dr. Kenneth Keniston, Dr. Donald J. Cohen, Director of the Yale Child Study Center, Robin Hardman of the Families and Work Institute, Dr. T. Berry Brazelton, Edward Tuck and the French-American Foundation, the staff of the National Center on Education and the Economy, Karen Johnson of the National Organization for Women, Tanya Russell of the Arkansas Headstart Association, Barbara Gilkey of the Arkansas Home Instruction for Preschool Youth Program, Ken Boehm of the National Legal and Policy Center, Dianne Williams of the Winthrop Rockefeller Foundation, John Robert Starr, Diane Ravitch and Dr. D. L. Cuddy.

Jim Robinson and the Free Republic website epitomize everything that Hillary Rodham Clinton finds dangerous about the Internet. Whether or not one agrees with the views of the Freepers, their site proved to be an invaluable source of daily news stories and contentious but often illuminating opinion.

Special thanks to my agent, Barbara Lowenstein, who gave her support for this project even though she can't have been pleased with all my conclusions, and to my unfailingly cheerful and patient editor, Henry Ferris.

Notes

ONE: THE FIRST VICTIM

1. Office of the Independent Counsel, Kenneth W. Starr, "Supplemental Materials to the Referral to the U.S. House of Representatives Pursuant to Title 28, United States Code 595c," House Document 105–316, September 29, 1998, Part II, pp. 2703, 2722–2727. (Hereafter, Starr Referral, Supplemental Materials.)

2. Ibid., pp. 2739–40.

3. Matthew Campbell, "Hillary in Retreat," London *Sunday Times,* March 8, 1998.

4. Ron Fournier, "Hillary Clinton Is a Key Advisor in Sex Controversies, Attorneys Contend," Associated Press, March 28, 1998.

5. Kenneth T. Walsh, "The Survivalist," *U.S. News & World Report,* May 11, 1998.

6. Francis X. Clines, "Once a Political Debit, Now a Powerful Force," *The New York Times,* November 22, 1998, p. A26.

TWO: RUGGED INDIVIDUALIST

1. Roger Morris, *Partners in Power: The Clintons and Their America* (New York: Henry Holt, 1996), p. 115. For Hillary Rodham's life in Park Ridge, see also Hillary Rodham Clinton, *It Takes a Village and Other Lessons Children Teach Us* (New York: Simon & Schuster/Touchstone, 1996); Donnie Radcliffe, *Hillary Rodham Clinton: A First Lady for Our Time* (New York: Warner Books, 1993); Martha Sherrill, "The Education of Hillary Rodham Clinton: Growing Up in a Chicago Suburb, A Good Girl Getting Better All the Time," *The Washington Post,* January 11, 1993; Judith Warner, *Hillary Clinton: The Inside Story* (New York: Penguin, 1993).

2. Interview with Penny Pullen.

3. Cynthia Hanson, "I Was a Teenage Republican," *Chicago Magazine,* September 1994.

4. Peter Baker, "First Lady Recalls Early Taste of Ethnic Tension," *The Washington Post,* December 10, 1997, p. A4.

5. Interview with Don Jones.

THREE: THE ART OF THE IMPOSSIBLE

Background on Hillary Rodham Clinton's years at Wellesley, including text of her commencement speech, is courtesy of Wellesley College; also Martha Sherrill, "The Education of Hillary Rodham Clinton"; Charles Kenney, "Hillary: The Wellesley Years," *The Boston Globe,* January 12, 1993, p. 65; and interviews.

1. Amy Bernstein, "Mother of Candidates," *U.S. News & World Report,* March 13, 1992, p. 20.

2. John Taft, *Mayday at Yale: A Case Study in Student Radicalism* (Boulder, Colo.: Westview Press, 1976), pp. 59–60. Quotations from student and faculty speeches in this section are drawn from Taft.

3. Miriam Horn, "The Group Twenty-five Years Ago," *U.S. News & World Report,* June 6, 1994; also Radcliffe, *Hillary Rodham Clinton,* p. 93.

4. Radcliffe, *Hillary Rodham Clinton,* p. 93.

5. Kenneth Keniston, *Youth and Dissent: The Rise of a New Opposition* (New York: Harcourt Brace Jovanovich, 1971), p. 316.

6. John A. Andrew III, *Lyndon Johnson and the Great Society* (Chicago: Ivan R. Dee, 1998), p. 16.

7. Ellen Condliffe Lagemann, *The Politics of Knowledge: The Carnegie Corporation, Philanthropy, and Public Policy* (Chicago: University of Chicago Press, 1989), p. 250.

8. Marian Wright Edelman, *The Measure of Our Success* (New York: HarperPerennial, 1992), p. 46. See also Robert V. Pambianco, "CDF's Radical Agenda and the Hillary Connection," *Human Events,* February 27, 1993, p. 10.

9. Reprinted in *Psychology Today,* November/December 1997, p. 58.

10. Interview with Penn Rhodeen.

11. Connie Bruck, "Hillary the Pol," *The New Yorker,* May 30, 1994, p. 61.

12. Interview with Robert Treuhaft; also Jessica Mitford, *A Fine Old Conflict* (New York: Vintage, 1978); David Rennie, "How Jessica Mitford's Son Became the Piano Tuner of Havana," London *Daily Telegraph,* June 28, 1997, p. 13; and for Mitford's visit to Arkansas, Jessica Mitford, "Appointment in Arkansas," *New West,* November 3, 1980.

13. David Maraniss, *First in His Class: A Biography of Bill Clinton* (New York: Simon & Schuster, 1995), p. 277.

14. Kenneth Keniston and the Carnegie Council on Children, *All Our Children: The American Family Under Pressure* (New York: Harcourt Brace Jovanovich, 1977), pp. 21–22.

15. *Library Journal,* September 15, 1977, p. 1830.

16. Interview with Faustina Solis.

17. Beatrice Siegel, *Marian Wright Edelman: The Making of a Crusader* (New York: Simon & Schuster, 1995), p. 98.

18. Jerry Zeifman, *Without Honor: The Impeachment of President Nixon and the Crimes of Camelot* (New York: Thunder's Mouth Press, 1995), pp. 151–52; additional material in this section is from interview with Jerry Zeifman.

19. Renata Adler, "Searching for the Real Nixon," *The Atlantic Monthly,* December 1976.

20. David A. Price, "First Lady's School for Scandal," *Investor's Business Daily,* December 6, 1996; interview with William Dixon.

FOUR: THE YANKEE GIRLFRIEND

1. Dolly Kyle Browning, *Purposes of the Heart* (Dallas: Direct Outstanding Creations, 1997), pp. 131–32.

2. Virginia Kelley with James Morgan, *Leading with My Heart* (New York: Simon & Schuster, 1994), p. 191.

3. Ibid., p. 75.

4. Interview by Peter Boyer, for "Once Upon a Time in Arkansas," *PBS/Frontline Online,* air date: October 7, 1997.

5. Charles F. Allen and Jonathan Portis, *The Comeback Kid: The Life and Career of Bill Clinton* (New York: Birch Lane Press, 1992), p. 45.

6. "Representative Is 'Out of Step,' Clinton Charges," *Arkansas Gazette,* August 8, 1974.

7. Maraniss, *First in His Class,* p. 335.

8. Norris Church talked about her dates with Clinton to Mailer biographer Peter Manso, but for an Arkansan take on the situation, see Paul Lake, "Why Norman Mailer's Wife Dumped Bill Clinton," *The Weekly Standard,* August 10/17, 1998, pp. 24–26.

9. Michael Kelly, "The Transition: Packaging the Candidate," *The New York Times,* November 14, 1992.

10. Maraniss, p. 336.

11. Ibid., p. 343.

12. Hillary Rodham Clinton, "Twenty Years of Marriage," *Talking It Over,* October 7, 1995, AOL Online.

FIVE: FIRST IN HER FIRM

1. Gregory Jaynes, "The Death of Hope," *Esquire,* November 1993, p. 85.

2. Interview with Webb Hubbell; see also Webb Hubbell, *Friends in High Places: Our Journey from Little Rock to Washington, D.C.* (New York: William Morrow, 1997).

3. Lisa Myers, "Clinton Accused of Raping (Julia) Juanita Broaddrick," MSNBC, March 28, 1998.

4. James B. Stewart, *Blood Sport: The President and His Adversaries* (New York: Simon & Schuster, 1996), p. 70.

5. Peter Boyer, *Frontline,* October 7, 1997.

6. Stewart, p. 70.

7. Jim McDougal and Curtis Wilkie, *Arkansas Mischief* (New York: Henry Holt, 1998), pp. 150–51.

8. Gennifer Flowers, *Passion and Betrayal* (Del Mar, Calif.: Emery Dalton, 1995), pp. 1–3, 40–44 and passim. The Flowers book is unreliable on dates. In televised interviews, Flowers states that Clinton had been married eighteen months when the affair began. The story of the abortion is from an appearance on *Hard Copy.*

9. *Arkansas Democrat,* November 8, 1978; *Arkansas Gazette,* December 2, 1978.

10. Meredith L. Oakley, *On the Make: The Rise of Bill Clinton* (Washington, D.C.: Regnery, 1994), pp. 205–9; Oakley is also a source for background on political initiatives of Clinton's first term, including SAWER.

11. Jeff Gerth, "Blair Led First Lady to $100,000 Windfall," *New York Times* syndication reprinted in *Arkansas Times-Gazette,* March 18, 1994, p. 1A.

12. Interview with William Harvey. See also David Brock, "Where Was Hillary?" *The American Spectator,* September 1994.

13. Caroline Baum and Victor Neiderhoffer, "Herd Instincts: Hillary's Cattle Profits," *The National Review,* February 20, 1995, p. 43; see also James K. Glassman, "Hillary's Cows," *The New Republic,* May 16, 1994; and Susan B. Garland, Dean Foust, et al., "Hillary Clinton, Go-Go-Getter," *Business Week,* April 18, 1994, p. 42.

14. Michael Kelly, "The President's Past," *The New York Times Magazine,* July 31, 1994, p. 20.

15. Bruck, "Hillary the Pol," p. 73.

SIX: SURVIVAL PLAN

1. Diane Blair, "Of Darkness and Light," in Ernest Dumas, ed., *The Clintons of Arkansas: An Introduction by Those Who Know Them Best* (Fayetteville: University of Arkansas Press, 1993), p. 60.

2. Rudy Moore, Jr., "They're Killing Me Out There," in Dumas, ed., *The Clintons of Arkansas,* p. 90.

3. Oakley, *On the Make,* p. 217.

4. David Gallen, *Bill Clinton as They Know Him: An Oral Biography* (New York: Marlowe & Co., 1996), p. 218.

5. David Brock, *The Seduction of Hillary Rodham Clinton* (New York: The Free Press, 1996), p. 82.

6. David Maraniss, "Lessons of Humbling Loss Guide Clinton's Journey," *The Washington Post*, July 14, 1992, p. A1.

7. Oakley, pp. 263–64.

8. Deposition of George Locke, "Investigation of Whitewater Development Corporation and Related Matters," Senate Report 104–869, Report of the Special Committee to Investigate Whitewater Development Corporation, administered by the Committee on Banking, Housing and Urban Affairs, 104 Cong. 1st Session (hereafter D'Amato Committee), Vol. XVII, pp. 5251–52.

9. "Lasater Maintains Distance," *Arkansas Democrat-Gazette*, April 9, 1995, p. 1B.

10. Oakley, p. 265.

11. Jeremiah Denton, "The LSC Survival Campaign," *Robber Barons of the Poor* (Washington, D.C.: Washington Legal Foundation, 1985), pp. 71–73.

12. Taylor Branch, "Closets of Power," *Harper's*, October 1982, pp. 35–50.

13. Phil Gailey, "Homosexual Takes Leave of a Job and of an Agency," *The New York Times*, March 31, 1982, p. A24.

14. Brock, pp. 104–5.

15. John A. Dooley and Alan W. Houseman, "Legal Services History," November 1985 (unpublished); this and other unpublished materials courtesy of Alan Houseman; additional materials courtesy of William Harvey.

16. Interview with William Harvey.

17. Gail Sheehy, "What Hillary Wants," *Vanity Fair*, May 1992.

18. Dick Morris, "Hillary: Life Saver, Power Grabber," *New York Post*, May 12, 1998.

19. Interview with Ivan Duda; for background on the Orsini case, see Gene Lyons, *Widow's Web* (New York: Ivy Books, 1993).

SEVEN: A WOMAN OF MANY PROJECTS

1. James Ring Adams, "What's Up in Jakarta?" *The American Spectator*, September 1995.

2. Daniel Wattenberg, "Love and Hate in Arkansas," *The American Spectator*, May/June 1994.

3. Ibid.

4. Audrey Duff, "Is a Rose a Rose?" *The American Lawyer*, July/August 1992.

5. Ibid.

6. Interview with Hermann Ivester; see also Doug Thompson, "Judge Comes Down Hard on Rose Firm in Aromatique Case," *Arkansas Democrat-Gazette*, July 12, 1994, p. 1D.

7. Roy Reed, "I Just Went to School in Arkansas," in Ernest Dumas, ed., *The Clintons of Arkansas*, pp. 148–60.

8. "Fulfilling the Promises of Reform, a Winthrop Rockefeller Foundation Policy Study" (Little Rock: Winthrop Rockefeller Foundation, 1988); see also Blant Hurt, "Mrs. Clinton's Czarist Past," *The Wall Street Journal,* March 19, 1993, p. A10.

9. Peter LaBarbera, "Clinton's 'Governor's School' Is Politically Correct," *Human Events,* September 12 and 26, 1992.

10. Interview with Barbara Gilkey.

11. Blant Hurt, "Dr. Elders' Record in Arkansas," *The Wall Street Journal,* July 22, 1993, p. A14.

12. Clinton discussed the subject with Carolyn Staley, among others. For more on Clinton's psychology, Paul M. Fick, *The Dysfunctional President: Inside the Mind of Bill Clinton* (New York: Citadel Press, 1996).

13. Allen and Portis, *The Comeback Kid,* p. 104.

14. Flowers, *Passion and Betrayal,* p. 65.

15. "Transcript of President's News Conference on the Whitewater Affair," *The Washington Post,* March 24, 1994, p. A18.

16. Stewart, *Blood Sport,* p. 133.

17. Deposition of Onie Elizabeth Wright, D'Amato Committee, Vol. XV, p. 1652.

EIGHT: THE GARY HART FACTOR

1. Oakley, *On the Make,* p. 348.

2. Gallen, *Bill Clinton as They Know Him,* pp. 190–91.

3. David Brock, "His Cheatin' Heart: Bill's Arkansas Bodyguards Tell the Story the Press Has Missed," *The American Spectator,* January 1994.

4. Ibid.

5. John Brummett, *High Wire: The Education of Bill Clinton* (New York: Hyperion, 1994), p. 114.

6. Murray Waas, "Clinton Takes the Offensive," *Salon,* August 18, 1998.

7. Judy Bachrach, "Clinton's Private Eye," *Vanity Fair,* September 1998.

8. Marlise Simons, "Child Care Sacred as France Cuts Back the Welfare State," *The New York Times,* December 31, 1997, p. A1.

9. Hillary Rodham Clinton, "In France, Day Care Is Every Child's Right," *The New York Times,* April 7, 1990.

10. Brummett, pp. 49–50.

11. Ira Magaziner and Hillary Rodham Clinton, "Will America Choose High Skills or Low Wages?" *Educational Leadership,* Vol. 49, No. 6, March 1992, pp. 10–14.

12. Rep. Ron Sunseri, "Certificates of Mastery vs. Diplomas," *Education Reporter,* June 1997. The article is the text of a speech given at a February 12, 1997, conference in Washington, D.C., entitled "What Goals 2000 Means to the States."

NINE: EVERYONE HAS A LIST

1. Landon Y. Jones, "Road Warriors," *People,* July 20, 1992, pp. 68ff.; Roxanne Roberts, "Hillary Clinton Gets Personal," *Redbook,* March 1992, pp. 86ff.; also Radcliffe, *Hillary Rodham Clinton,* p. 218.

2. Gallen, *Bill Clinton as They Know Him,* p. 201.

3. Oakley, *On the Make,* p. 452.

4. Jack Germond and Jules Witcover, *Mad as Hell* (New York: Random House, 1994), pp. 169–70.

5. Roberts, pp. 86ff.

6. Ruth Marcus, "Clinton's 'Captain of the Defense,' " *The Washington Post,* February 5, 1998, p. A12.

7. Brock, *The Seduction of Hillary Rodham Clinton,* p. 241.

8. Stewart, *Blood Sport,* pp. 177–78; McDougal and Wilkie, *Arkansas Mischief,* pp. 231–40.

9. Lucille Bolton was interviewed by Carl Limbacher of NewsMax.com; for the chronology of McIntosh's involvement, see Oakley, pp. 499–504; see also James Dalrymple, "Clinton's Black Son," London *Daily Mail,* June 1998, via Internet.

10. Jerry Seper, "After Pardon, Rumors Fly Like Pies," *Washington Times,* February 15, 1993, p. A1.

11. Interview with John B. Thompson. Thompson is a bitter critic of Attorney General Janet Reno and ran against her for Dade County attorney. He recounted his conversations with Jones in a sworn deposition for the Paula Jones lawsuit: Declaration of John B. Thompson, Civil Action LR-C-94–920, United States District Court for the Eastern District of Arkansas, Western Division, October 14, 1997. Jones then filed an affidavit denying that he ever said these things.

12. Steve Dunleavy, "I Was Victim of Clinton Reign of Terror," *New York Post,* September 27, 1998; Richard Johnson, "How Liz Gracen Got Her Big Break," *New York Post Online;* Steven Edwards, "Bill Clinton, Gennifer Flowers and Me," *Saturday Night,* May 1998, p. 34.

13. George Stephanopoulos, "Betrayal," *Newsweek,* August 31, 1998.

14. Jacob Weisberg, "Desperately Leaking Susan," *Vanity Fair,* June 1996, pp. 140ff.; Anonymous, *Primary Colors* (New York: Random House, 1996), p. 62; confidential interviews.

15. Stephanopoulos; Stewart, p. 212.

16. John King, "Political Atmosphere Is Right Neighborly in Clinton's Arkansas," Associated Press, April 19, 1992.

17. Barbara Matusow, "True Grit," *The Washingtonian,* January 1993.

18. Sheehy, "What Hillary Wants."

19. Murray Waas, "The Other Woman," *Salon Newsreal,* September 11, 1998.

20. Judy Bachrach, "Clinton's Private Eye," *Vanity Fair,* September 1998, pp. 192–208ff.; Nina Bernstein, "High-Tech Sleuths Find Private Facts On-line," *The New York Times,* September 17, 1997; Deposition of Terry Lenzner, Judicial Watch Online.

TEN: THE TRANSITION THAT WASN'T

1. Marc S. Tucker to Hillary Clinton, November 11, 1992.

2. Ronald Radosh, *Divided They Fell: The Demise of the Democratic Party, 1964–1996* (New York: The Free Press, 1996), pp. 221–23; for Lani Guinier, see also Jonathan Mahler, "Too Black Meant Too Scary to Let Speak: Lani Guinier Comes to Her Own Defense in 'Lift Every Voice' "; and Jonathan Mahler, "A Bizarre Footnote," *The Forward,* April 17, 1998.

3. Jacob Weisberg, "A Short History of Mr. Magaziner," *The New Republic,* January 24, 1994.

4. Haynes Johnson and David S. Broder, *The System: The American Way of Politics at the Breaking Point* (Boston: Little, Brown, 1997), p. 81; for background on discussions at the Governor's Mansion, see Bob Woodward, *The Agenda* (New York: Simon & Schuster, 1994).

5. Doug Thompson, "Turnover in Clinton's Secret Service Detail 'Highest That Anyone Can Remember,' " *Capitol Hill Blue,* July 17, 1998.

6. Ronald Kessler, *Inside the White House* (New York: Pocket Books, 1996), p. 234.

ELEVEN: RECLAIMING AMERICA

1. *The Wall Street Journal,* January 14, 1993, p. A18.

2. Gwen Ifill, "Executive Brief: The White House; Young Staff's First Steps Are Right Back in Time," *The New York Times,* January 28, 1993, p. 14.

3. Michael Duffy, "The Kids Down the Hall," *Time,* March 8, 1993, p. 44; also Michael Duffy, "Obstacle Course," *Time,* February 8, 1993, p. 26.

4. Interview with Diane Ravitch.

5. Elizabeth Drew, *On the Edge* (New York: Simon & Schuster/Touchstone, 1995), p. 103.

6. Margaret Carlson, "At the Center of Power," *Time,* May 10, 1993, p. 28.

7. Investigation into the White House and Department of Justice on Security of FBI Background Investigation Files, Interim Report, 104 Cong., H.R. 104–862; Nineteenth Report by the Committee on Government Reform and Oversight, 104 Cong., 2nd Session (popularly known as the "Clinger committee").

8. Johnson and Broder, *The System,* p. 114.

9. Joycelyn Elders and David Chanoff, *Joycelyn Elders, M.D.* (New York: William Morrow, 1996), pp. 294–95.

10. Martha Sherrill, "Hillary Clinton's Inner Politics," *The Washington Post,* May 6, 1993.

11. Remarks by the First Lady, Institute of Medicine Annual Meeting, The White House, Office of the Press Secretary, October 19, 1993.

12. Robert Pear, "Clinton Criticized as Too Ambitious with Vaccine Plan," *The New York Times,* May 30, 1994, p. A1.

13. General Accounting Office, "Vaccines for Children: Reexamination of Program Goals to Insure Implementation," June 1, 1995; for a more complete discussion of this episode, see also Stanley A. Renshon, *High Hopes: The Clinton Presidency and the Politics of Ambition* (New York: New York University Press, 1996), pp. 287–94.

14. Milt Freudenheim, "Drug Prices Overstated, GAO Says," *The New York Times,* May 30, 1995, p. D9.

15. Johnson and Broder, p. 123.

16. Henry Allen, "A New Phase at the White House," *The Washington Post,* June 9, 1992, p. D1; for Lerner's response, see "Hillary's Politics: My Meaning," *The Washington Post,* June 13, 1993, p. C1.

TWELVE: A BOX WITH NO WINDOWS

1. Rebecca Borders, "Bill's Cousin Cornelius?" *The American Spectator,* May 1996; see also Kim I. Eisler, "Fall Guy: Everybody Liked Billy Dale," *The Washingtonian,* February 1996; John Podesta, "White House Travel Office Management Review," The White House, July 2, 1993; and "Investigation of the White House Travel Office Firings and Related Matters," House Report 104–849, House Committee on Government Reform and Oversight, September 26, 1996.

2. Peter J. Boyer, "A Fever in the White House," *The New Yorker,* April 15, 1996, p. 72; for more on the Thomasons, see also Margy Rochlin, "The Prime of Linda Bloodworth-Thomason," *Los Angeles Times,* September 27, 1992; and Kim Masters, "Four Pals in Their Prime Time," *The Washington Post,* January 4, 1993.

3. Michael McMenamin and James Oliphant, "All the President's Fault," *Reason,* April 1996, pp. 39ff.

4. Stewart, *Blood Sport,* pp. 284–85.

5. D'Amato Committee, Vol. IV, pp. 2493, 2496.

6. Ibid., p. 2997.

7. Ibid., Vol. III, p. 846.

8. Starr Referral, Supplemental Materials, Part III, p. 4276.

9. Statement of Elizabeth Braden Foster, May 9, 1994, 29D-LR-35063: Federal Bureau of Investigation; also Office of the Independent Counsel, Kenneth W. Starr, "Report on the Death of Vincent Foster, Jr.," October 31, 1997. While the psychological profile of Foster's depression is compelling, the direct evidence is far from complete. The Park Police assumed from the beginning that Foster was a suicide and never looked for evidence to the contrary. There are no 35-millimeter crime scene photos (the camera inexplicably "went haywire," according to the Park Police lab technician); nor are there X rays of Foster's head wound (an omission attributed to another

malfunctioning machine). Foster's eyeglasses, bearing a trace of gunpowder, were found thirteen feet below his body; the only explanation is that they flew off his head with the gunshot impact and bounced down the heavily vegetated slope—which seems hard to imagine. Lisa Foster could not positively identify the .38 special found in her husband's hand, and three witnesses who were in the Fort Marcy parking lot at a time when Foster was supposedly already dead did not see his car. Nor, for that matter, did any trace of Deseryl show up in the autopsy.

10. D'Amato Committee, Final Report, pp. 88–89.

11. Dana Priest, "Hillary Clinton Parries Criticism of Health Plan in Hill Testimony," *The Washington Post,* September 30, 1993, p. A6.

12. Remarks by the First Lady to the American Medical Association, June 13, 1993, Office of the First Lady.

13. Johnson and Broder, *The System,* p. 319.

14. Ibid., p. 356.

THIRTEEN: LAWYERING UP

1. Stuart Taylor, Jr., "Her Case Against Clinton," *The American Lawyer,* November 1996, p. 57.

2. Information in this section comes from "Investigation into the White House and Department of Justice on Security of the FBI Investigation Files," House Report 104–862, 104th Congress, 2nd Session.

3. Howard Kurtz, *Spin Cycle* (New York: The Free Press, 1998), pp. 88–89; one reason Schmidt was unpopular in the White House was her ability to keep track of the discrepancies in the First Lady's statements; see especially David Maraniss and Susan Schmidt, "Hillary Clinton and the Whitewater Controversy: A Close-up," *The Washington Post,* June 22, 1996, p. A1.

4. "Investigation of the White House Travel Office Firings and Related Matters," report by the Committee on Government Reform and Oversight, House Report, 104–849.

5. Interview with Chuck Harder.

FOURTEEN: FROM THE WHITE HOUSE TO YOUR HOUSE

1. Gary Aldrich, *Unlimited Access: An FBI Agent Inside the Clinton White House* (Washington, D.C.: Regnery, 1998), pp. 96–97.

2. Sally Quinn, "Look Out, It's Superwoman," *Newsweek,* February 15, 1993, p. 24.

3. Martin Walker, "Star-Spangled Banners," *The Guardian,* November 30, 1995, p. T2.

4. Blanche Wiesen Cook, "Presidential Papers in Crisis: Some Thoughts on Lies, Secrecy and Silence," *Presidential Studies Quarterly,* Winter 1996; also Cook's review of *Previous Convictions: A Journey Through the 1950's,* in *The Nation,* January 22, 1996.

5. Confidential interviews; also Susan Schmidt, "First Lady Encouraged Plan to Share

Computer Data with DNC," *The Washington Post*, March 4, 1997; Paul M. Rodriguez, "Memos Suggest Obstruction Linked to Big Brother Database," *Insight*, December 1, 1997; on Peter Collins, Mark Levin, "Dancing Around the Law," *The Wall Street Journal*, January 8, 1998; and on the federal privacy act, "Judicial Watch Interim Report," Part I, Section II, Judicial Watch Online.

6. Barbara Matusow, "Be My Guest," *The Washingtonian*, October 1993; also Phyllis C. Richman, "Eschewing the Fat with Hillary: A New Point of 'Lite,' " *The Washington Post*, December 1, 1993, p. E1.

7. *Sean T. Haddon v. Gary J. Walters, Chief Usher*, case nos. 01940418 and 01954162, Social Security Administration, Office of Hearings and Appeals.

8. Paul Bedard, "Papers Disclose Plan to Force Out Chef Who Filed Bias Claim," *The Washington Times Weekly Edition*, September 22, 1997; see also additional reporting by Bedard in *The Washington Times*, September 8, 9, and 10, 1997.

9. Martha Sherrill, "First Lady Has Usher Dismissed," *The Washington Post*, March 13, 1994, p. A19. Before Emery was dismissed, Anthony Marceca of the White House Office of Personnel Security requisitioned his confidential file from the Internal Revenue Service even though he was not due for a background check for three more years; see "White House Used IRS File, Documents Say," *Los Angeles Times*, July 4, 1996.

10. Interview with Rodney Sweetland III.

11. "White House Holidays Feel Crunch of DNC Debt," Associated Press, December 4, 1997.

12. Elders and Chanoff, *Joycelyn Elders, M.D.*, p. 267.

13. Melissa Ludtke, *On Our Own: Unmarried Motherhood in America* (New York: Random House, 1998), pp. 396, 400.

14. "Remarks by the President and the First Lady at White House Conference on Early Child Development and Learning," April 17, 1997; also "Press Briefing by Mike McCurry," same date, and "White House Conference on Early Childhood Development Policy Announcements," undated, White House Online.

15. For a summary of the new brain research, see Rima Shore, *Rethinking the Brain* (New York: Families and Work Institute, 1997); for dissenting views on day care with references to other sources, see "What's the Matter with Day Care?" *American Enterprise*, May/June, 1998.

16. Christopher Lasch, "Hillary Clinton, Child Saver," *Harper's*, October 1992.

FIFTEEN: THE CHINA SYNDROME

1. Shelly H. Han, "The First Lady's Big Mistake," *The New York Times*, August 26, 1995, p. 19; William Neikirk, "Hillary Clinton Back in Limelight," *Chicago Tribune*, September 3, 1995.

2. Interview with Karen Johnson.

3. Myrna Blyth, "Women Together," *Ladies' Home Journal*, January 1996, p. 10.

4. Mary Ann Glendon, "Feminism and the Family: An Indissoluble Marriage," *Commonweal*, February 14, 1996.

5. Walter V. Robinson and Jill Zuckman, "Lawyer Says Clinton In-Law Peddled Access," *The Boston Globe*, January 15, 1997, p. A1. For Rodham's reply, see his letter to *The Boston Globe*, January 14, 1995, p. A1.

6. Jill Abramson et al., "Latin Connection: DNC Donor with Ties to Paraguay Presses His Case in the White House," *The Wall Street Journal*, February 20, 1997, p. A1; also confidential interview.

7. Jonathan Peterson, "Donor Discussed Paraguay Policy, Officials Say," *Los Angeles Times*, February 21, 1997, p. A1.

8. "Fund-Raising Calls from the White House," Investigation of Illegal or Improper Activities in Connection with 1996 Federal Election Campaign, Special Committee administered by the Senate Governmental Affairs Committee, Senate Report 105–167, on-line at senate.gov. (hereafter Thompson Committee).

9. Brian Duffy, "A Fund Raiser's Rise and Fall," *The Washington Post*, May 13, 1997, p. A1; for information on the Lums, see Peter J. Boyer, "American Guanxi," *The New Yorker*, April 14, 1997.

10. Thompson Committee, "John Huang at Commerce," p. 13.

11. For more details on this busy week, see Edward Timperlake and William C. Triplett, *The Year of the Rat* (New York: Regnery, 1998), pp. 36–38.

12. Thompson Committee, "John Huang at Commerce," see especially pp. 43–44.

13. Thompson Committee, p. 2507. Although the committee concluded that there was no "direct evidence" of Huang's espionage, the lack of an alternative explanation and the CIA's point of view on the matter are clear.

14. Affidavit of Nolanda Butler Hill; deposition of Nolanda Butler Hill, *Judicial Watch, Inc. v. United States Department of Commerce*, U.S. District Court for the District of Columbia, Civil Action No. 95–0133 (RCL); see also Hill's interview on *Prime Time Live*, ABC News, June 19, 1997.

15. Deposition of Maggie Williams in *Johnny Chung: His Unusual Access to the White House, His Political Donations, and Related Matters: Hearings Before the Committee on Government Reform and Oversight, House of Representatives*, 105th Cong., 1st Session, November 13 and 14, 1997, serial 105–69, pp. 12–13, 22, 26.

16. Ibid., p. 355.

SIXTEEN: THE MAKING OF AN ICON

1. Mike Seccombe and Jodie Brough, "How the First Lady Smuggled in Her Wish List," *Sydney Morning Herald*, November 23, 1996, p. 1; Bettina Arndt, "Trials of Being a First Lady," *Sydney Morning Herald*, November 22, 1996, p. 8; "That Secret Women's Business," *Canberra Times*, November 23, 1996.

2. Kurtz, *Spin Cycle*, pp. 78–82.

3. Alan Jolis, "Micro-Credit: A Weapon in Fighting Extremism," *International Herald-Tribune,* February 19, 1997, p. 9.

4. "World Bank Press Briefing," August 23, 1995, Federal News Service.

5. Hillary Rodham Clinton, speech of November 8, 1995, Federal Document Clearing House.

6. "Review of Management Practices at the Treasury Department's Community Development Financial Institutions Fund," Majority Staff Report, Subcommittee on General Oversight and Investigations, Committee on Banking and Financial Services, U.S. House of Representatives, 105th Cong., 2nd Session, June 1998, p. 24.

7. Kenneth T. Walsh, "Hillary's Resurrection," *U.S. News & World Report,* October 20, 1997, p. 26.

8. Thomas B. Edsall, "Clinton and Blair Envision a Third Way International Movement," *The Washington Post,* June 28, 1998, p. A24.

9. Michael Shnayerson, "Sid Pro Quo," *Vanity Fair,* May 1998.

10. Michael Isikoff, "A Twist in *Jones v. Clinton,*" *Newsweek,* August 11, 1997.

11. Gloria Borger, "Her Bill and Their Enemies," *U.S. News & World Report,* August 17, 1998.

12. Doug Ireland, "Of Closets and Clinton," *The Nation,* March 30, 1998.

13. Jesse Jackson, "Keeping Faith in a Storm," *Newsweek,* August 31, 1998.

14. Ruth Marcus, "Hillary Clinton 'Committed' to Her Marriage," *The Washington Post,* August 19, 1998.

15. Carolyn Skorneck, "Hillary Clinton, Already on the Road," AP Online, September 28, 1998.

EPILOGUE: THE HALF-LIFE OF A FIRST LADY

1. C. David Heymann, *RFK* (New York: Dutton, 1998), p. 400.

2. Ben Macintyre, "Nip and Tuck in Hillary's Senate Race," *The Times* (London), September 13, 1999, sect. 5L, p. 16.

3. Christopher Caldwell, "The Empress of the Empire State," *Jewish World Review,* July 19, 1999; see also Scott Christianson, "Hillary's Launch," *Empire State Report,* online edition.

4. Lucinda Franks, "The Intimate Hillary," *Talk,* September 1999.

5. Seth Gitell, "Meet Hillary Clinton's Grandmother," *Forward,* August 5, 1999.

6. Sergio Bustos, "Report: Gore, Not Hillary Clinton, Intended Beneficiary of FALN Clemency," Gannett News Service, November 11, 1999.

7. Jeff Gerth, "Friendship Counts: Clinton's Top Fund-Raiser Made Lots for Himself, Too," *The New York Times,* December 12, 1999, p. A1.

8. Matthew Miller, "Why Budgets Really Bloat: Wretched Excess," *New Republic,* November 1, 1999, pp. 17–20.

9. John Brummett, "Trying to Cope with Hillary," *Arkansas Democrat-Gazette,* November 28, 1999.

10. The White House, "Remarks by the President to the Little Rock Chamber of Commerce," December 13, 1999.

11. Clifford J. Levy, "Silence Turns to Boasts of Candidates' Fund-Raising Successes," *The New York Times,* December 31, 1999, p.1; Susan B. Glasser, "Clinton Taps Big Donors for Special N.Y. Account," *The Washington Post,* January 4, 2000, p. A1.

Index

464 INDEX